ACS SYMPOSIUM SERIES **712**

Combined Quantum Mechanical and Molecular Mechanical Methods

Jiali Gao, EDITOR
University of New York at Buffalo

Mark A. Thompson, EDITOR
WRQ, Inc.

American Chemical Society, Washington, DC

Library of Congress Cataloging-in-Publication Data

Combined quantum mechanical and molecular mechanical methods / Jiali Gao,
    Mark A. Thompson, editors.

   p.   cm.—(ACS symposium series ; ISSN 0097–6156 ; 712)

"Developed from a symposium sponsored by the Division of Computers in
Chemistry at the 214th National Meeting of the American Chemical Society,
Las Vegas, Nevada, September 7–11, 1997."

Includes bibliographical references and index.

ISBN 0–8412–3590–2

1. Quantum chemistry—Congresses.

   I. Gao, Jiali, 1962–   . II. Thompson, Mark A., 1957–   . III. American
Chemical Society. Division of Computers in Chemistry. IV. American
Chemical Society. Meeting (214th : 1997 : Las Vegas, Nevada) V. Series.

QD462.A1C63   1998
541.2′8—dc21   98–34174
                                                              CIP

The paper used in this publication meets the minimum requirements of American National Standard for
Information Sciences—Permanence of Paper for Printed Library Materials, ANSI Z39.48–1984.

PRINTED IN THE UNITED STATES OF AMERICA

Foreword

THE ACS SYMPOSIUM SERIES was first published in 1974 to provide a mechanism for publishing symposia quickly in book form. The purpose of the series is to publish timely, comprehensive books developed from ACS sponsored symposia based on current scientific research. Occasionally, books are developed from symposia sponsored by other organizations when the topic is of keen interest to the chemistry audience.

Before agreeing to publish a book, the proposed table of contents is reviewed for appropriate and comprehensive coverage and for interest to the audience. Some papers may be excluded in order to better focus the book; others may be added to provide comprehensiveness. When appropriate, overview or introductory chapters are added. Drafts of chapters are peer-reviewed prior to final acceptance or rejection, and manuscripts are prepared in camera-ready format.

As a rule, only original research papers and original review papers are included in the volumes. Verbatim reproductions of previously published papers are not accepted.

ACS BOOKS DEPARTMENT

Contents

Solvation

Biochemical Applications

Indexes

Preface

Quantum mechanical studies of chemical reactions in solution and in enzymes present a great challenge in computational chemistry because of the enormous size and complexity of these systems. Although molecular mechanical force fields have been used in modeling a wide range of problems, they are not adequate for the study of chemical phenomena involving bond forming and breaking, electron transfer, and electronic excited states. Yet, it is not practical in the near future to routinely carry out explicit ab initio molecular dynamics simulations of macromolecules in solution. The combination of quantum mechanics and molecular mechanics, by treating a small part of the system (e.g., the solute) with quantum mechanics and the remainder (e.g., the solvent) with classical force field, provides a tremendously powerful tool for studies of structure and reactivity.

This book was developed from the symposium on Hybrid Quantum Mechanical and Molecular Mechanical Methods. Speakers from Australia, Canada, England, France, Germany, Holland, Japan, Switzerland, and the United States presented their most recent results and engaged in stimulating discussions in this area. The breadth and scope of the symposium are clearly reflected by the diverse topics covered by this volume.

This book is divided into four sections, featuring a broad range of recent methodology developments and novel applications. Chapters 1 through 6 focus on methods for the treatment of the connection between quantum and classical regions. Chapters 7 through 10 discuss novel approaches in ab initio molecular dynamics. Chapters 11 through 14 feature models for the study of solvation effects. Chapters 15 through 19 highlight applications to biochemical problems. In all, this book provides a self-contained introduction to the state of the art in combined quantum mechanical and molecular mechanical (QM/MM) methods.

We take this opportunity to thank all of the participants of our symposium and we look forward to future gatherings on this subject.

This book was developed from a symposium sponsored by the Division of

Computers in Chemistry at the 214th National Meeting of the American Chemical Society, Las Vegas, Nevada, September 7–11, 1997.

Jiali Gao
Department of Chemistry
State University of New York at Buffalo
Buffalo, NY 14260

MARK A. THOMPSON
WRQ Inc.
1500 Dexter Avenue North
Seattle, WA 98109

MODEL DEVELOPMENT

Chapter 1

Quantum Mechanical–Molecular Mechanical Coupled Potentials

Kenneth M. Merz, Jr.

152 Davey Laboratory, Department of Chemistry, The Pennsylvania State University, University Park, PA 16802

The basic formulation of the quantum mechanical/molecular mechanical (QM/MM) methodology is presented. Critical issues related to running a QM/MM calculation are described and the future prospects of this methodology are discussed in detail.

Exceptional progress has been made in elucidating protein structures since the first crystal structure of a protein was determined by X-ray crystallography.(1) With the large numbers of structures now being obtained using NMR and X-ray techniques there is a growing expectation (hope?) that a solution to the protein folding and protein tertiary structure prediction problems may be obtained in the not too distant future.(2) A greater appreciation of how protein dynamics affects protein structure and function has lead to extensive NMR(3) and molecular dynamics (MD) studies(4,5) of protein dynamics and how these motions might affect protein activity.(6) Biochemists have provided detailed insights into protein function by utilizing numerous techniques to characterize the possible intermediates along a catalytic cycle as well as by identifying appropriate catalytic residues within a protein using site-directed mutagenesis as well as chemical modification.(7) Furthermore, detailed kinetic analyses provide insights into the energetic costs associated with the conversion of an enzyme substrate into its corresponding product(s).(7) Even with the tremendous progress that has been made in understanding protein structure, function and dynamics using a plethora of techniques it is safe to say that we still do not have a full appreciation of how protein structure and dynamics affects function. Indeed, in order to effectively accomplish this we need methodologies that can structurally and energetically identify transient intermediates and transition states along a catalytic pathway while simultaneously accounting for the dynamics of the surrounding environment in a realistic manner.

Theoretical chemistry techniques - especially quantum mechanics (QM) based methods can in principle be applied to address these issues at very high accuracy if only the appropriate calculations could be carried out.(8) Thus, QM methods can accurately represent the electronic and structural changes that occur along a reaction coordinate and, thereby, provide insights into the structure, energetics and dynamics of a given reactive process. However, in practice we are a long way from carrying out lengthy MD simulations (or, indeed, any other type of QM based simulations) using an accurate and purely *ab initio* Hamiltonian on a system as large as a typical protein in aqueous solution. This is due to the fact that these methods scale formally as $O(N^3)$ (where N is the number of basis functions) or greater, making them computationally intractable for large systems even on modern supercomputers. Thus, approximations have been made over the years to model a protein and its associated environment.(4,5) In particular, classical representations, (which formally scale computationally as $O(M^2)$ (where M is the number of atoms), which were originally applied to organic molecules(9-11) have found significant utility in the study of the structure and dynamics of biomolecules.(4,5) However, classical models, while particularly well suited to studying protein structure and dynamics, by their very nature are not able to realistically represent activated processes.(4,5,12) Thus, it has been clear for a number of years that a combination of the accuracy and bond-breaking ability of quantum mechanics with the speed and ability to model solvent and protein structure of classical methods (*a.k.a.*, molecular mechanical (MM) methods) would be required to effectively study enzyme structure, function and dynamics in a meaningful and relatively rigorous manner.(12,13) This has lead to the rapid development of the so-called QM/MM approach over the past decade in a number of groups worldwide.(14)

The Quantum Mechanical/Molecular Mechanical Method

Methodology. The basic strategy for this approach was laid out in a seminal paper by Levitt and Warshel(13) where a classical potential was combined with an early semiempirical quantum mechanical method (MINDO/2). Formally the method can be described as follows independent of the QM or MM potential function used: The Hamiltonian used within the QM/MM formulation is taken to be an effective Hamiltonian (H_{eff}) which operates on the wavefunction (ψ) of the system. The wavefunction is dependent on the position of the quantum mechanical nuclei, R_{QM}, the molecular mechanical nuclei, R_M, as well as the positions of the electrons, r.

$$H_{eff}\Psi\left(r, R_{QM}R_M\right) = E\left(R_{QM}R_M\right)\Psi\left(r, R_{QM}R_M\right) \qquad (1)$$

The effective Hamiltonian can be divided into three terms (equation 2), which are generated from the interactions which occur within and between the components of the system (see Scheme 1). The contributions considered here include the completely quantum mechanical, H_{QM} (QM in Scheme 1), the purely molecular mechanical interactions, H_{MM} (MM in Scheme 1) and the interactions between the QM and MM portions of the system, $H_{QM/MM}$ (indicated by the arrow in Scheme 1).

$$H_{eff} = H_{QM} + H_{MM} + H_{QM/MM} \qquad (2)$$

The total energy of the system can likewise be divided into three component parts.

$$E_{eff} = E_{QM} + E_{MM} + E_{QM/MM} \qquad (3)$$

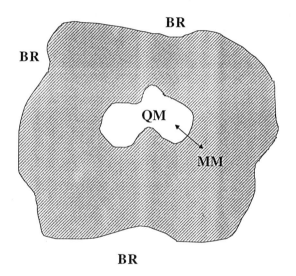

SCHEME 1

These component energies can be obtained by solving either the Roothaan-Hall equations(8) (associated with Hartree-Fock based methodologies) or the Kohn-Sham equations(15) (associated with density functional theory based methodologies) for H_{eff} (equation 1). Another way to express the total energy of the system is as the expectation value of H_{eff}. The purely MM term can be removed from the integral because it is independent of the electronic positions. Thus, the total energy of the system can be given as,

$$E_{eff} = \langle \Psi | H_{QM} + H_{QM/MM} | \Psi \rangle + E_{MM} \tag{4}$$

In equation 4, H_{QM} is the Hamiltonian given by either semiempirical(16), Hartree-Fock(8) or density functional theory(15), while E_{MM} is an energy obtained using a classical force field of some type (see below for more discussion regarding the choices).(9,11,17-19) The key remaining term is $H_{QM/MM}$. This term represents the interaction of the MM atom "cores" with the electron cloud of the QM atoms, as well as the repulsion between the MM and QM atomic cores. Finally, it was found to be necessary to add a Lennard-Jones term to the QM atoms to obtain good interaction energies as well as good geometries for intermolecular interactions.(20) The form of this term is:

$$H_{QM/MM} = -\sum_{iM} \frac{q_M}{|r_{iM}|} + \sum_{A} \frac{q_M Z_A}{|R_{AM}|} + \sum_{AM} \left(\frac{A_{AM}}{R_{AM}^{12}} - \frac{B_{AM}}{R_{AM}^{6}} \right) \tag{5}$$

Where, q_M is the atomic point charge on the MM atom, r_{iM} is the QM electron to MM atom distance, Z_A is the core charge of QM atom A, R_{AM} is the QM atom A to MM atom M distance and A_{AM} and B_{AM} are the Lennard-Jones parameters for QM atom A interacting with MM atom M. The critical term that allows the QM region to "see" the MM environment is the first term in equation 5 where the summation is over all interactions between MM atoms and QM electrons. This represent the core-electron interaction between MM and QM atoms and is incorporated into the QM Hamiltonian explicitly. Thus, the QM electronic structure can respond to its environment through the interaction of its electrons with the surrounding solvent/protein. The last two terms of equation 5 are added on to the total energy once the electronic energy has been determined by a self-consistent field (SCF) procedure and does not affect the electronic distribution of the system directly, but does affect the geometry of the system through the computed gradients and, hence, the resulting electronic energy on subsequent SCF cycles.

The original 1976 paper describing this method by Warshel and Levitt was clearly ahead of its time as this approach was not widely used again until the late 1980's and early 1990's when it was re-examined by several research groups using a number of different quantum mechanical methods, including: semiempirical (S)(20-26), density functional theory (DFT)(27-30) and Hartree-Fock (HF)*ab initio* methods.(31-33) Warshel has also pioneered an alternative strategy to study reactive processes in solution and in enzymes which he has termed the empirical valence bond (EVB) method.(12) This approach has been applied to a wide range of problems(12) and has some advantages and disadvantages relative to QM/MM methods that use molecular orbital or density functional QM methodologies.(12) We will not comment on this approach further, but the interested reader is directed towards Warshel's recent book on this subject.(12) In addition to the various potentials a number of different sampling techniques have been used in conjunction with QM/MM studies including energy minimizations(20,31), Monte Carlo (MC) simulations(21) and molecular dynamics (MD) simulations.(27,32)

Practical Aspects. In this section several of the practical aspects related to running a QM/MM calculations are described.

Choice of Sampling Methods. There are a number of decisions required when designing and carrying out a QM/MM study. The first is the way in which conformational sampling is to be done. In many ways the choice of sampling technique and choice of model Hamiltonian are interdependent. For example, using a purely *ab initio* approach on a large biomolecular system is too computationally expensive to effectively use a sampling technique like MD, but is more compatible with an energy minimization approach. However, it must be kept in mind that the energy minimization techniques, while less computationally demanding since only a few thousand energy and gradient evaluations are involved, are prone to being trapped in local minima that are near the initial starting configuration. Thus, care must be exercised when using energy minimization techniques when one is interested in sampling the very rich conformational space of a protein. In order to get effective sampling of a protein system a Monte Carlo (MC) or MD approach are preferred to minimization, and because of the generality of MD techniques they are more typically used in QM/MM studies. MC can be a very powerful tool to study solution phase systems(14), but they are not as powerful as MD techniques when it comes to the study of enzymatic systems(12) even though Jorgensen has made several recent advances in their application to protein systems.(34) MD and MC methods are much more computationally demanding then energy minimization because by their very nature they require the evaluation of the energy and forces (MD only) of a system many thousand of times. For example, in a protein system to sample 100ps using an MD simulation requires at least the evaluation of the energy and forces 100,000 times with a typical time step of 1fs. Indeed, we have found that in some cases a timestep of 0.5fs or less may be more appropriate in some QM/MM MD studies of protein systems which further increase the number of times the energy and gradient must be evaluated. Early studies using the basic QM/MM approach as applied to enzymes utilized energy minimization techniques(13,35), while more recent studies have begun to utilize MD simulation techniques.(36-38) However, in all cases the QM part of the system utilized a semiempirical method because of the expense associated with a fully *ab initio* or DFT method. Indeed, at the present time it is unlikely that an MD simulation using an *ab initio* or DFT Hamiltonian will be carried out on a protein given the inherent computational expense of this type of calculation. Thus, we are at present limited to MD simulations of proteins using semiempirical models and using energy minimization (or even so-called single point calculations where the energy is evaluated for a single geometric configuration) with *ab initio* or DFT Hamiltonians. This is discussed further below.

Choice of the Model Hamiltonian. The next necessary choice is the selection of the quantum mechanical level to be employed. The choice of MM model will be discussed briefly below, but it should be pointed out the QM portion of a QM/MM calculation so dominates the MM portion, in computational expense and energy, that the critical computational choice which will impact the speed of the algorithm has to do with selecting the QM potential and not the MM model. However it must be noted that recent efforts to incorporate a polarized MM model into a QM/MM calculation increases the expense of the MM portion of the calculation significantly.(24) Semiempirical QM models (*e.g.*, MNDO(39,40), AM1(16) and

PM3(41-43)) formally scale as $O(N^3)$ (where due to semiempirical approximations N is typically no greater than 4 for a "heavy" atom like carbon), while local density approximation(LDA) DFT methods also scale as $O(N^3)$(15), but N is typically much greater than that found in semiempirical theory (*e.g.*, for a typical good quality basis set like 6-31G** 17 contracted basis functions are used to represent a carbon atom(8)). Hartree-Fock *ab initio* methods formally scale $O(N^4)$ and correlated Hartree-Fock approaches can scale as $O(N^5)$ or higher.(8) Thus, in terms of computational expense semiempirical is the cheapest followed by LDA-DFT, Hartree-Fock which is followed by correlated Hartree-Fock. However, the accuracy of these methods follow roughly the reverse order. It is important to note, with the advent of new linear-scaling methodologies(44-46) which introduce divide and conquer algorithms into the Fock matrix generation and solution by including cutoffs for many of the three centered integrals, the expense of all of these QM methods, when applied to very large systems, is dropping dramatically and, hence, the scaling properties are changing, but for our purposes the above analysis will suffice.

While in principle one would prefer to use the most accurate level of theory in all cases this is not practical at this point in time. Thus, using contemporary computers and current numerical approaches it is only particle to carry out "long" (*e.g.*, 100ps or more) MD simulations using semiempirical Hamiltonians on a typical protein active site (*e.g.*, about 20-30 heavy atoms), while it is possible to carry out energy minimization studies using all QM levels except for highly correlated Hartree-Fock approaches.

Choice of the Molecular Mechanical Model. Typically a simple MM potential is used in QM/MM studies.(17-19,47,48) These potentials contain harmonic bonds and angles along with a truncated Fourier series expansion to represent the torsion potential. Non-bonded interactions (*i.e.*, those interactions beyond the 1 and 4 positions along a polymer chain) are typically represented with a Lennard-Jones "6-12" potential function, while electrostatic interactions are handled using atom centered charges. Examples of force fields of this type include OPLS(19), AMBER(18,47,48) and CHARMM.(17) This class of potential function has been quite successful in modeling protein structure and dynamics. The major deficiency in these models is the lack of explicit polarization. This leads to an unbalanced model in QM/MM studies typically because the QM region is polarized while the MM region is not. However, recent work has gone into incorporating polarization effects into the MM region of a QM/MM study (so-called QM/MMpol approach).(24,25) This is a promising approach, but for now will be confined to explicit solvent studies as opposed to studies of enzymatic systems.(24,25)

VDW Parameters for QM Atoms. One of the more difficult decisions which needs to be made when setting up a QM/MM simulation is the proper value for the van der Waals parameters to be used on the QM atoms for their interaction with MM atoms (term 3 of equation 5). While one of the strengths of QM/MM methods is their ability to accurately model a system without extensive reparameterization of either the QM or MM methodology, it has been noted empirically that a 5-10% scaling(20,27,32,49,50)of the VDW parameters normally used within a classical force field for those parameters applied to QM atoms can greatly improve the accuracy of the free energies of solvation calculated using the methodology. Some researchers will also scale the MM core charges seen by the QM partition of the system (terms 1 and 2 of equation 5) to improve calculated radial distribution functions and solvation free energies calculated for a solute. While researchers(14) have worked to establish definitively the VDW parameters to be used in QM/MM calculations through in depth parameterization on test systems it must be noted that the necessary scaling factor is highly method dependent (ie. S, DF, or HF) and even basis set dependent. This is unfortunate as it requires individual parameterization for any particular combination of MM and QM levels of theory or acceptance of the 5-10% scaling factor approximation. One method which neatly circumvents much of this problem is Warshel's use of EVB for the QM portion of his QM/MM calculations, as this requires parameterization to the system being studied independent of any concerns over the proper VDW parameters.(12)

Link Atoms in QM/MM Studies of Enzymes. In solution phase reactions the boundary between the solute (usually QM) and the solvent (usually MM) is very clear and typically it is possible to avoid introducing the QM/MM interface between atoms that are covalently linked. However, in the case of enzymes this is not the case and we have to introduce the concept of link atoms that covalently connect the QM and MM regions used in the representation of the protein. This interfacial region can be quite arbitrary and should be chosen with care. For example, consider modeling a glutamic acid residue within the QM/MM framework (see Scheme 2). The first consideration is where to make the QM and MM "cut" such that it does not adversely affect the electronic structure that we associate with a glutamate ion. If we treat the carboxylate as the formate anion this will greatly alter the pK_a relative to a typical carboxylate anion (4.0 *versus* 4.5!). Thus, while formate is computationally convenient it is electronically the incorrect choice. A better choice is to add an extra carbon atom to generate the acetate anion, which has a better pK_a match with the glutamate anion.

The next step necessary when the QM/MM interface falls at a covalent bond is the introduction of an extra QM atom, for example H_{QM} in Scheme 2. This is necessary to cap the exposed valence at the carbon atom such that a closed shell calculation is done. There are several reasons for doing this not the least of which being that the free radical of the carboxylate anion is not what we are attempting to model. Another import consideration is the fact that closed-shell calculations are from a technical aspect far easier to carry out than are open-shell calculations. This so-called link atom does not see any atoms within the MM region because it is an extra QM atom solely included to satisfy valence considerations that normally would not be present in the real system. The final MM carbon atom of the Glu residue is then attached to the final QM carbon atom by a MM type harmonic bond. This bond keeps the MM and QM regions at the appropriate distance apart. The remaining angles and dihedrals present across the QM/MM boundary are typically represented using MM terms (*i.e.*, an harmonic bond angle and a Fourier torsion potential) and care should be taken in selecting these such that the torsional properties around the bond at the QM/MM boundary are as accurate as possible (*e.g.*, based on the appropriate QM or MM calculations).

The concept of a link atom is an approximation inherent in the basic QM/MM approach that ones needs to be aware of when treating protein systems or any system that requires the use of a covalent QM/MM boundary region. However, through careful selection of the location of this region gross errors can be avoided. New and better ways need to be formulated to treat the link atom region and, indeed, some work has proceeded along these lines. In particular, the work of Rivail and co-workers(26,33) using localized orbitals to represent the QM/MM boundary region looks promising, but needs to be explored further to see how it works in a variety of situations.

Suggested QM/MM Terminology. Unfortunately, at this point there is no universally accepted nomenclature that can be used to describe QM/MM methods in a concise manner. Indeed, there are several competing nomenclatures present in the literature which tends to confuse the uninitiated about exactly the differences between the various methods. Herein we will use the nomenclature outlined in Table 1. This table also contains some approximate information regarding the accuracy of the various methods and also summarizes other abbreviations that we use consistently throughout this review.

Weaknesses of the Basic QM/MM Approach and Future Prospects

The QM/MM approach clearly shows great promise, but there a few fundamental problems associated with this method that will need to be addressed in the coming years. These problems, as we see them, can be summarized as follows:

(1) Issues relating to the boundary region: What effect does the QM/MM boundary have on the calculations? Is the link atom approach effective or should a localized orbital approach be utilized?

The boundary region and link atom issues are easily the most controversial in any application of a QM/MM potential as it lies at both the heart of the effectiveness and problems of such potentials. The question really comes down to the acceptability of the boundary region (QM/MM approximation) or link atoms for the system being investigated. A seemingly obvious but often over-looked point is that QM/MM potentials are never going to be exactly equivalent to the analogous full QM calculation. It is admirable to attempt to make QM/MM potentials mimic the fully *ab intio* and experimental results as closely as possible but the question which is more important is whether the potential is good enough to gain insight into relevant biological and chemical questions. We have found the answer to this question many times to be, yes, as the inclusion of environmental effects leads to a more accurate representation than currently feasible with any other method.

(2) Issues relating to sample sizes: How does a finite QM "sample" size affect the calculations? Will results dramatically change as more and more QM residues are introduced?

The issue of sample size is luckily one which can be addressed more and more easily with the ever increasing speeds of currently available computers. While it might be assumed that the largest possible region should be defined as QM, this is not always the case. Undoubtedly the more QM atoms which can be placed around a specific area of interest (be it a bond breaking/formation process or the bond length between two specific atoms) the less any spurious effects from the QM/MM boundary region are likely to effect the results of the calculation. Electronic polarization and charge fluctuation is also more accurately taken into account with a larger QM region. However, the benefits obtained from a larger number of QM atoms have to be weighed against the problems created by their inclusion. An obvious disadvantage to having a large number of QM atoms is the increased computational expense incurred at every energy and gradient determination. The number of conformations which can be considered is greatly reduced, and depending on the size of the accessible region of phase space the accuracy of any properties determined can be greatly affected. A good example of this problem occurred in studies of ribonuclease A we conducted where the two catalytic histidines and a portion of the RNA to be cleaved were used as the QM region. While it might be desirable for the model to include several additional active site residues this would make it nearly impossible to adequately sample the extremely flat free energy surface present during the formation of the penta-coordinate intermediate. This is in contrast to the studies of carbonic anhydrase presented in

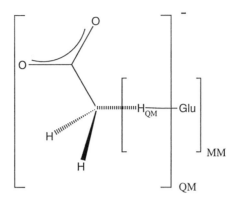

SCHEME 2

Table 1. Summary of QM/MM Methods: Definition of Abbreviations and the Approximate Accuracy.

Abbreviation	Meaning	Accuracy[1]
QM/MM	Quantum Mechanical/Molecular Mechanical. Term used to broadly define this class of methods.	N/A
MM	Molecular Mechanical. Term used to broadly define force fields based on classical mechanics. Only three widely used examples are given below.	N/A
AMBER	Assisted Model Building and Energy Refinement. Optimized for proteins and DNA.	Depends on system, but generally good.
OPLS[2]	Optimized Potentials for Liquid Simulations. Optimized for proteins and various liquids.	Depends on system, but generally good.
CHARMm	Chemistry at Harvard Molecular Mechanics. Optimized for proteins and DNA.	Depends on system, but generally good.
S/MM	Semiempirical/Molecular Mechanical. Term used to generically define the class of semiempirical QM/MM methods.	N/A
MNDO/MM	Modified Neglect of Differential Overlap/Molecular Mechanical. MNDO is a semiempirical Hamiltonian. Hydrogen bonding poorly handled, therefore, not generally useful.	Depends on system ±3-±7 kcal/mol
AM1/MM	Austin Model 1/Molecular Mechanical. MNDO is a semiempirical Hamiltonian. Hydrogen bonding treated, but geometric details poorly reproduced.	Depends on system ±3-±6 kcal/mol
PM3/MM	Parametric Model 3/Molecular Mechanical. PM3 is a semiempirical Hamiltonian. Best method for hydrogen bonding, therefore, this is the most versatile method.	Depends on system ±3-±5 kcal/mol
DFT/MM	Density Functional Theory/Molecular Mechanical. Term used to generically define the class of density functional theory QM/MM methods.	N/A
LDA/Basis/MM	Local Density Approximation/Basis/Molecular Mechanical.	Depends on system ±3-±6 kcal/mol
GC/Basis/MM	Gradient Corrected/Basis/Molecular Mechanical.	Depends on system ±2-±4 kcal/mol
HF/MM	Hartree Fock/Molecular Mechanical. Term used to generically define the class of Hartree-Fock QM/MM methods.	N/A
HF/Basis/MM	Hartree Fock/Basis/Molecular Mechanical.	Depends on system ±2-±5 kcal/mol
MPX/Basis/MM	Møller PlessetX(X=2 or 4 typically)/Basis/Molecular Mechanical.	Depends on system ±1-±3 kcal/mol

1) These numbers qualitatively define the accuracy of the methods.
2) The OPLS term also includes the TIP3P and TIP4P water models of Jorgensen.

this review where there is very little conformational flexibility around the active site and the reaction mechanism is dependent on proton transfers within a hydrogen bonding network. A less obvious difficulty encountered with increased numbers of QM atoms within QM/MM MD simulations is accurately controlling the temperature and dynamics of the QM atoms. The harmonic nature of classical force fields is much more tolerant of velocity "spikes" on particular atoms caused by bad contacts. QM potentials are not nearly so tolerant and velocity spikes can easily result in bond cleavage, thus requiring a robust temperature coupling algorithm along with a short time step.

(3) Issues relating to representational details: In QM/MM calculations an unbalanced model is employed. Thus, the QM region is polarized, *etc.*, while the MM region uses a fixed charge model. How does this effect calculated results using a QM/MM approach in general?

Two methods have been used to avoid the problem of having an unbalanced model with respect to polarization in the standard implementation of QM/MM methodology. The first was to explicitly include polarization by using a polarizable MM potential.(24,25) While this method, for example, was able to successfully reproduce spectral shifts, it was also extremely computational expensive.(24) The second method which has been used to address polarization issues approached the question from a different point of view. Instead of worrying about the polarization of the MM atoms, the concern was for the changing polarization of a solute during an MM simulation. To account for this a QM/MM potential was used to recalculate the charges on the solute at every point along a defined reaction pathway.(51) It is our opinion however that in systems where polarization is expected to have a significant effect QM/MM potentials are probably not adequate for the simulation unless the QM region is extended significantly or an explicitly polarizable MM potential is used. A better choice in such situations might be the use of a wholly QM representation using a divide and conquer algorithm.(45,52)

(4) Issues relating to the cutoff used: How does the use of a cutoff for QM/MM interactions effect the calculated results? Should all QM---MM long-range interactions be included?

As with classical force fields there are many opinions regarding the issue of long range interactions and cutoffs in QM/MM calculations. Unfortunately, the cutoff length used tends to have a much larger affect on the electronic energy of the QM region than is normally encountered from the additional energy associated with long range electrostatic effects in classical force fields. As an example of this the electronic energy using a PM3 potential for CA system discussed in section 3 of this review using a 10Å cutoff is ~-50kcal/mole. Where as when a 20Å cutoff is used the electronic energy can drop to ~-500kcal/mole. Clearly we are only interested in relative energies but with effects of this magnitude problems can be encountered if residues move into or out of the cutoff region during the simulation. In practice we have found that considering the QM region as one residue for the purpose of calculating the non-bonded interaction list is important, in this way all of the QM atoms see an equivalent environment. Also helpful to the stability of the simulation is the use of a dual cutoff with the QM atoms having a much larger cutoff than the MM atoms. In this way the computational expense of using a very large cutoff is partially mitigated. Even using such precautions though certain protein systems can remain extremely sensitive to the cutoff approximation.

(5) Issues relating to sampling: Can we use energy minimization techniques effectively to map out reaction profiles for enzymes or must we always resort to

MD simulations? How long should MD studies be and what is the most appropriate timestep to use?

The problem of adequate conformational sampling is not one restricted to QM/MM methods but rather becomes an issue in any simulation of a protein or solvated system. The potential energy surfaces of such systems are notoriously complex with many local minima and extremely flat regions of the potential energy surface which are extremely difficult to sample. QM/MM potentials add unique difficulties to the sampling question which have not been adequately addressed to this point. One question which is still a point of open debate is whether minimization is adequate to map out the reactive mechanism of an enzyme. While some researchers have used minimization effectively to examine active site mechanisms(35) others strongly insist that the methodology is inadequate to describe the processes when the system contains kinetic energy.(12) While computationally attractive due to the relatively few conformations which need to be considered minimization is statistically difficult to justify in a protein system. To adequately sample a phase space as complicated as that of a protein active site it is generally agreed that a simulation of many hundreds of pico-seconds may be necessary and even then depending on the starting conditions of the simulation, and the sampling technique used, important conformations may not be sampled. The expense of doing QM/MM simulations even when using a semiempirical Hamiltonian can be quite daunting. Added to this problem is the necessity of using extremely small time steps (on the order of >0.5 fs) due to the inclusion of hydrogen motions and the sensitivity of the QM portion of the potential to temperature spikes. Currently it only seems feasible to use semiempirical or EVB methods for the QM potential when large QM partitions are being used or long simulations are necessary to sample conformational space. Some researchers have attempted to avoid this problem by using a multiple time step approach in which the QM portion of the system is held fixed for several time steps in which the MM atoms are allowed to move between each MD step of the QM partition.(29) While this allows for greater equilibration/sampling of specific conformations of QM atoms it would be undesireable to use the energies from this approach for the calculation of, for example, reaction free energies using free energy perturbation methodologies. Another method proposed to allow for the use of higher level QM potential functions in QM/MM studies is to sample conformational space using MD and a semiempirical Hamiltonian as a first step and then calibrate the semiempirical calculations through single point calculations and minimizations using higher level QM theory on those conformations deemed to be chemically significant.

(6) Issues relating to the potential functions used: Can we trust semiempirical calculations and what level of *ab initio* or DFT calculation must be carried out to obtain reliable results. Should specific parameterizations of semiempirical Hamiltonians be used instead of existing semiempirical models, *etc.*. What is the best MM model to use?

The issue of which potential function is appropriate to a particular problem is highly dependent on the chemical system being considered. Recent books by Hehre can provide a helpful guide in making this always difficult decision.(8) Semiempirical potentials have proven effective for use with the organic reactions they were originally parameterized for, but less accurate in their application to protein systems (*e.g.*, methods such as AM1 and PM3 contain a classical term specifically designed to improve their ability to model peptide bonds). An example of the inaccuracies encountered with semiempirical representation can be seen in model system calculations done by us for triose phosphate isomerase.(53) For this system PM3 is seen to give energies tens of kilocalories different from the HF values for species along the reaction pathway. In particular semiempirical methods can be inappropriate for examination of transition states and strained intermediates because these types of structures where not considered in the parameterizations of many of these methods. We refer the reader back to the answer of the last question we posed in this section of the review (on conformational sampling) for a possible use for semiempirical methods for protein simulations. The question of which classical force field is more appropriate is more difficult to address and luckily probably less important.

These issues and many more regarding this method will be addressed in the coming years. Even with these potential deficiencies the QM/MM approach can now be added to the theoretical chemists bag of tricks to help him or her to solve chemical reactivity problems in fields ranging from organic chemistry to biochemistry.

Summary and Conclusions

In this brief review we describe the basic QM/MM approach and how it can be applied effectively to study enzymatic processes. We also address the strengths and weaknesses of this approach and point to future directions to be followed in order to fully appreciate the power of this methodology.

Acknowledgments

We would like to thank the NIH for supporting this research through Grant GM44974. We also thank the Pittsburgh Supercomputer Center and the Cornell Theory Center for generous allocations of supercomputer time.

References

1. Blundell, T. L.; Johnson, L. N. *Protein Crystallography*; Academic Press: New York, 1976.
2. *The Protein Folding Problem and Tertiary Structure Prediction*; Legrand, S.; Merz, K. M., Jr., Eds.; Birkhaüser: Boston, Massachusetts, 1994.
3. Wüthrich, K. *NMR of Proteins and Nucleic Acids*; John Wiley & Sons: New York, 1986.
4. McCammon, J. A.; Harvey, S. C. *Dynamics of Proteins and Nucleic Acids*; Cambridge University Press: New York, 1987.
5. Brooks, C. L., III; Karplus, M.; Pettitt, B. M. *Proteins: A Theoretical Perspective of Dynamics, Structure, and Thermodynamics*; John Wiley and Sons: New York, 1988; Vol. LXXI.
6. Xiang, S.; Short, S. A.; Wolfenden, R.; Carter, C. W., Jr. *Biochemistry* **1996**, *35*, 1335-1341.
7. Fersht, A. R. *Enzyme Structure and Mechanism*; W. H. Freeman and Co.: New York, 1985.
8. Hehre, W. J.; Radom, L.; Schleyer, P. v. R.; Pople, J. A. *Ab Initio Molecular Orbital Theory*; John Wiley & Sons: New York, 1986.
9. Burkhert, U.; Allinger, N. L. *Molecular Mechanics*; American Chemical Society: Washington, D.C., 1982.
10. Allinger, N. L.; Yuh, Y. H.; Lii, J.-H. *J. Am. Chem. Soc.* **1989**, *111*, 8551-8566.

14

11. Bowen, J. P.; Allinger, N. L. *Reviews in Computational Chemistry* **1991**, *2*, 81-98.
12. Warshel, A. *Computer Modelling of Chemical Reactions in Enzymes and Solutions*; John Wiley & Sons, Inc.: New York, 1991.
13. Warshel, A.; Levitt, M. *J. Mol. Biol.* **1976**, *103*, 227-249.
14. Gao, J. In *Reviews in Computational Chemistry*; Lipkowitz, K. B., Boyd, D. B., Eds.; VCH Press: New York, 1995; Vol. 6.
15. Parr, R. G. In *Annual Review of Physical Chemistry*; Rabinovitch, B. S., Schurr, J. M., Strauss, H. L., Eds.; Annual Reviews Inc.: Palo Alto, 1983; pp 631.
16. Dewar, M. J. S.; Zoebisch, E. G.; Healy, E. F.; Stewart, J. J. P. *J. Am. Chem. Soc.* **1985**, *107*, 3902-3909.
17. Brooks, B. R.; Bruccoleri, R. E.; Olafson, B. D.; States, D. J.; Swaminathan, S.; Karplus, M. *J. Comput. Chem.* **1983**, *4*, 187-217.
18. Cornell, W. D.; Cieplak, P.; Bayly, C. I.; Gould, I. R.; Merz, K. M., Jr.; Ferguson, D. M.; Spellmeyer, D. C.; Fox, T.; Caldwell, J. W.; Kollman, P. A. *J. Am. Chem. Soc.* **1995**, *117*, 5179-5197.
19. Jorgensen, W. L.; Tirado-Rives, J. *J. Am. Chem. Soc.* **1988**, *110*, 1657-1666.
20. Field, M. J.; Bash, P. A.; Karplus, M. *J. Comp. Chem.* **1990**, *11*, 700-733.
21. Gao, J.; Xio, X. *Science* **1992**, *258*, 631-635.
22. Luzhkov, V.; Warshel, A. *J. Comput. Chem.* **1992**, *13*, 199-213.
23. Thompson, M. A. *J. Am. Chem. Soc.* **1995**, *117*, 11341-11344.
24. Thompson, M. A. *J. Phys. Chem.* **1996**, *100*, 14492-14507.
25. Bakowies, D.; Thiel, W. *J. Phys. Chem.* **1996**, *100*, 10580-10594.
26. Monard, G.; Loos, M.; Théry, V.; Baka, K.; Rivail, J.-L. *Int. J. Quant. Chem.* **1996**, *58*, 153-159.
27. Stanton, R. V.; Hartsough, D. S.; Merz, K. M., Jr. *J. Comput. Chem.* **1994**, *16*, 113-128.
28. Wesolowski, T. A.; Warshel, A. *J. Phys. Chem.* **1993**, *97*, 8050-8053.
29. Wei, D.; Salahub, D. R. *J. Chem. Phys.* **1994**, *101*, 7633-7642.
30. Tuñón, I.; Martins-Costa, M. T. C.; Milloy, C.; Ruiz-López, M. F.; Rivail, J.-L. *J. Comput. Chem.* **1996**, *17*, 19-29.
31. Singh, U. C.; Kollman, P. A. *J. Comput. Chem.* **1986**, *7*, 718-730.
32. Stanton, R. V.; Little, L. R.; Merz, K. M., Jr. *J. Phys. Chem.* **1995**, *99*, 17344-17348.
33. Assfeld, X.; Rivail, J.-L. *Chem. Phys. Lett.* **1996**, *263*, 100-106.
34. Jorgensen, W. L.; Duffy, E. M.; Severance, D. L.; Blake, J. F.; Jones-Hertzog, D. K.; Lamb, M. L.; Tirado-Rives, J. In *Biomolecular Structure and Dynamics: Recent Experimental and Theoretical Advances*; Vergoten, G., Ed.; to be published in NATO ASI Ser., 1997.
35. Bash, P. A.; Field, M. J.; Davenport, R. C.; Petsko, G. A.; Ringe, D.; Karplus, M. *Biochemistry* **1991**, *30*, 5826-5832.
36. Merz, K. M., Jr.; Banci, L. *J. Phys. Chem.* **1996**, *100*, 17414-17420.
37. Merz, K. M., Jr.; Banci, L. *J. Am. Chem. Soc.* **1997**, *119*, 863-871.
38. Hartsough, D. S.; Merz, K. M., Jr. *J. Phys. Chem.* **1995**, *99*, 11266-11275.
39. Dewar, M. J. S.; Thiel, W. *J. Am. Chem. Soc.* **1977**, *99*, 4899-4907.
40. Dewar, M. J. S.; Thiel, W. *J. Am. Chem. Soc.* **1977**, *99*, 4907-4917.
41. Stewart, J. J. P. *J. Comp. Chem.* **1989**, *10*, 209-220.
42. Stewart, J. J. P. *J. Comp. Chem.* **1989**, *10*, 221-264.
43. Stewart, J. J. P. *J. Comp. Chem.* **1991**, *12*, 320-341.
44. Yang, W.; Lee, T.-S. *J. Chem. Phys.* **1995**, *103*, 5674-5678.
45. Dixon, S. L.; Merz, K. M., Jr. *J. Chem. Phys.* **1996**, *104*, 6643-6649.

46. White, C. A.; Johnson, B. G.; Gill, P. M. W.; Head-Gordon, M. *Chem. Phys. Lett.* **1994**, *230*, 8-16.
47. Weiner, S. J.; Kollman, P. A.; Nguyen, D. T.; Case, D. A. *J. Comput. Chem.* **1986**, *7*, 230-252.
48. Weiner, S. J.; Kollman, P. A.; Case, D. A.; Singh, U. C.; Ghio, C.; Alagona, G.; Profeta, S.; Weiner, P. *J. Am. Chem. Soc.* **1984**, *106*, 765-784.
49. Gao, J. *J. Phys. Chem.* **1992**, *96*, 537-540.
50. Stanton, R. V.; Hartsough, D. S.; Merz, K. M., Jr. *J. Phys. Chem.* **1993**, *97*, 11868-11870.
51. Marrone, T. J.; Hartsough, D. S.; Merz, K. M., Jr. *J. Phys. Chem.* **1994**, *98*, 1341-1343.
52. Dixon, S. L.; Merz, K. M., Jr. *J. Chem. Phys.* **1997**, *107*, 879-893.
53. Alagona, G.; Ghio, C.; Kollman, P. A. *J. Am. Chem. Soc.* **1995**, *117*, 98 55-9862.

Chapter 2

Quantum Mechanical–Molecular Mechanical Approaches for Studying Chemical Reactions in Proteins and Solution

Jörg Bentzien, Jan Florián, Timothy M. Glennon, and Arieh Warshel

Department of Chemistry, University of Southern California, Los Angeles, CA 90089-1062

In this paper we address the challenge of developing reliable approaches for quantum mechanical calculations of chemical processes in proteins and solution. The requirements for obtaining meaningful results are discussed emphasizing the importance to include the entire system and the problems associated with rigorous treatments of small parts of the system are presented. With this philosophy in mind we consider several hybrid QM/MM approaches for modeling reactions in proteins and solutions. Special emphasis is given to two approaches recently developed in our group. The first is an ab initio quantum mechanical/Langevin dipole (QM(ai))/LD) approach that allows to obtain reliable solvation free energies and to gain insight into relevant biochemical reactions such as amide and phosphate hydrolysis. The second method is a new ab initio QM/MM (QM(ai)/MM) approach for calculating chemical reactions in enzymes. This approach uses an empirical valence bond (EVB) surface as a reference potential for evaluating ab initio free energies. The potential and limitations of this approach are examined by analyzing the nucleophilic attack step in the catalytic reaction of subtilisin. This study, which is probably the first consistent evaluation of ab initio activation free energies of an enzymatic reaction, provides a major support for the use of the EVB approach in studies of enzymatic reactions. That is, it is shown that the EVB electrostatic free energy is very similar to the corresponding ab initio free energy, yet the EVB calculations are much simpler and faster.

Reliable studies of enzymatic reactions involve major conceptual and technical challenges. Here, one deals with a very complex system where simple rigorous approaches are not always the most effective strategies. For example, applying high level ab initio calculations to the substrate in the gas phase is not going to tell us very much about the effect of the enzyme on the substrate. In our opinion it is essential to start by modeling the complete system consisting of the substrate (the solute), the enzyme and the surrounding solvent in order to obtain meaningful insight into enzymatic reactions. This often forces one to represent the entire system in an approximate way rather than to examine a smaller part in the most rigorous way. Once the complete solute-protein-solvent model starts to reproduce observable results (e.g. pK_a values and asymptotic energies in solution), it is useful to begin with a gradual improvement of the components of the model, such as improving the level of theory describing the solute. We emphasize this point since the natural tendency in science is to move in an incremental way building part of the puzzle step by step. However, such an approach is almost guaranteed to lead to blind spots in understanding what the important factors are in enzyme catalysis. An excellent example is the common assumption that enzymes work like gas phase [1,2] or low dielectric solvents [3]. The fundamental problem with this hypothesis can easily be deduced by considering the complete system [4].

With the philosophy that it is essential to represent the entire system one is forced to design effective and practical strategies as is now starting to be widely recognized. Here, we consider such hybrid quantum mechanical/molecular mechanical (QM/MM) approaches for modeling reactions in solution and enzymes [5,6]. We will emphasize the importance of reliable estimates of the energetics of chemical reactions in solution as a prerequisite for studies of enzymatic reactions. Thus in the first part of this work we will focus on a recent implementation of the quantum mechanical/Langevin dipole (QM/LD) model that allows one to obtain reliable ab initio results for chemical reactions in solution. The second part will consider different QM/MM approaches for modeling enzymatic reactions. It will be pointed out that at present only approaches that are calibrated on the relevant energetics in solution, for example the empirical valence bond (EVB) approach, provide reliable results for enzymatic reactions. A new approach for calculating free energies of enzymatic reactions by hybrid ab initio/MM potential surfaces will be outlined. The potential and limitation of this approach will be illustrated.

Simulation of Chemical Reactions in Solution

General Background. Enzyme catalysis is best defined by considering the given reaction in the enzyme relative to the corresponding reaction in solution. Typically enzymes provide rate accelerations of 10^7 or more over the rates in aqueous solution [7,8]. By comparing reaction profiles in aqueous solvent to those in the protein we can identify the essential elements of the protein responsible for the rate acceleration. At the same time, the similarity of protein and aqueous solvent energetics and charge distribution (as opposed to the corresponding values in the gas phase, which are qualitatively different) insures that our comparison is meaningful. Furthermore, it is frequently easier to obtain experimental information about solution reactions than about gas phase processes.

In the following section we consider briefly different methods for modeling chemical reactions in solution. At present the most effective methods for modeling chemical reactions in solution are the so-called hybrid quantum-classical methods, which include QM/MM approaches and simplified solvent models such as quantum mechanical/continuum dielectric (QM/CD) and quantum mechanical/Langevin dipole (QM/LD) techniques. The basic idea of these models is to represent the reacting region quantum mechanically and the surroundings classically. Table I gives a concise overview of early contributions in the field, and below we consider the main developments.

TABLE I. Early Hybrid QM/Classical Studies of Reactions in Solution[a]

Reaction	Solvent	Solute	Averaging	Ref.
PT/GA	dipolar	MINDO	Average over EM	[9-11]
PT	reaction field	INDO	—	[12]
PT/GA	all-atom	EVB	FEP/MD	[13,14]
S_N2	all-atom	ab initio[b]	FEP/MC	[15]
NA	all-atom	ab initio	EM	[16]
S_N2	all-atom	ab initio	EM	[17]
S_N2	dipolar	ab initio	EM	[18]
S_N2	all-atom	AM1/MNDO	FEP/MD[c]	[19]
S_N2	all-atom	EVB	FEP/MD	[20]
S_N2	reaction field	AM1/MNDO	—	[21]

[a]PT, GA, NA denote proton transfer, general acid catalysis and nucleophilic attack, respectively. FEP denotes free energy perturbation. EM, MD, MC denote energy minimization molecular dynamics and Monte Carlo, respectively. [b]No consistent solute-solvent coupling. [c]Used gas phase structures in the mapping procedure.

The earliest treatment of actual chemical reactions by hybrid models dates back to the work of Warshel and Levitt [6], who introduced QM/MM approaches to reactions in condensed phases; similar ideas were implemented in subsequent studies [10,11]. The early use of hybrid QM/CD models for studies of chemical reactions involved a non-quantitative reaction field model (e.g. ref. [12]). Quantitative QM/CD results started to appear with the emergence of discretized continuum models [22-24] and the recognition that the van der Waals or effective Born radius should be parameterized to reproduce observed solvation energies [11]. Recent years saw a renaissance in hybrid QM/MM approaches [19,25-39], with attempts to implement such approaches with more rigorous microscopic solvation models and to use them in free energy perturbation (FEP) calculations. The earliest attempts to progress in this direction were done using combined Valence Bond/MM methods within the empirical valence bond (EVB) formulations, which have been reviewed extensively elsewhere [40,41]. At present, the implementation of molecular orbital (MO) approaches in FEP calculations of reaction profiles in solution is problematic (see discussion in ref. [42]) and only sophisticated techniques of controlling the progress of the reaction from reactants to products, such as using the EVB surface as a mapping potential, allow one to rigorously determine activation free energies. Note in this respect that the interesting work of Bash et al. [19] simulated an S_N2 reaction using a mapping potential based on the ground and transition state structures in the gas phase. Such an approach does not provide a general method for FEP studies since it is not expected to work in heterolytic bond cleavage reactions when the gas phase transition state structure is drastically different from the corresponding solution structures.

Density functional theory (DFT) treatments of chemical reactions in solution are now starting to emerge. These include combined DFT/MM approaches [43] and attempts to treat the entire solvent quantum mechanically, either with the direct approach of Parrinello and coworkers [44], who try to treat the entire solute-solvent system on the same level of DFT, or our frozen DFT (FDFT) approaches [45-47] or constrained DFT (CDFT)[48] where the density of the solvent molecules is constrained, achieving significant savings in computer time. The DFT based approaches were reviewed recently [49] and will not be considered here any further.

Simplified Solvent Models offer an Effective Alternative to QM/MM (all atom) Approaches. After this brief review it is useful to establish the applicability and potential of current QM/MM approaches. The accuracy of our FEP approach and our Langevin dipole (LD) approach has been discussed previously [50,51]. We have also developed more rigorous and more expensive approaches [42,52,53]. Here, we illustrate the current applicability of these methods in terms of the trade-off between the computational efficiency versus stability and quantitative accuracy of the obtained results. First we introduce the recent version of what is our simplest and, in fact, oldest strategy of combined QM/Langevin dipole (LD) solvent models [6,9,11,31]. Later we compare the hydration free energies of nucleic acid bases obtained by using different computational methods. Here, as well as in a general case of solvation energetics for neutral and single charged solutes [54], the performance of various approaches is comparable and depends to a large extent on the effort invested in their parameterization. Finally, more challenging issues involving the quantitative evaluation of the hydration free energies of multiple charged phosphates, and contributions of solute polarization to the activation barriers in model phosphate and peptide bond hydrolysis reactions are presented.

In its current version designed for aqueous solution [54,55], the Langevin dipole (LD) solvation model approximates the solute charge distribution by electrostatic potential-derived atomic charges that are evaluated at the Hartree-Fock (HF) level using the 6-31G* basis set. (More rigorous approaches are currently being implemented in the Gaussian 94 and QChem quantum chemical programs.) We denote this method QM(ai)/LD to indicate that the Langevin dipoles solvate the charges which are obtained from ab initio quantum mechanics. These charges are used to evaluate the electrostatic part of the solvation free energy, ΔG_{ES}, as well as the potential-dependent hydrophobic surface of the solute, ΔG_{phob}, a van der Waals term (ΔG_{vdW}), a term approximating the polarization (relaxation) energy of the solute, ΔG_{relax}, and the Born correction, ΔG_{Born}, that accounts for the contribution of the bulk solvent lying outside the region filled with the Langevin dipoles. Thus, the total solvation free energy is given by

$$\Delta G_{solv} = \Delta G_{ES} + \Delta G_{phob} + \Delta G_{vdW} + \Delta G_{relax} + \Delta G_{Born}. \qquad (1)$$

In order to obtain a quantitative agreement with experimental solvation free energies for both neutral and ionic solutes, ab initio calculated parameters must be combined with a set of empirically adjusted van der Waals (vdW) radii for the solute atoms. In general, these vdW radii need to be hybridization dependent. A standard set of vdW and other parameters for our QM(ai)/LD method have been recently developed for the training set encompassing 44 neutral and 39 ionic solutes that contained C, O, N, P, S, F and Cl atoms[54]. In addition, LD results for this set of molecules were compared with the results obtained with the polarized continuum model (PCM) of Tomasi and coworkers [22,56], taking into account that the PCM model used by us employed hybridization independent vdW parameters and did not include the hydrophobic (cavitation) and vdW contributions to the hydration free energy. The LD and PCM solvation models have shown comparable accuracy and generally similar trends for series of structurally related solutes, as long as we compared neutral and mono-ionic solutes [54].

The performance of a wider set of computational methodologies can be compared for nucleic acid bases (Table II). Note, that only relative hydration free energies are given in this table, since this approach enables us to directly compare results obtained for methylated and unmethylated bases and to include studies that did not evaluate absolute energies. Since only the order of hydration free energies of nucleobases (guanine > cytosine > thymine > adenine) [57] is known from solution experiments, theoretical methods have been the only source of quantitative data for this biologically important property. The first quantitative evaluation of the hydration energy of cytosine was done by Warshel [11] as part of a study of the stacking interactions in the cytosine dimer in aqueous solution. Interestingly, the predicted solvation energy of -16.0 kcal/mol is in excellent agreement with the results of the current most sophisticated computational methodologies. (Note that the data in the last row of Table VIII of ref. [11] correspond to the two infinitely separated cytosine monomers.) The classical free energy perturbation (FEP) calculations of Kollman and coworkers [25,58] that used the AMBER force field resulted in the incorrect order of solvation free energies. Later studies that employed similar methodologies [59,60] have shown that this disagreement was caused primarily by insufficient sampling of the configuration space in the molecular dynamics calculation. Nevertheless, significantly different solvation free energy values of -12.0 [60] and -16.3 kcal/mol [61] were obtained for adenine even when comparable simulation times and identical force fields were used. This point illustrates that despite the large demands on computer resources the classical FEP calculations are still rather unstable. On the other hand, the computational efficiency and stability of results are the main virtues of modern QM(ai)/LD and QM(ai)/PCM methods, which use realistic solute-solvent boundaries. Our LD and PCM calculations resulted not only in the correct ordering of hydration free energies, but the results obtained by the iterative LD calculation fell within 1 kcal/mol from the most accurate FEP calculations by Miller and Kollman [60] and by Elcock and Richards [59] for all studied bases. In addition, the combined QM(ai)/LD approach provided the equilibrium constant for the keto-iminol tautomerization of protonated cytosine in aqueous solution that agreed well with the available experimental estimates [62].

Table II. Relative[a] Hydration Free Energy (kcal/mol) of
Nucleic Acid Bases and Base Pairs

Method	cytosine	guanine	thymine	uracil	AT[b]	GC[b]
QM(ai)/LD [c]	-7.2	-9.6	-1.8	-2.8	-8.6	-14.3
QM(ai)/NLD [d]	-4.7	-8.9	-2.0	-3.0	-8.0	-12.9
QM(ai)PCM [e]	-4.6	-5.5	-2.4	-2.4	-7.1	-9.1
TI/AMBER [f,g]	-6.4	-10.4	-0.4	-2.0		
FEP/OPLS [f,h]	-8.5	-10.1	-1.5	-1.2		
FEP/QM(AM1)/MM [i]	-11.2	-8.4	-3.4	-4.8		
FEP/AMBER [j]	-0.1	-7.0	5.1		-5.8	-13.8
QM(AM1)/PCM [k]	-3.1	-6.8	2.7			
QM(ai)/SCRF [f,l]	-6.5	-9.6	-2.1	-3.4		
FDPB [m]	-6.4	-9.3	0.0			
AM1-SM2 [f,n]	2.2	-3.4	7.6	6.1		

[a]The hydration free energy of adenine was taken as a reference point. [b]Adenine-thymine and guanine-cytosine Watson-Crick base pairs. The HF/6-31G** geometries of base pairs were used [63]. [c]Iterative LD calculation using ChemSol 1.0 [55] and a standard parameterization [54]. The calculated solvation free energy for adenine (ΔG_{solv} (Ade)) is -10.8 kcal/mol. [d]Noniterative LD calculation using ChemSol 1.0 [55] and a standard parameterization [54]. ΔG_{solv} (Ade) = -9.5 kcal/mol. [e]This work, HF/6-31G* calculation using Gaussian 94 with the default van der Waals radii scaled by 1.2. ΔG_{solv} (Ade) = -10.5 kcal/mol. [f]Data for 9-Me adenine, 9-Me guanine, 1-Me cytosine, 1-Me thymine and 1-Me uracil. [g]Thermodynamic integration using the AMBER force field, 400-1000 ps simulation length [60]. ΔG_{solv} (Ade) = -12.0 kcal/mol. [h]Free energy perturbation method, OPLS force field, 500 ps simulation length [59]. [i]Free energy perturbation method, AMBER force field, 60 ps simulation length [58]. ΔG_{solv} (Ade) = -12.6 kcal/mol. [j]Reference [64], ΔG_{solv} (Ade) = -5.1 kcal/mol. [k]Reference [65], ΔG_{solv} (Ade) =

-11.3 kcal/mol. [l]Self-consistent reaction field (SCRF) calculation at the HF/6-31G(d,p) level using the eliptic solute cavity and the multipole expansion (l=7) of the solute charge distribution [66], ΔG_{solv} (Ade) = -6.5 kcal/mol. [m]Finite-difference Poisson-Boltzmann calculations [67], ΔG_{solv} (Ade) = -10.4 kcal/mol. [n]Reference [68], ΔG_{solv} (Ade) = -20.1 kcal/mol.

The solute polarization contribution, ΔG_{relax}, to the total solvation free energy represents a particularly valuable outcome of QM/MM studies since this quantity is usually neglected in classical MM force fields [69]. Therefore we compared in Table III the magnitudes of the electrostatic contributions, ΔG_{ES}, (calculated from the gas phase solute charge distribution) and the solute polarization contributions, ΔG_{relax}, to the hydration free energies of nucleobases and base-pairs. Here, the FEP/QM(AM1)/MM calculations of Gao [70] resulted in ΔG_{relax} terms that were significantly larger than the values calculated by using the QM(ai)/LD method. The large differences between our results and QM(AM1)/MM results of Gao, are further magnified when the ratio of ΔG_{relax} to ΔG_{ES}, is considered. Thus, although reasonable values of relative solvation free energies for nucleic acid bases are available, the extent of the solute polarization contribution to these energies remains to be determined with a higher degree of certainty.

Table III. Electrostatic and Solute Polarization Contributions to the Total Free Energy of Hydration (kcal/mol)

Solute	QM(ai)/LD[a]		FEP/QM(AM1)/MM[b]	
	ΔG_{ES}	ΔG_{relax}	ΔG_{ES}	ΔG_{relax}
adenine	-6.6	-0.7	-2.4	-3.1
thymine	-7.6	-1.5	-5.9	-3.4
guanine	-13.5	-2.2	-6.0	-7.3
cytosine	-11.3	-2.5	-10.4	-6.4
AT[c]	-10.4	-1.6		
GC[c]	-15.0	-2.1		

[a]Iterative LD calculation. This calculation was based on the change in the solute charge distribution upon going from the gas phase to aqueous solution evaluated by using the PCM model (Pauling's vdW radii scaled by 1.2 and a dielectric constant of $\varepsilon = 80$ were used for the PCM calculation). [b]Combined Monte Carlo QM(AM1)/MM calculation [30]. [c]Adenine-thymine and guanine-cytosine Watson-Crick base pairs.

The solute polarization term becomes particularly important for accurate modeling of chemical reactions in solution. This is because the changes in this term upon going from the ground to the transition state of a reaction may affect significantly the calculated activation barriers in solution. In our opinion, useful insights into the polarization properties of transition states can be obtained by comparing magnitudes of the ΔG_{relax} term between the transition states, reactants, and products. Thus, in Table IV we examine the behavior of the solute polarization term for three biologically important reactions in aqueous solution:

$$MeO^- + MeC(O)NH_2 \rightleftharpoons MeOC(O)(NH_2)Me^- \tag{2}$$

$$H_2O + MePO_4H^- \rightleftharpoons MePO_5H_3^- \tag{3}$$

$$MePO_4H^- \rightleftharpoons MeOH + PO_3^- \tag{4}$$

Reaction 2 corresponds to the cleavage of the peptide bond and considers the nucleophilic attack of a methoxide ion on the sp^2 carbon of acetamide, resulting in the tetrahedral intermediate. Reactions 3 and 4 are related to the associative and dissociative mechanisms of the hydrolysis of methyl phosphate. The calculated magnitudes of the solvation and solute polarization terms to the activation barriers for these reactions show a regular pattern in so far that the solute polarization contributes always about 10% of the total solvation free energy difference ($\Delta\Delta G_{solv}^{\neq}$) between the reactants and transition states. Similar regularity occurs for the solvation and solute relaxation contributions to the reaction free energies. Thus, our LD calculations indicate that the solute polarization contributions during chemical reactions are quite small and can be reasonably estimated by scaling the total solvation energy by a factor of 0.1.

Table IV. Solvation and Solute Polarization Contributions[a] (kcal/mol)
to the Activation Barriers for Reactions 2-4

#[b]	ΔG_{solv}			ΔG_{relax}			$\Delta\Delta G_{solv}^{\neq}$	$\Delta\Delta G_{relax}^{\neq}$
	Reactants	TS	Products	Reactants	TS	Products		
2	-99.1,-8.9	-80.7	-88.7	-6.8,-1.3	-5.5	-7.3	27.3	2.6
3	-7.4,-71.2	-70.4	-66.6	-0.7,-1.8	-1.8	-1.2	8.2	0.7
4	-70.4	-69.2	-66.7,-60.8	-1.8.	-1.7	-0.2,-0.6	1.2	0.1

[a]Iterative LD calculations. For the complete energetics of these reactions and related structural data see references in[71] and[52]. [b]# denotes the reaction number.

Another challenging problem for biomolecular modeling is the calculation of solvation free energies for solutes with high charge densities. For such systems (e.g. phosphate groups in nucleotide mono-, di- and triphosphates) accurate evaluation of long range electrostatic interactions by the all atom solvent models becomes computationally very expensive. Thus, approximate treatments of solute-solvent interactions often represent the only practical way for quantitative evaluation of solvation energies. Here, the Langevin dipole model seems to provide more accurate results than the PCM model (Table V). One might tend to attribute the good performance of the LD model to the possibility of saturation of solute-dipole interactions for large solute electrostatic fields due to the presence of the Langevin function. (Note, that the nonlinear dependence of induced dipoles upon the magnitude of the solute electrostatic field is expressed in terms of the Langevin function in LD-based approaches). Such a saturation effect is missing in the current PCM and DC solvation models. However, the lack of the saturation in the PCM model could be offset by the outlying charge effect [22]. This cancellation of errors might arise as a consequence of the fact that only the solute electron density enclosed inside the cavity contributes to the polarization of the reaction field represented by the charges placed on the surface of the cavity. As the negative charge on the solute increases, its electron density becomes more diffuse and, if the atomic vdW radii of the solute atoms are kept constant, a larger proportion of the electron density appears outside the solute cavity.

The balance between the outlying charge and saturation effects is studied in Table V by changing the scaling factor of Pauling's atomic vdW radii in the PCM calculation from 1.0 to 1.5. One can see that the use of the unscaled vdW radii results in overestimated solvation energies regardless of the phosphate charges. The increase in the vdW radii by a factor of 1.2 results in a large decrease of ΔG_{solv} for the neutral phosphate, but also in an unexpectedly small drop in the ΔG_{solv} for negatively charged phosphates. A further increase in the vdW radii results in an unphysically low solvation energy for the neutral phosphate, while the magnitudes of solvation energies for the charged phosphates remain substantially larger than the corresponding experimental values. On the other hand, the noniterative LD model significantly underestimates the ΔG_{solv} values for the 2- and 3- ions. Thus, it seems that an accurate prediction of the solvation properties of solutes with high charge densities requires that both the saturation of the solute-solvent interactions and the explicit dipole-dipole interactions of the solvent molecules are taken into account in the construction of simplified solvent models.

Table V. Comparison of the LD,[a] PCM[b] and Experimental Hydration Free Energies (kcal/mol) of Multiple Charged Phosphate Groups

Solute	NLD	LD	PCM/1.0	PCM/1.2	PCM/1.4	PCM/1.5[c]	experiment[d]
H_3PO_4	-10	-12	-22	-13	-5.7	-1.9	
$H_2PO_4^-$	-71	-70	-86	-81	-74	-74	-68
HPO_4^{2-}	-225	-247	-278	-275	-269	-268	-245
PO_4^{3-}	-439	-536	-601	-594	-582	-572	-536

[a]The results of both the iterative and noniterative LD calculations [54] are presented. In the iterative LD method (LD), the electrostatic field at a given Langevin dipole is given as the sum of contributions from the solute and other Langevin dipoles, evaluated iteratively. In the noniterative LD method (NLD), this electrostatic field is generated by the solute charge distribution using a distance-dependent screening constant. [b]PCM HF/6-31G* calculations were carried out using Gaussian 94 [72]. [c]An increase of the scaling factor above 1.5 results in a numerical error during the evaluation of the electrostatic field and leads to the termination of the PCM calculation. [d]see ref. [54].

The most important conclusion from the results presented in this section is the fact that simplified solvation models like the combined QM(ai)/LD method are starting to give sufficiently reliable information on chemical reactions in solution, and can be used in calibrating approaches for studies of enzymatic reactions.

Computer Simulations of Enzymatic Reactions

The Reliability of Current QM/MM Approaches. The challenge of understanding enzymatic reactions has led to significant theoretical activity in recent years [25,41,53,73,74]. However, it appears that despite this impressive progress many studies are still overlooking the major factors and focusing on quantum mechanical features of the reacting fragments that essentially cancel out in the evaluation of catalytic effects. This sometimes leads to focus on the exact intramolecular potential of the reacting substrate while paying little attention to the crucial interaction with the surroundings which is the essence of enzyme catalysis. Bersuker *et al.* [75], for example, have developed a QM/MM approach which focuses on the importance of charge transfer (CT) effects in biochemical metallo-complexes. Though an explicit quantum mechanical treatment of these effects is necessary when studying spectroscopic details the importance of CT for *catalysis* is not clear. It is quite likely that similar CT effects occur in the enzyme and in the reference reaction in solution using the same ligands, so that these contributions cancel out. Thus, it is sometimes preferable to treat a metal ion as a point charge with proper calibration of its effect in solution, rather than to include CT to this metal without accurate treatment of solvation effects. With the above points in mind we consider below several strategies for calculations of enzymatic reactions and discuss their scope and limitations. We will conclude by discussing our new ab initio QM(ai)/MM method which presents, in principle, a way of obtaining ab initio based activation free energies for enzymatic reactions [52].

We begin our consideration of techniques for simulating chemical reactions in enzymes by noting that attempts to consider the substrate and several enzyme residues in the gas phase (e.g. refs. [76,77]) are potentially problematic and may even be deceiving because the missing solvation effects can be enormous. For example, recent calculations that consider ionized residues in the reaction of ribonuclease [78] or related studies [77] can be considered as a part of an incremental attempt to systematically understand the reacting system. However, such calculations overlook enormous solvation contributions. These contributions generally change drastically between the ground and transition states. Even inclusion of the entire enzyme and some solvent molecules (in a QM/MM fashion) does not guarantee correct results owing to variations in how the surrounding environment is treated. This is true even in the case of recent QM/MM methods that formally include the entire system except the bulk solvent around the protein region (e.g. refs. [26,79,80]). Most of these methods are not yet able to provide accurate free energies and generally only involve energy minimization. Furthermore, they are ineffective in providing the proper solvent reorganization in response to the formation of charges. Finally, the approaches used do not involve proper electrostatic boundary conditions (e.g. surface polarization and bulk effects which are discussed elsewhere [50,81]) and therefore cannot reproduce the correct compensation for charge-charge interaction. The importance of these points may be seen from the examples given below. These examples are not meant to detract from significant progress in the field but to point out major traps.

One apparent problem with modeling enzyme reactions using a solute and several residues in the gas phase is associated with the relative contribution of ionized residues [26,80]. According to these calculations, ionized groups 15 Å from the reacting system can provide a major contribution to catalysis (up to 10 kcal/mol). However, observations from mutation experiments [82] and more consistent calculations [50] show that charge-charge interactions more than 8 Å away from the reacting region can never contribute more than 1-2 kcal/mol. This issue is discussed in great length in many of our papers (e.g. see ref. [81]). Apparently, properly handling electrostatic effects presents a major problem in QM/MM calculations. It is not enough to simply include the residual charges of the solvent and the protein in the calculation; the proper reorientation upon charging must also be consistently reproduced.

An additional problem with most current QM/MM approaches is the fact that the quantum mechanical method used is not accurate enough. In many cases, this oversight may contribute to very large errors. Since ab initio approaches are often too expensive to be used to describe an enzyme active site, one is forced to use AM1-type methods that are not expected to give accurate results even when the environment is treated properly. This is demonstrated by the findings of Lyne *et al.* [80] where the difference between the AM1 and ab initio barrier for the reaction of chorismate mutase is shown to be around 20 kcal/mol. To further illustrate our point, we refer the reader to two sets of calculations of the catalytic reaction of triosephosphate isomerase, the combined AM1/MM calculation of Bash *et al.* [26] and the recent EVB calculation of Åqvist and Fothergill [83]. The AM1/MM calculation does not reproduce quantitative results since the observed rate determining step for the reaction is the first step with an observed barrier of about 12 kcal/mol, whereas this calculation gave a rate determining barrier of 20 kcal/mol at the third step. On the other hand the calculations of Åqvist and Fothergill reproduce the experimental trend with impressive accuracy and without adjusting any parameters in moving from water to the enzyme active site.

At present the most accurate way of obtaining activation free energies for enzymatic reactions is provided by the EVB method. Previous EVB studies are reviewed in ref. [53]. More recent examinations include the catalytic mechanism of DNA polymerase [84], of ras p21 [85], the mechanism of triosephosphate isomerase [83], acetyl cholinesterase [86], and the proton transfer step in the reaction of carbonic anhydrase [53] as well as the effect of tunneling in this step [87]. Most of these studies capture the catalytic effect of the enzyme in a quantitative or semiquantitative way.

The reader might question how the seemingly simple EVB method achieves correct results where other methods fail. The answer is largely in the physical insight of this method. Many theoreticians assume that one can simply take gas phase calculations, add the enzyme, and eventually reproduce the correct results. But, even with the emergence of faster computers this approach will not work, when the electrostatic and long range forces are not treated correctly. On the other hand, the EVB technique reflects the realization that enzyme catalysis involves the difference between the reaction in the enzyme and the corresponding reference reaction in water. Enzyme catalysis is due to environmental effects and is not due to the solute potential surface. The EVB method avoids the need to evaluate the exact free energy surface in the enzyme site by calibrating it on experimental information in solution or fitting it to high level ab initio gas phase calculations [20]. The advance of the QM(ai)/LD techniques described in the first part of this paper makes it possible to calibrate EVB parameters using ab initio calculations in solution. Most importantly, the EVB method focuses on the difference between the potential surfaces in enzyme and solution, thus errors in the intramolecular contributions simply cancel out.

Evaluating the ab initio Free Energy of Enzymatic Reactions. Even with the scope and utility of the EVB method, it is still very important to go beyond this level to try to obtain results for enzymatic reactions by "first principle" approaches. This section concludes with mentioning our progress in this direction.

We have modified our ab initio hybrid QM/MM technique [42] to model chemical reactions in proteins [52]. This method uses a molecular mechanics force field to represent the solvent, and an ab initio quantum mechanical technique that incorporates the electrostatic field from the solvent and/or protein in its Hamiltonian to represent the environment. The preliminary results, reported here, were obtained using the HF/4-31G approximation. The EVB mapping potential is used as a reference potential for free energy perturbation (FEP) calculations of the potential surface. A mapping parameter drives the reaction from the reactants to the products and molecular dynamics trajectories are calculated along the course of the reaction [41]. The reactions considered in ref. [42] were simple enough to fit by hand EVB parameters that reproduced the ab initio surface. This fitting process becomes more involved for larger and more varied substrates. We have consequently developed an automated procedure that constructs an EVB gas phase potential surface whose minima are as close to the corresponding minima obtained by gas phase ab initio calculations as possible. This automated procedure results in an EVB surface that accurately approximates the ab initio surfaces of the reactant, product, and crucial intermediates. This potential surface may be further refined by using the values of ab initio energies generated during the calculations of the reference solution reaction. Next we use a perturbation approach to "mutate" the system from the EVB mapping potential to the ab initio free energy surface. The automated procedure that forces the EVB surface to reproduce the ab initio minima significantly reduces the amount of sampling needed to obtain the ab initio free energy surface.

As in previous hybrid QM/MM procedures (e.g. refs. [6,26,33]) we have to properly connect the quantum and classical regions. Warshel and Levitt introduced an excellent hybrid orbital treatment [6], which was followed recently by the approach of Rivail and coworkers [33]. Here we use the much simpler approach of link atoms (LA) [17,88]. These atoms are inserted along the bonds where the quantum mechanical and the classical regions are connected. The link atom replaces the original atom, the so called link atom host (LAH). In this study hydrogen atoms are used as link atoms, but other options are possible. To obtain stable results the charges on groups bonded to the quantum region are set to zero. We note that errors introduced by the connecting atoms are likely to be the same in solution and in the protein. We parameterize our method so that the overall free energy curve reproduces experimental results in solution, and use the same parameters for the reaction in the enzyme.

As a preliminary check of our approach we simulated the nucleophilic attack step of subtilisin. This step involves the attack of a deprotonated serine residue on the carbonyl carbon of a peptide substrate, as shown in Scheme I.

We calculated the complete FEP/EVB free energy surface for the reaction and applied our QM(ai)/MM method to the reactant, product and transition state. The results of the study are summarized in Figures 1 and 2.

The most encouraging finding that emerges from our study is the fact that the ab initio electrostatic free energy obtained by our procedure is almost identical to the corresponding EVB free energies. That is, although the EVB charge distribution of the reactant and product are taken as the corresponding gas phase ab initio charges there is no guarantee (except in the inherent physics of the EVB method) that the charge distribution along the reaction coordinate will be similar in both methods. Remarkably, it is evident from Figure 1 that the ab initio electrostatic free energy reaches very reasonable convergence and practically coincides with the EVB free energy surface. This has several consequences. First, we obtain for the first time converging results for QM/MM ab initio "solvation" free energies in an enzyme. This is fundamentally different than results that would be obtained using Jorgensen's approach [15] where one uses inconsistently the gas phase charges along the reaction without considering the effect of the environment on the solute Hamiltonian. Second, establishing that the EVB method can be used to reproduce the ab initio solvation free energies provides further validation to strategies that use the EVB method in quantitative studies of enzymatic reactions. This is obviously quite important since the EVB method is faster by several orders of magnitude than the alternative ab initio strategies.

The free energy profiles of the full QM(ai)/MM potential (Figure 2) present less encouraging results although the average values of the free energies are reasonable. The corresponding free energy surfaces involve a somewhat unsatisfactory convergence. The problem is associated with fluctuations of the intramolecular ab initio potential which is partially due to the deviation between the EVB mapping potential and the ab initio surfaces. This deviation can be reduced by further refinement of the EVB potential using the ab initio energies generated during simulation of the reaction in solution.

The above results indicate clearly that QM/MM calculations that only consider one or a few configurations of the protein are rather meaningless and a proper averaging procedure is essential. Furthermore, the superior convergence of the intermolecular free energies indicate that these quantities should be used (if possible) instead of the total free energy. As emphasized above and in the spirit of the EVB philosophy it is not unreasonable to estimate the catalytic effect of the enzyme by considering only the change in ab initio electrostatic (solvation) free energy while assuming that the intramolecular contribution is the same in the protein and in solution. This approach is clearly a viable approximation as long as the convergence of the intramolecular contribution is not satisfactory.

Concluding Remarks

This report considered different strategies for modeling chemical reactions in enzymes and solution, discussed problems associated with some of these approaches and the potential of others. We have also summarized our recent progress in attempts to use hybrid ab initio QM/MM techniques to simulate enzymatic reactions. Special attention has been given to two QM/MM strategies developed and refined in our Laboratory. The first, a hybrid QM(ai)/LD approach, reproduces the accuracy of dielectric continuum

Scheme I

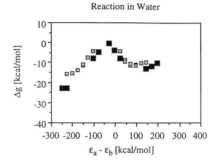

Reaction in Water

Reaction in Protein

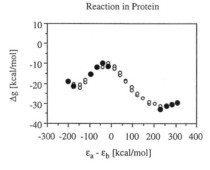

Figure 1. Electrostatic QM(ai)/MM free energy surfaces for the reaction in water (top) and the reaction in protein (bottom). The reaction coordinate is the energy difference between the reactant and product EVB potentials. EVB points in water are marked by dotted squares ▫, ab initio points are marked by filled squares ■. The EVB points for the reaction in protein are marked with open circles ○, ab initio points with filled circles ●.

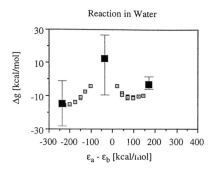

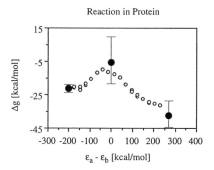

Figure 2. Full QM(ai)/MM free energy surfaces for the reaction in water (top) and the reaction in protein (bottom). The reaction coordinate is the energy difference between the reactant and product EVB potentials. EVB points in water are marked by dotted squares ⊡, ab initio points are marked by filled squares ■. The EVB points for the reaction in protein are marked with open circles o, ab initio points with filled circles ●. I represents the maximum deviation from the average ab initio value.

models, while providing a more realistic physical picture. Moreover, the QM(ai)/LD method is also quite promising because its speed and accuracy allow calibration of EVB parameters in solution rather than in the gas phase. Furthermore, this approach allows one to elucidate the mechanism of solution reactions whose understanding is crucial for further progress in studies of key biological systems. One of the best examples is our ability to start probing the nature of phosphate hydrolysis in solution and the relative importance of associative and dissociative mechanisms [71,89].

The second approach considered here is a hybrid QM(ai)/MM approach that uses EVB potential surfaces as a reference potential for evaluating activation free energies of enzymatic reactions. This approach was demonstrated by considering the nucleophilic attack step in the catalytic reaction of subtilisin.

The simulation produced is, to the best of our knowledge, the first QM/MM ab initio free energy surface for an enzymatic reaction. In particular, we find it significant that the ab initio electrostatic free energies provide converging estimates for the catalytic effect of the enzyme. The intramolecular ab initio free energies provide less encouraging convergence although the average values appear to be reasonable. Improving the convergence of the overall ab initio free energy is clearly an important challenge.

Regardless of the difficulties in obtaining converging ab initio free energy surfaces, we believe that the present results provide strong support for the EVB philosophy. That is, the effect of the enzyme is mainly associated with the difference in solvation energy upon moving from solution to the enzyme active site and the EVB intermolecular free energy surfaces are found to accurately reproduce the corresponding ab initio surfaces. Thus, one can use the EVB approach and reproduce the ab initio catalytic effect with much less computer time and with much less convergence problems.

Acknowledgments

This work was supported by Grant GM24492 from the National Institutes of Health. J.B. acknowledges the support by a grant of the Deutsche Forschungsgemeinschaft (DFG).

References

(1)   Lightstone, F. C.; Zheng, Y.-J.; Maulitz, A. H.; Bruice, T. C. Non-enzymatic and enzymatic hydrolysis of alkyl halides: A haloalkane dehalogenase enzyme evolved to stabilize the gas-phase transition state of an S_N2 displacement reaction, *Proc. Natl. Acad. Sci. USA* **1997**, *94*, 8417.

(2)   Dewar, M. J. S.; Storch, D. M. Alternative View of Enzyme Reactions, *Proc. Natl. Acad. Sci. USA* **1985**, *82*, 2225.

(3)   Lee, J. K.; Houk, K. N. A Proficient Enzyme Revisited: The Predicted Mechanism for Orotidine Monophosphate Decarboxylase, *Science* **1997**, *276*, 942.

(4)   Warshel, A.; Åqvist, J.; Creighton, S. Enzymes Work by Solvation Substitution rather than by Desolvation, *Proc. Natl. Acad. Sci. USA* **1989**, *86*, 5820.

(5)   Gao, J. *Reviews in Computational Chemistry*; VCH: New York, 1995; Vol. 7, pp 119.

(6)   Warshel, A.; Levitt, M. Theoretical Studies of Enzymatic Reactions: Dielectric, Electrostatic and Steric Stabilization of the Carbonium Ion in the Reaction of Lysozyme, *J. Mol. Biol.* **1976**, *103*, 227.

(7)   Fersht, A. R. *Enzyme Structure and Mechanism*; W.H. Freeman and Company: New York, 1985.

(8)   Radzicka, A.; Wolfenden, R. A Proficient Enzyme, *Science* **1995**, *267*, 90.

30

(9)   Warshel, A. A Microscopic Dielectric Model for Reactions in Water, *Phil. Trans. R. Soc. Lond. B* **1977**, *278*, 111.

(10)   Warshel, A. A Microscopic Model for Calculations of Chemical Processes in Aqueous Solutions, *Chem. Phys. Lett.* **1978**, *55*, 454.

(11)   Warshel, A. Calculations of Chemical Processes in Solutions, *J. Phys. Chem.* **1979**, *83*, 1640.

(12)   Tapia, O.; Johannin, G. An Inhomogeneous Self-Consistent Reaction Field Theory of Protein Core Effects. Towards a Quantum Scheme for Describing Enzyme Reactions, *J. Chem. Phys.* **1981**, *75*, 3624.

(13)   Warshel, A. Dynamics of Reactions in Polar Solvents. Semiclassical Trajectory Studies of Electron-Transfer and Proton-Transfer Reactions, *J. Phys. Chem.* **1982**, *86*, 2218.

(14)   Warshel, A. Simulating the Energetics and Dynamics of Enzymatic Reactions, *Pontif. Acad. Sci. Scr. Var.* **1983**, *55*, 59.

(15)   Chandrasekhar, J.; Smith, S. F.; Jorgensen, W. L. Theoretical Examination of the S_N2 Reaction Involving Chloride Ion and Methyl Chloride in the Gas Phase and Aqueous Solution, *J. Am. Chem. Soc.* **1985**, *107*, 154.

(16)   Weiner, S. J.; Singh, U. C.; Kollman, P. A. Simulation of Formamide Hydrolysis by Hydroxide Ion in the Gas Phase and in Aqueous Solution, *J. Am. Chem. Soc.* **1985**, *107*, 2219.

(17)   Singh, U. C.; Kollman, P. A. A Combined *Ab Initio* Quantum Mechanical and Molecular Mechanical Method for Carrying out Simulations on Complex Molecular Systems: Applications to the $CH_3Cl + Cl^-$ Exchange Reaction and Gas Phase Protonation of Polyethers, *J. Comput. Chem.* **1986**, *7*, 718.

(18)   Kong, Y. S.; Jhon, M. S. Solvent Effects on S_N2 Reactions, *Theor. Chim. Acta* **1986**, *70*, 123.

(19)   Bash, P. A.; Field, M. J.; Karplus, M. Free Energy Perturbation Method for Chemical Reactions in the Condensed Phase: A Dynamical Approach Based on a Combined Quantum and Molecular Mechanics Potential, *J. Am. Chem. Soc.* **1987**, *109*, 8092.

(20)   Hwang, J.-K.; King, G.; Creighton, S.; Warshel, A. Simulation of Free Energy Relationships and Dynamics of S_N2 Reactions in Aqueous Solution, *J. Am. Chem. Soc.* **1988**, *110*, 5297.

(21)   Ford, G. P.; Wang, B. Incorporation of Hydration Effects within the Semiempirical Molecular Orbital Framework. AM1 and MNDO Results for Neutral Molecules, Cations, Anions, and Reacting Systems, *J. Am. Chem. Soc.* **1992**, *114*, 10563.

(22)   Miertus, S.; Scrocco, E.; Tomasi, J. Electrostatic Interaction of a Solute with a Continuum. A Direct Utilization of Ab initio Molecular Potentials for the Prevision of Solvent Effects, *Chem. Phys.* **1981**, *55*, 117.

(23)   Honig, B.; Sharp, K.; Yang, A.-S. Macroscopic Models of Aqueous Solutions: Biological and Chemical Applications, *J. Phys. Chem.* **1993**, *97*, 1101.

(24)   Rashin, A. A.; Bukatin, M. A.; Andzelm, J.; Hagler, A. T. Incorporation of Reaction Field Effects into Density Functional Calculations for Molecules of Arbitrary Shape in Solution, *Biophys. Chem.* **1994**, *51*, 375.

(25)   Bash, P. A.; Singh, U. C.; Langridge, R.; Kollman, P. A. Free Energy Calculations by Computer Simulation, *Science* **1987**, *236*, 564.

(26)   Bash, P. A.; Field, M. J.; Davenport, R. C.; Petsko, G. A.; Ringe, D.; Karplus, M. Computer Simulation and Analysis of the Reaction Pathway of Triosephosphate Isomerase, *Biochemistry* **1991**, *30*, 5826.

(27)   Field, M. J.; Bash, P. A.; Karplus, M. A Combined Quantum Mechanical and Molecular Mechanical Potential for Molecular Dynamics Simulations, *J. Comp. Chem.* **1990**, *11*, 700.

(28)   Gao, J. Absolute Free Energy of Solvation From Monte Carlo Simulations Using Combined Quantum and Molecular Mechanical Potential., *J. Phys. Chem.* **1992**, *96*, 537.

(29)   Gao, J.; Xia, X. A Priori Evaluation of Aqueous Polarization Effects through Monte Carlo QM-MM Simulations, *Science* **1992**, *258*, 631.

(30)   Gao, J. Combined QM/MM Simulation Study of the Claisen Rearrangement of Allyl Vinyl Ether in Aqueous Solution, *J. Am. Chem. Soc.* **1994**, *116*, 1563.

(31)   Luzhkov, V.; Warshel, A. Microscopic Models for Quantum Mechanical Calculations of Chemical Processes in Solutions: LD/AMPAC and SCAAS/AMPAC Calculations of Solvation Energies, *J. Comp. Chem.* **1992**, *13*, 199.

(32)   Théry, V.; Rinaldi, D.; Rivail, J.-L.; Maigret, B.; Ferenczy, G. G. Quantum Mechanical Computations on Very Large Molecular Systems: The Local Self-Consistent Field Method, *J. Comp. Chem.* **1994**, *15*, 269.

(33)   Ferenczy, G. G.; Rivail, J.-L.; Surján, P. R.; Náray-Szabó, G. NDDO Fragment Self-Consistent Field Approximation for Large Electronic Systems, *J. Comp. Chem.* **1992**, *13*, 830.

(34)   Stanton, R. V.; Dixon, S. L.; Merz, K. M., Jr. General Formulation for a Quantum Free Energy Perturbation Study, *J. Phys. Chem.* **1995**, *99*, 10701.

(35)   Stanton, R. V.; Hartsough, D. S.; Merz, K. M., Jr. Calculation of Solvation Free Energies Using a Density Functional/Molecular Dynamics Coupled Potential, *J. Phys. Chem.* **1993**, *97*, 11868.

(36)   Stanton, R. V.; Little, L. R.; Merz, K. M., Jr. An Examination of a Hartree-Fock/Molecular Mechanical Coupled Potential, *J. Phys. Chem.* **1995**, *99*, 17344.

(37)   Hwang, J.-K.; Warshel, A. A Quantized Classical Path Approach for Calculations of Quantum Mechanical Rate Constants, *J. Phys. Chem.* **1993**, *97*, 10053.

(38)   Thompson, M. A. Hybrid Quantum Mechanical/Molecular Mechanical Force Field Development for Large Flexible Molecules: A Molecular Dynamics Study of 18-Crown-6, *J. Phys. Chem.* **1995**, *99*, 4794.

(39)   Thompson, M. A.; Glendening, E. D.; Feller, D. The Nature of K^+/Crown Ether Interactions: a Hybrid Quantum Mechanical Molecular Mechanical Study, *J. Phys. Chem.* **1994**, *98*, 10465.

(40)   Warshel, A.; Weiss, R. M. An Empirical Valence Bond Approach for Comparing Reactions in Solutions and in Enzymes, *J. Am. Chem. Soc.* **1980**, *102*, 6218.

(41)   Warshel, A. *Computer Modeling of Chemical Reactions in Enzymes and Solutions*; John Wiley & Sons: New York, 1991.

(42)   Muller, R. P.; Warshel, A. Ab Initio Calculations of Free Energy Barriers for Chemical Reactions in Solution, *J. Phys. Chem.* **1995**, *99*, 17516.

(43)   Wei, D.; Salahub, D. R. A Combined Density Functional and Molecular Dynamics Simulation of a Quantum Water Molecule in Aqueous Solution, *Chem. Phys. Lett.* **1994**, *24*, 291.

(44)   Tuckerman, M.; Laasonen, K.; Sprik, M.; Parrinello, M. Ab Initio Molecular Dynamics Simulation of the Solvation and Transport of H_3O^+ and OH⁻ Ions in Water, *J. Phys. Chem.* **1995**, *99*, 5749.

(45)   Wesolowski, T. A.; Warshel, A. Frozen Density Functional Approach for *Ab Initio* Calculations of Solvated Molecules, *J. Phys. Chem.* **1993**, *97*, 8050.

(46)   Wesolowski, T.; Warshel, A. *Ab Initio* Free Energy Perturbation Calculations of Solvation Free Energy Using the Frozen Density Functional Approach, *J. Phys. Chem.* **1994**, *98*, 5183.

(47)   Wesolowski, T.; Muller, R. P.; Warshel, A. *Ab Initio* Frozen Density Functional Calculations of Proton Transfer Reactions in Solution, *J. Phys. Chem.* **1996**, *100*, 15444.

(48)   Wesolowski, T. A.; Weber, J. Kohn-Sham Equations with Constrained Electron Density: An Iterative Evaluation of the Ground-State Electron Density of Interacting Molecules, *Chem. Phys. Lett.* **1996**, *248*, 71.

(49)   Muller, R. P.; Wesolowski, T.; Warshel, A. Calculations of Chemical Processes in Solution by Density-functional and other Quantum-Mechanical

32

Techniques, In *Density-Functional Methods: In Chemistry and Materials Science*; M. Springborg, Ed.; John Wiley & Sons: Chichester, 1997.

(50)   Lee, F. S.; Chu, Z. T.; Warshel, A. Microscopic and Semimicroscopic Calculations of Electrostatic Energies in Proteins by the POLARIS and ENZYMIX Programs, *J. Comp. Chem.* **1993**, *14*, 161.

(51)   Warshel, A.; Chu, Z. T. Calculations of Solvation Free Energies in Chemistry and Biology, In *ACS Symposium Series: Structure and Reactivity in Aqueous Solution. Characterization of Chemical and Biological Systems*; C. J. Cramer and D. G. Truhlar, Ed.; 1994.

(52)   Bentzien, J.; Muller, R. P.; Florián, J.; Warshel, A. Hybrid ab initio Quantum Mechanics/Molecular Mechanics (QM(ai)/MM) Calculations of Free Energy Surfaces for Enzymatic Reactions: The Nucleophilic Attack in Subtilisin, *J. Phys. Chem. B* **1997**, submitted.

(53)   Åqvist, J.; Warshel, A. Simulation of Enzyme Reactions Using Valence Bond Force Fields and Other Hybrid Quantum/Classical Approaches, *Chem. Rev.* **1993**, *93*, 2523.

(54)   Florián, J.; Warshel, A. Langevin Dipoles Model for Ab Initio Calculations of Chemical Processes in Solution: Parameterization and Application to Hydration Free Energies of Neutral and Ionic Solutes, and Conformational Analysis in Aqueous Solution, *J. Phys. Chem B* **1997**, *101*, 5583.

(55)   Florián, J.; Warshel, A. *ChemSol*; University of Southern California: Los Angeles, 1997.

(56)   Miertus, S.; Tomasi, J. Approximate Evaluations of the Electrostatic Free Energy and Internal Energy Changes in Solution Processes, *Chem. Phys.* **1982**, *65*, 239.

(57)   Cullis, P. M.; Wolfenden, R. Affinities of Nucleic Acid Bases for Solvent Water, *Biochemistry* **1981**, *20*, 3024.

(58)   Cieplak, P.; Kollman, P. A. Calculation of the free energy of association of nucleic acid bases in vacuo and water solution, *J. Am. Chem. Soc.* **1988**, *110*, 3334.

(59)   Elcock, A. H.; Richards, W. G. Relative hydration free energies of nucleic acid bases, *J. Am. Chem. Soc.* **1993**, *115*, 7930.

(60)   Miller, J. L.; Kollman, P. A. Solvation Free Energies of the Nucleic Acid Bases, *J. Phys. Chem.* **1996**, *100*, 8587.

(61)   Cornell, W. D.; Cieplak, P.; Bayly, C. I.; Gould, I. R.; Merz, K. M., Jr.; Ferguson, D. M.; Spellmeyer, D. C.; Fox, T.; Caldwell, J. W.; Kollman, P. A. A Second Generation Force Field for the Simulation Of Proteins, Nucleic-Acids, And Organic-Molecules, *J. Am. Chem. Soc.* **1995**, *117*, 5179.

(62)   Florián, J.; Baumruk, V.; Leszczynski, J. IR and Raman Spectra, Tautomeric Stabilities, and Scaled Quantum Mechanical Force Fields of Protonated Cytosine, *J. Phys. Chem.* **1996**, *100*, 5578.

(63)   Sponer, J.; Leszczynski, J.; Hobza, P. Structures and Energies of Hydrogen-Bonded DNA Base Pairs. A Nonempirical Study with Inclusion of Electron Correlation, *J. Phys. Chem.* **1996**, *100*, 1965.

(64)   Gao, J. The hydration and solvent polarization effects of nucleotide bases, *Biophys. Chem.* **1994**, *51*, 253.

(65)   Orozco, M.; Luque, F. J. SRF computation of the Reactive Characteristics of DNA bases in water, *Biopolymers* **1993**, *33*, 1851.

(66)   Young, P. E.; Hillier, I. J. Hydration Free Energies of Nucleic Acid Bases using an ab initio Continuum Model, *Chem. Phys. Lett.* **1993**, *215*, 405.

(67)   Mohan, V.; Davis, M. E.; McCammon, J. A.; Pettit, B. M. Contiuum Model Calculations of Solvation Free Energies - Accurate Evaluation of Electrostatic Contributions, *J. Phys. Chem.* **1992**, *96*, 6428.

(68)   Cramer, C. J.; Truhlar, D. G. Polarization of the Nucleic-Acid Bases in Aqueous Solution, *Chem. Phys. Lett.* **1992**, *198*, 74.

(69) Gao, J. Energy Components of Aqueous Solution - Insight from Hybrid QM/MM Simulations using a Polarizable Solvent Model, *J. Comput. Chem.* **1997**, *18*, 1061.

(70) Gao, J.; Kuczera, K.; Tidor, B.; Karplus, M. Hidden Thermodynamics of Mutant Proteins: A Molecular Dynamics Analysis, *Science* **1989**, *244*, 1069.

(71) Florián, J.; Warshel, A. On the Nature of Phosphate Ester Hydrolysis in Aqueous Solution: Associative Versus Dissociative Mechanisms, *J. Phys. Chem.* **1997**, submitted.

(72) Frisch, M. J.; Trucks, G. W.; Schlegel, H. B.; Gill, P. M. W.; Johnson, B. G.; Robb, M. A.; Cheeseman, J. R.; Keith, T.; Petersson, G. A.; Montgomery, J. A.; Raghavachari, K.; Al-Laham, M. A.; Zakrzewski, V. G.; Ortiz, J. V.; Foresman, J. B.; Cioslowski, J.; Stefanov, B. B.; Nanayakkara, A.; Challacombe, M.; Peng, C. Y.; Ayala, P. Y.; Chen, W.; Wong, M. W.; Andres, J. L.; Replogle, E. S.; Gomperts, R.; Martin, R. L.; Fox, D. J.; Binkley, J. S.; Defrees, D. J.; Baker, J.; Stewart, J. P.; Head-Gordon, M.; Gonzalez, C.; Pople, J. A. *Gaussian 94, Revision D.2*; Gaussian, Inc.: Pittsburgh PA, 1995.

(73) Náray-Szabó, G.; Fuxreiter, M.; Warshel, A. Electrostatic Basis of Enzyme Catalysis, In *Computational Approaches to Biochemical Reactivity*; A. Warshel and G. Náray-Szabó, Ed.; Kluwer Academic Publishers: Dordrecht, Boston, London, 1997.

(74) Kollman, P. Free Energy Calculations: Applications to Chemical and Biochemical Phenomena, *Chem. Rev.* **1993**, *93*, 2395.

(75) Bersuker, I. B.; Leong, M. K.; Boggs, J. E.; Pearlman, R. S. A Method of Combined Quantum Mechanical (QM)/Molecular Mechanics (MM) Treatment of Large Polyatomic Systems with Charge Transfer Between the QM and MM Fragments, *J. Comp. Chem.* **1997**, *63*, 1051.

(76) Krauss, M.; Garmer, D. R. Assignment of the Spectra of Protein Radicals in Cytochrome c Peroxidase, **1993**, *J. Phys. Chem.*, 831.

(77) Nakagawa, S.; Umeyama, H. Ab Initio Molecular Orbital Study on the Effects of Zinc Ion, its Ligands and Ionic Amino Acid Residues for Proton Transfer Energetics between Glu 270 and Zn Co-ordinated Water Molecule in Carboxypeptidase A, *J. theor. Biol.* **1982**, *96*, 473.

(78) Wladowski, B. D.; Krauss, M.; Stevens, W. J. Ribonuclease A Catalyzed Transphosphorylation: An *ab Initio* Theoretical Study, *J. Phys. Chem.* **1995**, *99*, 6273.

(79) Waszkowycz, B.; Hillier, I. H.; Gensmantel, N.; Payling, D. W. A Combined Quantum Mechanical/Molecular Mechanical Model of the Potential Energy Surface of Ester Hydrolysis by the Enzyme Phospholipase A$_2$, *J. Chem. Soc. Perkin Trans. 2* **1991**, 225.

(80) Lyne, P. D.; Mulholland, A. J.; Richards, W. G. Insights into Chorismate Mutase Catalysis From a Combined QM/MM Simulation of the Enzyme Reaction, *J. Am. Chem. Soc.* **1995**, *117*, 11345.

(81) Alden, R. G.; Parson, W. W.; Chu, Z. T.; Warshel, A. Calculations of Electrostatic Energies in Photosynthetic Reaction Centers, *J. Am. Chem. Soc.* **1995**, *117*, 12284.

(82) Russel, A. J.; Fersht, A. R. Rational modification of enzyme catalysis by engineering surface charge, *Nature* **1987**, *328*, 496.

(83) Åqvist, J.; Fothergill, M. Computer Simulation of the Triosephosphate Isomerase Catalyzed Reaction, *J. Biol. Chem.* **1996**, *271*, 10010.

(84) Fothergill, M.; Goodman, M. F.; Petruska, J.; Warshel, A. Structure-Energy Analysis of the Role of Metal Ions in Phosphodiester Bond Hydrolysis by DNA Polymerase I, *J. Am. Chem. Soc.* **1995**, *117*, 11619.

(85) Schweins, T.; Langen, R.; Warshel, A. Why Have Mutagenesis Studies Not Located the General Base in ras p21, *Nature Struct. Biol.* **1994**, *1*, 476.

(86)   Fuxreiter, M.; Warshel, A. On the Origin of the Catalytic Power of Acetylcholinesterase: Computer Simulation Studies, *J. Am. Chem. Soc.* **1997**, in press.

(87)   Hwang, J.-K.; Warshel, A. How Important are Quantum Mechanical Nuclear Motions in Enzyme Catalysis?, *J. Am. Chem. Soc.* **1996**, *118*, 11745.

(88)   Eurenius, K. P.; Chatfield, D. C.; Brooks, B. R.; Hodoscek, M. Enzyme Mechanisms with Hybrid Quantum and Molecular Mechanical Potentials. 1. Theoretical Considerations, *Int. J. Quant. Chem.* **1996**, *60*, 1189.

(89)   Florián, J.; Warshel, A. A Fundamental Assumption About OH$^-$ Attack in Phosphate Hydrolysis Is Not Fully Justified, *J. Am. Chem. Soc.* **1997**, *119*, 5473.

Chapter 3

The Geometry of Water in Liquid Water from Hybrid Ab Initio-Monte Carlo and Density Functional-Molecular Dynamics Simulations

Cristobal Alhambra, Kyoungrim Byun, and Jiali Gao[1]

Department of Chemistry, State University of New York at Buffalo, Buffalo, NY 14260

The geometry of the water molecule in liquid water is computed using a hybrid ab initio molecular orbital-Monte Carlo (MC) simulation method at the HF/3-21G and HF/6-31G* level, and using a hybrid density functional-molecular dynamics (MD) technique with nonlocal exchange and correlation functionals. The OH bond length is calculated to be elongated by 0.005 to 0.008 Å in the liquid compared with the gas phase value, whereas the change in the average HOH angle is minimal with a predicted decrease of 0.2 to 1.3 degrees. The HOH angle exhibits large fluctuations, ranging from ca. 97 to 110 degrees. The condensed phase effect is further characterized by computing the vibrational frequencies through Fourier transform of the normal coordinate autocorrelation functions, and by computing the electric polarization of a QM water in liquid water. The present high-level ab initio MC and MD simulation results should be of use for the development of empirical flexible models for liquid water.

Solvent effects, in particular, aqueous solvation, play an important role in determining the structure and reactivity of molecules of biological interest (1). Until recently, computer simulation studies of aqueous solvation have been primarily carried out using pairwise additive intermolecular potentials with rigid water geometries. These models are illustrated by the simple point charge (SPC) model (2), and the transferrable intermolecular potentials (TIP3P and TIP4P) for water (3), in which the structure of water is held rigid either at an idealized geometry with an OH bond length of 1 Å and an HOH angle of 109.47^{o}, or at a gas phase experimental geometry (R(OH) = 0.9572 Å; θ(HOH) = 104.52^{o}). In recent years, there has been a growing interest in developing many-body nonadditive potentials for liquid water, which incorporate both

[1]Corresponding author.

intramolecular bond flexibility and/or electric polarization (*4-11*). The new development can potentially further extend the applicability of molecular dynamics and Monte Carlo simulations to phenomena that involve large environmental changes, such as water near interfaces, supercritical fluids, and salt solutions. An important property charactering a flexible water model is the average geometry of water in the liquid state. However, there is a tremendous ambiguity from various flexible water models (Table I).

Table I. The Gas to Liquid Geometrical Changes of Water from Various Flexible Models

Model	$\Delta R(OH)$, Å	$\Delta \theta(HOH)$, deg
TIPS(DP)[4]	0.015	NR
TJE[5]	0.016	-4.6
WT[5]	0.017	-4.6
ZSR[8]	0.006	-2.2
ZYZR[8]	0.009	-3.5
MCYL[7]	0.018	-1.0

The predicted average bond angle in liquid water ranges from 101.0° to 105°, representing decreases from the corresponding gas phase values by 1.0° to 4.6° (*4-9*). On the experimental side, neutron diffraction experiments indicate that there is an increase in the OH bond length, and only small geometrical variations in the bond angle, although the prediction of the water geometry is complicated by inelastic scattering and data analyses (*12,13*). Therefore, it is of interest to obtain a computational estimate of the average geometrical parameters for water in liquid water.

    The development of hybrid quantum mechanical/molecular mechanical (QM/MM) methods makes it possible to study solvent effects on the structure and chemical reactivity of solutes using modern electronic structure methods (*14-22*). In this article, we employ combined ab initio molecular orbital/molecular mechanical (AI/MM) and density functional theory/molecular mechanical (DFT/MM) potentials along with Monte Carlo (MC) and molecular dynamics (MD) simulation techniques to examine the condensed phase effect on the structure and vibrational frequencies of water in liquid water. In what follows, we will first present the computational details of the combined QM/MM ab initio MC and MD simulation methods. These approaches are then illustrated by fluid simulations of aqueous solution.

Hybrid AI/MM and DFT/MM Potentials

Hybrid Ab Initio/Molecular Mechanical Potential. The basic assumption in a hybrid quantum mechanical/molecular mechanical potential is that a molecular system can be partitioned into a quantum mechanical region consisting of the solute molecule and a classical region containing the rest of the solvent system (*14*). Consequently, the effective Hamiltonian of the system may be written as follows:

$$\hat{H}_{eff} = \hat{H}_X + \hat{H}_{Xs} + \hat{H}_{ss} + \hat{H}_{Xs}^{vdW} \tag{1}$$

where $\hat{H}_X$ is the Hamiltonian of the solute molecule, $\hat{H}_{ss}$ is the potential energy of the solvent, and $\hat{H}_{Xs}$ describes solute-solvent interactions. If the solute molecule in the QM region is treated by ab initio Hartree-Fock molecular orbital theory, the potential energy of the system, which is a function of both the solute (R_X) and solvent (R_s) nuclear coordinates, is determined by the expectation value of the solute molecular wave function Ψ:

$$E_{tot}(R_X, R_s) = <\Psi|\hat{H}_X + \hat{H}_{Xs}|\Psi> + E_{ss} + E_{Xs}^{vdW} \tag{2}$$

Implementation of the procedure in Hartree-Fock MO calculations is particular simple since it only involves the inclusion of the one-electron integrals $I_{\mu\nu}$ resulting from the QM/MM interaction Hamiltonian $\hat{H}_{Xs}$ into the gas phase Fock matrix for X (*14,17*). Recently, Marek Freindorf incorporated the GAMESS program into a standard Metropolis MC simulation program (BOSS) in our laboratory (*23-25*), and carried out a combined AI/MM MC simulation study of the hydration of N-methylacetamide in aqueous solution, making use of the HF/3-21G method for the QM region (*26*). The same procedure is employed in the present study.

Hybrid Density Functional/Molecular Mechanical Potential. In the framework of Kohn-Sham density functional theory (*27,28*), the total energy of a hybrid QM/MM system can be computed analogously:

$$E_{tot}(R_X, R_s) = E_X[R_X, \rho(r)] + E_{Xs}[R_X, R_s, \rho(r)] + E_{ss}(R_s) + E_{Xs}^{vdW} \tag{3}$$

where $\rho(r)$ is the electron density of the QM solute molecule, which is obtained from the electronic wave functions by solving the self-consistent equation (equation 4):

$$[-\frac{1}{2}\nabla_i^2 + \int dr' \frac{\rho(r')}{|r-r'|} + \frac{\delta E_{xc}}{\delta\rho(r)} - \sum_{m=1}^{M} \frac{eZ_m}{r_{i...}} + v_i(r)]\psi_i(r) = \epsilon_i\psi_i(r) \tag{4}$$

In equation 4, ψ_i is the molecular orbital for electron i, Z_m is the nuclear charge of atom m in the solute molecule, M is the total number of nuclei in the QM region, E_{xc} includes the exchange and correlation contributions to the electronic energy for the solute, and $v_i(\mathbf{r})$ is the external potential due to the solvent molecules.

$$\hat{H}_{Xs} = \sum_{i=1}^{2N} v_i(\mathbf{r}) = -\sum_{i=1}^{2N} \sum_{s=1}^{S} \frac{e q_s}{r_{is}} \tag{5}$$

where q_s is the partial atomic charge of a solvent site, S is the total number of solvent atoms, and 2N is the total number of electrons in the QM region. Having obtained the electron density, the energy terms involving electronic degrees of freedom in equation 3 can be determined as follows:

$$
\begin{aligned}
E_X[\mathbf{R}_X, \rho(r)] = & -\frac{1}{2}\int dr \sum_i \psi_i(r)\nabla^2\psi_i(r) - \int dr \sum_m \frac{Z_m}{|\mathbf{R}_m - r|}\rho(r) \\
& + \int\int dr dr' \frac{\rho(r)\rho(r')}{|r - r'|} + E_{xc} + \sum_{n>m} \frac{Z_n Z_m}{R_{nm}}
\end{aligned}
\tag{6}
$$

and

$$E_{Xs}[\mathbf{R}_X, \mathbf{R}_s, \rho(r)] = \int dr \sum_i v_i(r)\rho(r) + \sum_{s=1}^{S}\sum_{m=1}^{M} \frac{q_s Z_m}{R_{sm}} \tag{7}$$

In equation 6, the term E_{xc} is the exchange-correlation energy, for which we have used Perdew's correlation (29) and Becke's exchange (30) nonlocal density functionals in the present study.

The potential energy determined either using equation 2 from the combined AI/MM method, or using equation 3 from the combined DFT/MM method can be directly used in Metropolis Monte Carlo simulations, which typically generate several million configurations to obtain adequate ensemble averages. Alternatively, molecular dynamics simulations can be carried out using the forces derived from the hybrid QM/MM Born-Oppenheimer potential energy surface. The related Newton's equations of motion are

$$-\frac{\partial E_{tot}(\mathbf{R}_X, \mathbf{R}_s)}{\partial \mathbf{R}_j} = m_j \frac{\partial^2 \mathbf{R}_j}{\partial t^2} \tag{8}$$

where j specifies either a solute or a solvent atom, and m_j are the atomic masses.

The DFT program developed by St-Amant (31) has been used for the Kohn-Sham DFT calculations. The implementation features expansion of the wave functions and charge density in terms of a linear combination of Gaussian type orbitals (LCGTO) (32). Following the method developed by Fournier, Andzelm, and Salahub (33), analytical gradients of the potential energy with respect to the nuclear coordinates were obtained by including contributions from the orbital basis correction and density-fit basis correction terms in addition to the Hellman-Feynman force. The DFT code has been conveniently modified and interfaced with the CHARMM package for MD simulations (*34*).

Computational Details

To demonstrate the utility of the combined AI/MM and DFT/MM method for the study of solvent effects, and the computational accuracy on the predicted geometry and electronic polarization for water in aqueous solution, we have carried out both statistical mechanical Monte Carlo and molecular dynamics simulations. In the first set of studies, Metropolis Monte Carlo simulations were performed for a cubic box containing 260 classical water molecules plus one "QM" water using the isothermal-isobaric ensemble at 25 °C and 1 atm. The solvent water molecules are approximated by Jorgensen's TIP3P model (*3*), whereas the "solute" water molecule is treated at the ab initio HF/3-21G and HF/6-31G* levels of theory (*35*). In all computations, the solvent geometry was held rigid as the empirical potential function was developed. To obtain the average bond distance and bond angle of water, the QM water molecule is flexible during the fluid simulations. Thus, in addition to the standard translational and rotational moves, the internal degrees of freedom are also varied in the range of ±0.05 Å for bond distances and ± 0.5° for the HOH bond angle in each move of the QM molecule. In computing intermolecular interactions, periodic boundary conditions are used, along with a spherical cutoff distance of 9 Å. The Monte Carlo simulations included at 2 x 10^6 configurations of equilibration followed by 4 x 10^6 configurations of averaging for each calculation.

Subsequently, stochastic boundary molecular dynamics (SBMD) simulations (*36*) were carried out in a 20 Å sphere of 1024 TIP3P water molecules using the DFT/TIP3P potential. The QM water molecule is approximately located near the center of the sphere, and its geometry is fully relaxed throughout the simulation. The orbital basis set used in this study is of triple-ζ quality plus polarization for oxygen (7111/411/1) and hydrogen (311/1) (*32*). Auxiliary basis sets used in the fitting of the charge density and the exchange-correlation potential have the contribution patterns: O(4, 4; 4,4), and H(3, 1; 3, 1). A fine grid of randomly oriented points were deployed to fit the auxiliary density (832 points per atom) and to numerically integrate the exchange-correlation potential (2968 points per atom). The Becke exchange and Perdew correlation nonlocal functionals are adopted in the present calculation (*29,30*). It was found that these functionals can yield reasonable results for the geometries and vibrational frequencies for water in comparison with experiment and ab initio results (*38*).

The equations of motion of the hybrid DFT/MM system are integrated using the Verlet algorithm and the boundary potentials described by Brooks and Karplus (*36*). Although the water geometry in the MM region was kept rigid with the SHAKE method, in order to ensure that time correlation functions are accurately evaluated for the hydrogen atoms in the QM region, the time step is taken to be 0.2 fs. The system is first equilibrated using the TIP3P potential for a sphere of water for 100 ps. The coordinate for a water molecule near the center of the sphere is chosen for the QM molecule. The system is further equilibrated for 5 ps using the combined DFT/TIP3P potential, followed by 4.5 ps for data collection. Coordinates for the QM atoms are saved on each dynamics integration step, from which normal coordinate autocorrelation functions are determined to yield the vibrational frequencies.

The Lennard-Jones parameters needed for the QM atoms (last terms in equations 2 and 3) have been developed for the combined AI/TIP3P potential at the HF/3-21G level, and have been shown to yield excellent results for the water dimer both when the QM molecule is treated as a hydrogen bond donor and an acceptor (23). The performance is also good using the HF/6-31G* method for the QM region. These parameters are employed here without further modification. For the DFT/TIP3P potential the parameters listed in Table II are used, which give reasonable hydrogen bond distances and energies for the water dimer complex.

Table II. Van der Waals Parameters for the QM Atoms, and Dimer Interaction Energies for Water Used in the Combined AI/TIP3P and DFT/TIP3P Potential

	AI/TIP3P		DFT/TIP3P	
	σ (Å)	ϵ (kcal/mol)	R_{min} (Å)	ϵ (kcal/mol)
O	3.60	0.15	3.84	0.1521
H	1.30	0.10	1.50	0.0360
		water dimer		
(QM)...(MM)	R(OO), Å	ΔE (kcal/mol)	R(OO), Å	ΔE (kcal/mol)
HOH...OH$_2$	3.00	-5.48	2.94	-5.50
H$_2$O...HOH	2.96	-5.67	3.01	-5.66
Expt[a]	2.98	-5.4	2.98	-5.4

[a]Curtiss, L.A.; Frurip, D.J.; Blaunder M. *J. Chem. Phys.* **1979**, *71*, 2703.

Results and Discussion

The Geometry of Water in Liquid Water. Key results of the present study are listed in Table III, which compares the average geometrical parameters of water obtained from various levels of theory and simulation protocols. In comparing the computed values, it should be kept in mind that the gas phase geometries of water from different theoretical models are somewhat different. Thus, only the change in geometry due to the solvent effect is comparable. The results from all simulations indicate that the OH bond length of water increases by about 0.005 to 0.008 Å in the liquid state in comparison with the gas phase value. This is accompanied by a decrease in the HOH bond angle, ranging from $-1.3 \pm 0.5°$ from hybrid HF 6-31G*/TIP3P MC simulations to $-0.5 \pm 3°$ from DFT/TIP3P MD calculations. The combined HF 3-21G/TIP3P MC results fall into these ranges. Figure 1 shows the histogram of the bond angle fluctuations from our HF 6-31G*/TIP3P MC and DFT BP/TIP3P MD simulations. Clearly, the HOH angle is rather flexible in the liquid phase, which varies from ca. 97° to 113°. It is interesting to note that both the HF and DFT results show similar behaviors in the HOH angle flexibility.

In addition to the errors due to statistical fluctuation, another source of error of the present study is due to the use of fixed classical water geometry. We have examined this effect by carrying out DFT/MM MD simulation using a flexible water model developed by Dang and Pettitt (4). Essentially the same structural changes are obtained, suggesting the effect from the flexible model is minimal. Thus, combining the gas phase experimental data (1) and the computed condensed phase shifts, our best estimates for the geometry of water in liquid water are R(OH) = 0.965 ± 0.002 Å, and θ(HOH) = 103.4 ± 0.5° at the HF/6-31G* level, and R(OH) = 0.962 ± 0.010 Å, and θ(HOH) = 104.0 ± 4.0° from DFT-BP/TZVP MD simulations.

For comparison, early neutron diffraction experiments of Thiessen and Narten predicted an average OH distance of 0.966 Å and an HOH bond angle of 102.8°, or ΔR(OH) = -0.009 Å and $\Delta\theta$(HOH) = -1.7° (12), in accord with the present simulation results. However, a recent neutron diffraction study of D_2O yielded an R(OD) of 0.970 ± 0.005 Å and a DOD angle of 106.06 ± 1.8° (13). The latter, in contrast to the Thiessen-Narten results, is about 1.6° *greater* than the gas phase value. The zero-point vibrational contributions for D_2O have been estimated to be ΔR(OD) = 0.013 Å, which is of similar magnitude of the experimental shift, and $\Delta\theta$(DOD) = -0.02° (37). So, the zero-point vibrational effect further complicates a direct comparison of the computed results with the latest experiment on D_2O. To resolve the experimental discrepancies, additional neutron diffraction studies along with improved data analyses would be desirable.

Quantum chemical calculations have been mostly carried out on water clusters, which show only small structural changes with an increase in the OH bond length and a widening of the HOH angle. Cluster calculations clearly suffer from a lack of the long range dielectric effect, and of the dynamic, vibrational averaging. Hybrid QM/MM simulations of a QM water in liquid water have also been carried out in the past (38,39). These previous studies have mainly focused on the solvent polarization effect. Two recent studies have been aimed at the prediction of water geometry in water. Moriaty and Karlstrom carried out a QM/MM MC simulation of a rigid QM water in water represented by the polarizable NEMO model (40). The average deformation potential energy surface for the QM water molecule was determined from the classical water positions that have been saved during the MC calculation. It was found from geometry optimization using the deformation potential that the OH bond is elongated by 0.01 Å in liquid water, whereas the HOH bond angle increases 4.05 degrees (40). In another study, Nymand and Åstrand used a perturbation approach along with ensemble averages of the electric fields and field-gradients from a MD simulation of water using the NEMO potential (41). Utilizing an atomic natural orbital (ANO) basis set, the predicted geometrical changes in the liquid range from 0.0108 to 0.0304 Å for the OH bond, and -0.59 to +1.93 degrees for the HOH angle at the HF and CAS level of theory with the use of different perturbation terms. Surprisingly, the main contribution to the geometrical changes is from the van der Waals dispersion term (41), without which the

Table III. Computed Geometrical Parameters for Water in Liquid Water from ab initio Monte Carlo and Density Functional-Molecular Dynamics Simulations

	R (OH), Å			θ (HOH), deg		
	gas	liquid	Δ	gas	liquid	Δ
HF/3-21G	0.966	0.972 ± 0.001	0.006	107.7	107.5 ± 0.5	-0.2
HF/6-31G*	0.947	0.955 ± 0.002	0.008	105.5	104.2 ± 0.5	-1.3
BP/TZVP	0.973	0.978 ± 0.010	0.005	105.5	105.0 ± 3.5	-0.5
BP/TZVP[d]	0.973	0.979 ± 0.020	0.006	105.5	104.9 ± 4.5	-0.6
Expt	0.9572[a]	0.970 ± 0.005[b] 0.966 ± 0.006[c]	0.013 0.009	104.52[a]	106.1 ± 1.8[b] 102.4 ± 2.0[c]	+1.6 -2.1

[a]Ref 1. [b]Ref 13. [c]Ref 12. [d]Dang and Pettitt's flexible model for water is used for the liquid.

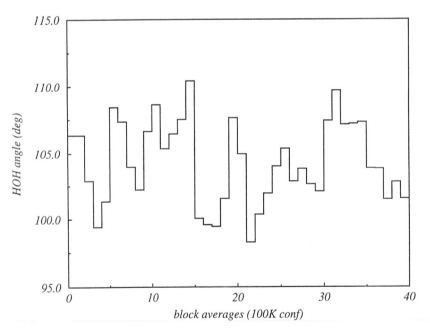

Figure 1. The HOH angle profile as a function of the simulation time from the hybrid HF 6-31G*/TIP3P Monte Carlo (a) and hybrid DFT/TIP3P molecular dynamics (b) simulations.

bond length and angle changes would be 0.0174 Å and -0.11° at the HF level, and 0.0180 Å and 0.47° at the CAS level. The origin of the difference in the predicted HOH angle changes between these and our simulation results is not clear at this point.

Solvent Effects on the Vibrational Frequencies of Water. The change in water geometry is echoed by the predicted shifts in intramolecular vibrational frequencies. Table IV summarizes the computed vibrational frequencies for water in the gas phase and in water at the DFT BP/TZVP level. Gas phase vibrational frequencies are conveniently computed by diagonalization of the Hessian or the second derivative matrix, whereas the aqueous results are obtained from Fourier transforms of the normal coordinate autocorrelation functions (NCAF) for the three normal modes. Thus, the condensed phase results also include possible anharmonic contributions. Wang et al. have used Fourier transforms of the NCAF to computed the vibrational frequency shifts of methanol in liquid methanol (*42*). The normal mode coordinates for water are defined as usual: $R_s = (R_1(OH) + R_2(OH))/2^{1/2}$; $R_a = (R_1(OH) - R_2(OH))/2^{1/2}$; $R_b = \theta(HOH)$.

Table IV. Computed vibrational frequencies (cm^{-1}) for water at the DFT BP/TZVP level

	calculated			experimental		
	gas	liquid	Δ	gas[a]	liquid	Δ
ν_a	3833	3747	-86	3756	3480[b] 3558[c]	-286
ν_s	3709	3666	-43	3657	3385[b]	-282
ν_b	1542	1629	87	1595	1669[c]	74

[a]Ref 43. [b]Ref 45. [c]Ref 46.

Figure 2 and 3 depict the three NCAF and the corresponding Fourier transforms for the "DFT" water molecule obtained from a 5 ps MD trajectory at an integration time-step of 0.2 fs to ascertain that the fast hydrogen vibrations are adequately treated in the MD simulation. At the BP/TZVP level, the computed harmonic vibrational frequencies in the gas phase are in good accord with experiment. In the liquid, both the symmetric (ν_s) and asymmetric (ν_a) stretching frequencies are predicted to be red-shifted by -43 and -86 cm^{-1}, respectively. The bending frequency (ν_b) exhibits a blue-shift of +87 cm^{-1} in water. The computational trends are in agreement with experiment (Table IV), although the calculated stretching frequency shifts are quantitatively too small (*1,43-46*).

We have also recorded the hydrogen velocity autocorrelation function for the QM water, of which the power spectrum obtained by the Fourier transform is shown in Figure 4. The high frequency stretching bands located at 3747 and 3666 cm^{-1} are in exact accord with the NCAF analyses (Table IV), while the shift in the HOH bending mode is determined to be 1629 cm^{-1} from the velocity autocorrelation function. Figure 4 also displays a broad band centered at ca. 500 cm^{-1}, corresponding to the liberation

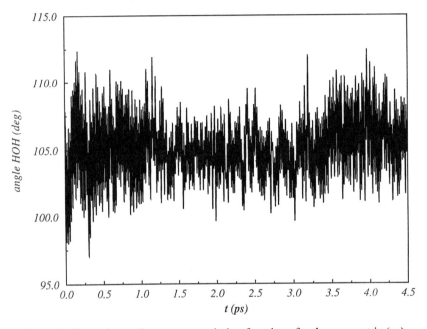

Figure 2. Normal coordinate autocorrelation functions for the symmetric (v_s) and asymmetric (v_a) stretching and the bending (v_b) vibrations for water. Ordinates have been shifted by 1.5 units for successive plots.

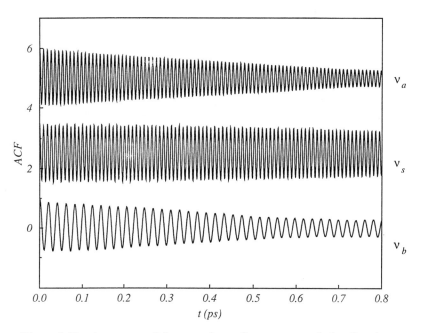

Figure 3. Fourier spectra of the normal coordinate autocorrelation functions displayed in Figure 2 for water. Spectrum intensities are given in arbitrary units.

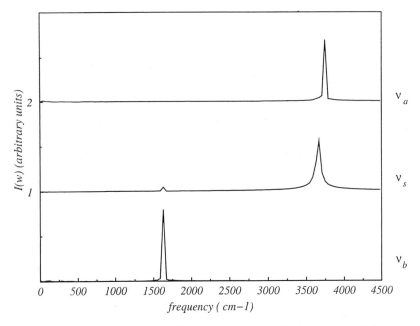

Figure 4. Power spectrum of the hydrogen atom velocity autocorrelation function for water.

frequencies in the liquid, arising from intermolecular interactions. These features are in accord with findings obtained using empirical flexible water models and with experiment (*4-10,45*).

Solvent Polarization Effects. The aqueous solvent polarization effect on water has been thoroughly investigated previously using combined semiempirical AM1/MM and DFT/MM methods. Here, we only wish to echo similar findings for water from combined AI/MM and DFT/MM simulations (*19,22,38,39*). In particular, we found that the molecular dipole moment of water increases from 2.15 D in the gas phase to 2.55 ± 0.01 D in water at the HF/6-31G* level, giving rise to an induced dipole moment of 0.40 D. Similarly, the change at the DFT BP/TZVP level from MD simulations is 0.38 ± 0.09 D. At the local density approximation using the TZVP basis set, Wei and Salahub obtained an induced dipole of 0.66 ± 0.05 D for water, and inclusion of the Perdew nonlocal correlation functional yielded an identical result (*38*). The computed Mulliken population charges from HF 6-31G*/TIP3P MC simulations, and the electrostatic potential fitted charges from DFT BP/TZVP/TIP3P MD calculations are listed in Table V. Both calculations predicted significant increases in atomic partial charge for water in going from the gas phase into the liquid phase.

Table V. Computed Dipole Moment (D) and Partial Charges (e)

	HF 6-31G*/TP3P		DFT BP-TZVP/TIP3P	
	gas	liquid	gas	liquid
dipole	2.15	2.55 ± 0.01	2.16	2.54 ± 0.09
q(O)	-0.868	-0.949 ± 0.007	-0.782	-0.900 ± 0.026
q(H)	0.434	0.475 ± 0.002	0.391	0.450 ± 0.018

Conclusions

Combined quantum and statistical mechanical simulations of a quantum-mechanically treated water in liquid water have been carried out, making use of hybrid AI/MM, and hybrid DFT/MM potentials. The geometry of water in liquid water has been determined both at the ab initio Hartree-Fock level and at the DFT level using gradient corrected exchange and correlation functionals. We found that the OH bond distance increases by about 0.005-0.008 Å, while the HOH angle decreases by 0.2-1.3° in going from the gas phase into the liquid phase. The predicted increase in the OH bond length is in agreement with experiment and previous computational results. However, both experimental and computational results on the HOH bond angle change are inconclusive in previous investigations. Molecular dynamics and Monte Carlo simulations employing empirical flexible models predicted a decrease as large as 5° in the HOH angle, whereas two recent molecular orbital calculations, which included the average solvent potentials, indicated an increase in the HOH angle. The present investigation, however, represents a direct ab initio QM/MM simulation study to yield the average geometrical parameters of water in the liquid phase, and thus, should be more reliable. The predicted structural changes are reflected in the computed vibrational frequencies of water in liquid water, which are derived from the normal coordinate autocorrelation functions. The trends of the condensed phase vibrational frequency shifts are found to be consistent with experiment.

Acknowledgments. This work has been partially supported by the National Institutes of Health and by Praxair, Inc. Discussions with Professor William L. Jorgensen are also useful.

Literature Cited

(1)   Eisenberg, D.; Kauzmann, W. *The Structure and Properties of Water*; Clarendon: Oxford, 1969.

(2)   Berendsen, H.; Potsma, J.; van Gunsteren, W.; Hermans, J. *Intermolecular Forces*; D. Reidel Publishing Co.; Dordrecht, Netherlands, 1981, pp 331.

(3)   Jorgensen, W. L.; Chandrasekhar, J.; Madura, J. D.; Impey, R.; Klein, M. *J. Chem. Phys.* **1983**, *79*, 926.

(4)   Dang, L.; Pettitt, B. M. *J. Phys. Chem.* **1987**, *91*, 3349.

(5)   Teleman, O.; Jonsson, B.; Engstrom, S. *Mol. Phys.* **1987**, *60*, 193. Wallqvist, A.; Teleman, O. *Mol. Phys.* **1991**, *74*, 515.

(6)   Toukan, K.; Rahman, A. *Phys. Rev. B.* **1985**, *31*, 2643.

(7)   Lie, C. G.; Clementi, E. *Phys. Rev. A*, **1986**, *33*, 2679.

(8)   Zhu, S.-B.; Singh, S.; Robinson, G. W. *J. Chem. Phys.* **1991**, *95*, 2791.

(9)   Mizan, T. I.; Savage, P. E.; Ziff, R. M. *J. Phys. Chem.* **1994**, *98*, 13067. Mizan, T. I.; Savage, P. E.; Ziff, R. M. *J. Comput. Chem.* **1996**, *17*, 1757.

(10)   Silvestrelli, P. L.; Bernasconi, M.; Parrinello, M. *Chem. Phys. Lett.* **1997**, *277*, 478. Sprik, M.; Hutter, J.; Parrinello, M. *J. Chem. Phys.* **1996**, *105*, 1142.

(11)   Stuart, S. J.; Berne, B. J. *J. Phys. Chem.* **1996**, *100*, 11934. Gao, J. *J. Chem. Phys.* **1998**, in press.

(12)   Thiessen, W. E.; Narten, A. M. *J. Chem. Phys.* **1982**, *77*, 2656.

(13)   Ichikawa, K.; Kameda, Y.; Yamaguchi, T.; Wakita, H.; Misawa, M. *Mol. Phys.* **1991**, *97*, 79.

(14)   Gao, J. *Acc. Chem. Res.* **1996**, *29*, 298. Gao, J. In *Reviews in Computational Chemistry*; Lipkowitz, K.; Boyd, D. B. Eds., VCH: New York, Vol. 7, 1995, pp 119-185.

(15)   Warshel, A.; Levitt, M. *J. Mol. Biol.* **1976**, *103*, 227.

(16)   Singh, U. C.; Kollman, P. A. *J. Comput. Chem.* **1986**, *7*, 718.

(17)   Field, M. J.; Bash, P. A.; Karplus, M. *J. Comput. Chem.* **1990**, *11*, 700.

(18)   Gao, J. *J. Phys. Chem.* **1992**, *96*, 537.

(19)   Gao, J.; Xia, X. *Science* **1992**, *258*, 631.

(20)   Stanton, R. V.; Hartsough, D. S.; Merz, K. M. *J. Phys. Chem.* **1993**, *97*, 11868.

(21)   Thompson, M. A.; Schenter, G. K. *J. Phys. Chem.* **1995**, *99*, 6374.

(22)   Tunon, I.; Martins-Costa, M. T. C.; Millot, C.; Ruiz-Lopez, M. F. *J. Chem. Phys.* **1997**, *106*, 3633. Strnad, M.; Martins-Costa, M. T. C.; Millot, C.; Tunon, I.; Ruiz-Lopez, M. F.; Rivail J. L. *J. Chem. Phys.* **1997**, *106*, 3643.

(23)   Freindorf, M.; Gao, J. *J. Comput. Chem.* **1996**, *17*, 386.

(24)   Schmidt, M. W.; Baldridge, K. K.; Boatz, J. A.; Elbert, S. T.; Gordon, M. S.; Jensen, J. H.; Koseki, S.; Matsunaga, N.; Nguyen, K. A.; Su, S. J.; Windus, T. L.; Dupuis, M.; Montgomery, J. S. *J. Comput. Chem.* **1993**, *14*, 1347.

(25)   Gao, J. *MCQUB*; Department of Chemistry, SUNY at Buffalo, 1996. Jorgensen, W. L. *BOSS*; Version 2.9, Department of Chemistry, Yale University, 1990.

(26)   Gao, J.; Freindorf, M. *J. Phys. Chem.* **1997**, *101*, 3182.

(27)   Parr, R. G.; Yang, W. *Density Functional Theory of Atoms and Molecules*; Oxford University: London, 1989.

(28)   Kohn, W.; Sham, L. J. *Phys. Rev.* **1965**, *A140*, 1133.

(29)   Perdew, J. P. *Phys. Rev. B.* **1986**, *33*, 8822. *34*, 7406.

(30)   Becke, A. D. *Phys. Rev. A.* **1988**, *38*, 3098.

(31)   St-Amant, A. Department of Chemistry, University of Ottawa, 1997.

(32)   Godbout, N.; Salahub, D. R.; Andzelm, J.; Wimmer, E. *Can. J. Chem.* **1992**, *70*, 560.

(33)   Fournier, R.; Andzelm, J.; Salahub, D. R. *J. Chem. Phys.* **1989**, *90*, 6371.

(34)   Brooks, B. R.; Bruccoleri, R. E.; Olafson, B. D.; States, D. J.; Swaminathan, S.; Karplus, M. *J. Comput. Chem.* **1983**, *4*, 187.

(35)   Hehre, W. J.; Radom, L.; Schleyer, P. v. R.; Pople, J. A. *Ab initio Molecular Orbital Theory*; Wiley: New York, 1986.

(36)   Brooks, C. L.; Karplus, M. *J. Chem. Phys.* **1983**, *79*, 6312.

(37)   Fowler, P. W.; Paynes, W. T. *Mol. Phys.* **1981**, *43*, 65.

(38)   Wei, D. Q.; Salahub, D. R. *Chem. Phys. Lett.* **1994**, *224*, 291.

(39)   Moriarty, N. W.; Karlstrom, G. *J. Phys. Chem.* **1996**, *100*, 17791.

(40)   Moriarty, N. W.; Karlstrom, G. *J. Chem. Phys.* **1997**, *106*, 6470.

(41)   Nymand, T. M.; Astrand, P.-O. *J. Phys. Chem.* **1997**, *101*, 10039.

(42)   Wang, J.; Boyd, R. J.; Laaksonen, A. *J. Chem. Phys.* **1996**, *104*, 7261.

(43)   Benedict, W. S.; Gailar, N.; Plyler, E. K. *J. Chem. Phys.* **1956**, *24*, 1139.

(44)   Bayly, J. G.; Kartha, V. B.; Stevens, W. H. *Infrared Phys.* **1967**, *47*, 116.

(45)   Ford, T. A.; Falk, M. *Can. J. Chem.* **1968**, *46*, 3579.

(46)   Chen, S. H.; Toukan, K.; Loong, C. K.; Price, D. L.; Teixeira, J. *Phys. Rev. Lett.* **1984**, *53*, 1360. Walfren, G. E. *J. Chem. Phys.* **1967**, *47*, 116.

Chapter 4

On the Treatment of Link Atoms in Hybrid Methods

Iris Antes and Walter Thiel

Organisch-chemisches Institut, Universität Zürich, Winterthurerstrasse 190, CH-8057 Zürich, Switzerland

Several coupling models and link atom options are investigated for combined quantum-chemical and classical approaches. Ab initio, density functional, and semiempirical methods are used for the quantum-chemical region, whereas the classical region is described by the AMBER force field. Numerical results are reported for the proton affinities of alcohols and ethers. Recommendations for link atoms are given based on these results and theoretical analysis.

Hybrid methods (1,2) treat the electronically important part of large systems by quantum mechanics (QM), and the remainder by molecular mechanics (MM). In many applications, there is a natural division of a large system into weakly interacting QM and MM regions (e.g. into a QM solute and MM solvent). Generally, however, this division may involve the cutting of covalent bonds (e.g. in enzymatic reactions) which causes a major perturbation of the electronic structure. Two approaches are commonly used to deal with this problem at the QM level: the electronic density at the bond being cut can be represented by frozen hybrid orbitals (1,3), or the free valencies can be saturated by link atoms (typically hydrogen atoms) (4,5). Both approaches have merits and drawbacks, and both introduce errors which may determine and limit the overall accuracy of QM/MM calculations. The present contribution reports a systematic study of some of the choices that need to be made when using link atoms.

Theoretical Background

In our previous work (6,7) we have defined a hierarchy of four QM/MM coupling models which may be characterized briefly as follows:

A Mechanical embedding of the QM subsystem, QM/MM interactions treated according to molecular mechanics.

B Electronic embedding of the QM subsystem, QM calculation in the presence of MM point charges (resulting in QM polarization), QM/MM Coulomb interactions evaluated from the QM electrostatic potential and MM atomic charges.

C Model B plus MM polarization evaluated from the QM electric field and MM polarizabilities using a published dipole interaction model (8).

D Model C with a self-consistent treatment of MM polarization (for similar recent approaches see (9-11)).

Each of the coupling schemes A-D contains a global link atom correction (E_{LINK}) which removes QM energy terms involving link atoms from the total energy expression. This correction has been discussed in detail (6,7) and will thus not be considered here. In our approach the total energy of the system (X-Y) can generally be written as

$$E_{tot}(X-Y) = E_{MM}(X) + E_{QM}(Y-L) + E_{QM/MM}(X,Y) + E_{LINK} \tag{1}$$

where $E_{MM}(X)$ denotes the MM energy of the MM subsystem X, $E_{QM}(Y-L)$ the QM energy of the QM subsystem Y saturated with link atoms L, and $E_{QM/MM}(X,Y)$ the interaction energy between the subsystems. Coupling schemes A-D differ only in the definition of the interaction energy.

In the case of mechanical embedding (model A), the QM/MM interaction is treated purely classically, without any contributions from the link atoms:

$$E_{QM/MM}^{A}(X,Y) = E_{MM}(X,Y) \tag{2}$$

Model A is therefore quite robust with regard to the link atoms. This is also true for other approaches with mechanical embedding (*12,13*) which adopt some slightly different link atom conventions (e.g. concerning positional constraints of link atoms or gradient evaluation).

Electronic embedding (model B and beyond) is appropriate when the MM environment contains strongly polar or charged groups that influence the electron distribution in the QM region. This effect can be taken into account by including the MM atomic charges q_J into the one-electron core hamiltonian:

$$\tilde{h}_{\mu\nu} = h_{\mu\nu} - \sum_J^X q_J V_{\mu\nu}^J \tag{3}$$

The resulting Coulomb (electrostatic) energy is given by:

$$E^{coul}(X,Y) = \sum_J^X q_J \Phi^J \tag{4}$$

$$\Phi^J = -\sum_\mu^{Y-L} \sum_\nu^{Y-L} P_{\mu\nu} V_{\mu\nu}^J + \sum_A^{Y-L} Z_A V^{AJ} \tag{5}$$

where $P_{\mu\nu}$ and Z_A are density matrix elements and nuclear charges, respectively. $V_{\mu\nu}^J$ and V^{AJ} denote nuclear attraction integrals and Coulomb repulsion terms which describe the interaction between a unit charge at MM atom J and an electron or nucleus A in the QM region, respectively. Φ^J is the electrostatic potential generated by the QM region at the position of MM atom J.

Using the modified core hamiltonian $\tilde{h}$ in the SCF procedure yields a perturbed density matrix $\tilde{P}$ and, consequently, a different SCF energy $\tilde{E}_{QM}(Y-L)$. This induction (polarization) in the QM region is accompanied by a net energy gain of

$$E^{ind}(Y) = \tilde{E}_{QM}(Y-L) - E_{QM}(Y-L) - E^{coul}(X,Y) \tag{6}$$

The QM/MM interaction energy for model B is

$$E_{QM/MM}^{B}(X,Y) = E^{coul}(X,Y) + E^{ind}(Y) + E_{MM}^{bonded}(X,Y) + E_{MM}^{vdW}(X,Y) \qquad (7)$$

where the last two terms represent the bonded and van-der-Waals interactions between the QM and MM subsystems (evaluated from the force field, as in model A).

Since the link atoms L are not part of the real system X-Y, it seems reasonable to require (5) that they do not contribute to the QM/MM interation energy. At the level of model B, this can be accomplished (5-7) by the following integral approximation:

$$L1: V_{\mu\nu}^{J} = 0 \text{ if } \mu \text{ or } \nu \text{ belongs to L, and } V^{AJ} = 0 \text{ if } A \in L \qquad (8)$$

which efficiently restricts the range of summation in eq. (5) to Y instead of Y-L (6,7). This seemingly natural option L1 might have two drawbacks: First, the QM polarization may become unrealistic when the polarizing influence of MM atoms at intermediate distances is felt only by the QM atoms Y and not by the link atoms L (see. eq. (3)). Secondly, the electrostatic potential generated by the QM region may become unbalanced when the contributions from the link atoms are omitted (see eq. (5)). To explore these effects, we have investigated many other link atom options which have in common that the link atoms "see" the MM atoms at intermediate and large distances and that integral approximations are applied only at short range, i.e. they involve only the MM atoms of the first neighboring neutral charge group (14) at the bond being cut (N_{group}) or are restricted to the MM atom being replaced by the link atom (N_1). From a large number of options studied, the following ones are considered here:

$$L2: V_{\mu\nu}^{J} = 0 \text{ and } V^{AJ} = 0 \text{ if } J \in N_{group} \qquad (9)$$

$$L3: V_{\mu\nu}^{J} = 0 \text{ if } \mu \text{ or } \nu \text{ belongs to L and } J = N_1, \text{ and}$$

$$V^{AJ} = 0 \text{ if } A \in L \text{ and } J = N_1 \qquad (10)$$

Option L2 implies that the atoms from the first neighboring MM charge group are electrostatically invisible to all QM and link atoms in Y-L. Option L3 includes all interactions except those involving the pair N_1-L (where L replaces N_1). Incorporating the N_1-L interactions has no conceptual foundation and leads, in numerical tests, to severely distorted unphysical wavefunctions for the QM subsystem. Hence, option L3 is the closest reasonable approach to the full interaction limit (X-L).

The electronic embedding (model B) may be refined by accounting for MM polarization in a non-iterative (model C) or iterative (model D) manner (6,7). The chosen classical treatment which is based on an induced dipole interaction scheme (8) requires as input the electronic field F_α^J (components α = x,y,z) generated by the QM region at the positions of the MM atoms J:

$$F_\alpha^J = \sum_\mu^{Y-L} \sum_\nu^{Y-L} \tilde{P}_{\mu\nu} \nabla_\alpha V_{\mu\nu}^J - \sum_A^{Y-L} Z_A \nabla_\alpha V^{AJ} \tag{11}$$

The link atom options L1-L3 are again defined by the integral approximations (8)-(10), in complete analogy to model B.

Up to this point, the development has been general, and no distinction has been made between different QM methods. When using ab initio or density functional theory (DFT) as QM component in hybrid approaches, the basic interactions $V_{\mu\nu}^J$ and V^{AJ} are defined (in the usual notation) as:

$$V_{\mu\nu}^J = \langle \mu(1) | r_{1J}^{-1} | \nu(1) \rangle \tag{12}$$

$$V^{AJ} = R_{AJ}^{-1} \tag{13}$$

When applying semiempirical QM methods, the appropriate semiempirical approximations must be used for $V_{\mu\nu}^J$ and V^{AJ} (6,7,15). It is of interest to check whether and how these integral approximations affect options L1-L3. For this purpose we consider the contribution of the link atoms to the electrostatic potential (Φ_{LINK}^J) for the standard case that a single hydrogen atom serves as a link atom. At the semiempirical level, hydrogen carries a single s orbital ($\mu = \nu = s$) so that eq. (5) simplifies to:

$$\Phi_{LINK}^{J} = -P_{ss}V_{ss}^{J} + V^{AJ} \qquad \text{(semiempirical)} \qquad (14)$$

Since we have typically $P_{ss} \approx 1$ (H is almost neutral in capping C-H bonds) and $V_{ss}^{J} \approx V^{AJ}$ (semiempirical neglect of penetration), we expect $\Phi_{LINK}^{J} \approx 0$ due to effective cancellation of attractive and repulsive terms. At the ab initio or density functional level, these arguments do not hold: In particular, eq. (5) contains attractive three-center contributions $V_{\mu\nu}^{J}$ (neglected in semiempirical methods) which have no direct repulsive counterparts. In addition, hydrogen is normally less neutral, and $V_{\mu\nu}^{J} < V^{AJ}$ (for analytically evaluated integrals). As a consequence, the cancellation of attractive and repulsive terms is much less effective at the ab initio or density functional level which implies much larger link atom contributions to Φ_J. An analogous conclusion holds for F_{α}^{J} (see eq. (11)).

These considerations suggest that the differences between the link atom options L1-L3 will be less pronounced at the semiempirical level (compared with ab initio or DFT).

Computational Methods

Our previous QM/MM work (6,7) employed an integrated program combining semiempirical QM methods (MNDO, AM1) with the MM3 force field. We have now developed a flexible modular program that allows the combination of different QM and MM codes using the coupling models A-C. Coupling scheme D has not yet been implemented because it had not been cost-effective in the previous MNDO/MM3 studies (6,7).

The present work is based on combining the following programs: CADPAC5.2 (16) (ab initio and DFT), MNDO95 (17) (semiempirical), and AMBER4.1 (18) (force field). Standard QM methods have been used: Ab initio restricted Hartree-Fock (RHF) and second order Møller-Plesset perturbation theory (MP2) with standard basis sets (STO-3G, 3-21G, 3-21G*, 6-31G, 6-31G*) (*19*), gradient corrected BLYP density functional theory (*20*), and the semiempirical AM1 method (*21*). Standard AMBER4.1 force field parameters (*22*) have been employed along with the following atomic charges (e) for the neutral charge groups in the MM region: CH_2 (H +0.09, C -0.18), CH_3 (H +0.09, C -0.27), OH (H +0.40, C -0.40), NH_2 (H +0.30, N -0.60), $C'H_2OC''H_3$ (H +0.09, C' -0.10, O -0.23, C'' -0.12).

Geometries have been optimized at the QM level and at the QM/MM level using coupling model B. Single-point QM/MM energies for coupling model C have been determined at the corresponding QM/MM/B geometries. Proton affinities have been evaluated from the computed total energies (QM = ab initio, DFT) corrected for the zero point energy and thermal contributions (*19*) or from heats of formation (QM = AM1) following previously described conventions (*6*).

Results

We only report numerical QM/MM results for coupling models B and C since model A is comparatively insensitive to details of the link atom treatment. In analogy to previous work (*5-7*) we consider proton affinities because the addition of a proton should lead to pronounced electrostatic and polarization effects which are modeled by coupling schemes B and C. Generally we focus on relative proton affinities or, equivalently, trends in absolute proton affinities which should be reproduced by reasonable QM/MM approaches.

Table I lists semiempirical, ab initio, and DFT results for simple alcohols. The calculated (QM) absolute proton affinities of methanol cover a wide range and are fairly basis set dependent at the RHF level, but they are close to the experimental value at the highest levels applied (RHF, BLYP, and MP2 with the 6-31G* basis set). The calculated (QM) proton affinities of the other alcohols *relative to methanol* are much more uniform and show little basis set variation, and they generally reproduce the increase of the proton affinities with increasing alkyl substitution quite well. The QM/MM calculations with model B (option L1, QM region: methanol) also show this increase, but underestimate its magnitude significantly. Inclusion of MM polarization

Table I. Proton affinities (kcal/mol): Absolute values for methanol, values relative to methanol for the other molecules[a]

	exp.[b]	AM1	RHF STO3G	RHF 3-21G	RHF 3-21G*	RHF 6-31G	RHF 6-31G*	BLYP 6-31G*	MP2 6-31G*
methanol									
QM	181.9	170.4	230.2	196.6	188.4	190.7	182.0	180.6	179.5
ethanol	6.4								
QM		7.8	6.2	4.4	4.9	4.8	4.8	6.8	4.8
QM/MM/B		4.1	3.5	2.4	2.2	2.1	2.3	2.3	3.4
QM/MM/C		7.0	10.1	10.9	8.8	9.7	9.6	9.7	10.5
propanol	8.9								
QM		7.1	7.1	4.7	6.1	5.9	6.2	8.4	6.3
QM/MM/B		3.7	3.6	2.7	2.6	2.9	2.8	3.0	5.0
QM/MM/C		6.2	11.4	11.2	11.4	13.3	12.8	12.9	14.4
iso-propanol	9.3								
QM		12.3	11.4	8.5	8.7	9.9	9.4	12.1	8.8
QM/MM/B		5.7	6.3	4.0	3.9	4.7	4.5	5.2	6.7
QM/MM/C		10.9	20.3	19.4	20.6	22.9	23.3	24.5	25.1
tert-butanol	11.8								
QM		15.6	18.2	14.0	14.7	17.0	16.2	19.4	15.0
QM/MM/B		5.3	6.5	3.3	2.7	4.7	4.3	5.9	6.5
QM/MM/C		12.7	27.4	30.3	30.8	36.4	36.5	39.8	38.1

(a) The QM/MM calculations employ models B and C with link atom option L1. The QM/MM division involves cutting the CC bond(s) next to oxygen.

(b) From ref. 23.

with model C yields reasonable relative proton affinities at the semiempirical AM1 level (6,7), but overestimates the experimental values strongly at the ab initio and DFT levels. This latter deficiency has been traced to excessively high electric fields generated by the QM region which are imbalanced due to the omission of the link atom contributions in option L1 (see the preceding analysis in the theoretical section). The numerical data thus confirm that the semiempirical methods are less sensitive in this regard (as expected theoretically). Analogous numerical results have been obtained for amines and thiols (not shown). We conclude that model C with link atom option L1 should only be applied with semiempirical QM methods.

Link atom options L2 and L3 have been introduced in an attempt to overcome imbalance problems with option L1 (see theoretical section). Tables II and III contain the corresponding AM1, RHF/6-31G*, and BLYP/6-31G* results for alcohols and ethers, respectively. Options L1 and L2 tend to give similar relative proton affinities with model B, except for the limiting case that the MM regions consist of single methyl groups (as in ethanol, iso-propanol, tert-butanol, and diethylether). In this case, the electrostatic effects with option L2 must be small by definition (see eq.(9)) which is indeed found (values below 1 kcal/mol, see Tables II and III). On the other hand, inclusion of MM polarization by model C and option L2 generally leads to reasonable results both for the trends and the actual values of the relative proton affinities (with a slight tendency for overestimation); this is true for all three QM approaches (AM1, RHF, BLYP). Concerning option L3, the results with model B seem inconsistent while those with model C are improved, but still less satisfactory (e.g. when compared with C/L2).

It is intuitively obvious that link atom effects should be reduced by moving the QM/MM boundary farther away from the electronically important site. This is documented in Table IV for n-propanol and n-butanol. In most cases, the results do approach the limiting QM value when cutting more distant CC bonds (as expected) but the variations are more systematic for options L1 and L2 than for L3. Taking n-butanol as an example, the differences between QM and QM/MM proton affinities are often below 2.5 (1.0) kcal/mol when cutting at the second (third) CC bond from the protonation site; this holds for all QM methods considered (see Table IV). Concomitantly, the differences between the results for models B and C diminish.

Table II. Proton affinities (kcal/mol): Absolute values for methanol, values relative to methanol for the other molecules[a]

	exp.[b]	QM	L1 model B	L1 model C	L2 model B	L2 model C	L3 model B	L3 model C
AM1								
methanol	181.9	170.4						
ethanol	6.4	7.8	4.1	7.0	0.5	6.3	1.8	4.9
propanol	8.9	7.1	3.7	6.2	3.4	9.4	1.8	6.2
iso-propanol	9.3	12.3	5.7	10.9	0.5	10.6	-3.5	3.3
butanol	9.2	7.5	3.6	6.6	2.9	10.0	1.2	6.3
iso-butanol	-	8.0	4.1	9.3	7.2	15.8	4.3	11.7
tert-butanol	11.8	15.6	5.3	12.7	1.9	14.0	-8.2	6.0
*RHF/6-31G**								
methanol	181.9	182.0						
ethanol	6.4	4.8	2.3	9.6	0.0	5.4	-1.6	2.5
propanol	8.9	6.2	2.8	12.8	3.6	10.6	1.2	6.5
iso-propanol	9.3	9.4	4.5	23.3	0.8	11.5	-2.9	6.2
butanol	9.2	7.4	2.6	13.9	3.2	10.9	0.5	6.6
iso-butanol	-	7.6	3.7	14.6	6.5	14.6	4.3	11.2
tert-butanol	11.8	16.2	4.3	36.5	0.7	16.4	-5.6	9.2
*BLYP/6-31G**								
methanol	181.9	180.6						
ethanol	6.4	6.8	2.3	9.7	0.0	5.6	-1.6	2.8
propanol	8.9	8.4	3.0	12.9	3.7	10.6	1.3	6.6
iso-propanol	9.3	12.1	5.2	24.5	0.8	11.9	-2.0	7.4
butanol	9.2	9.5	2.7	14.0	3.3	11.1	0.7	6.7
iso-butanol	-	9.8	4.0	16.8	6.6	14.9	4.7	11.9
tert-butanol	11.8	19.4	5.9	39.7	0.7	16.8	-3.8	11.4

(a) The QM/MM calculations employ models B and C with link atom options L1-L3. The QM/MM division involves cutting the CC bond(s) next to oxygen. (b) From ref. 23.

Table III. Proton affinities (kcal/mol): Absolute values for dimethylether, values relative to dimethylether for the other molecules[a]

	exp.[b]	QM	L1 model B	L1 model C	L2 model B	L2 model C	L3 model B	L3 model C
AM1								
dimethylether	192.1	175.9						
diethylether	8.1	11.3	6.3	9.6	0.1	8.8	-1.6	3.2
dimethoxyetane	12.8	11.0	7.6	10.2	7.7	14.1	9.2	16.9
di(methoxyethyl)ether	27.3	20.3	14.1	18.4	15.3	27.5	19.5	35.5
RHF/6-31G*								
dimethylether	192.1	190.5						
diethylether	8.1	7.4	3.4	18.0	0.2	9.7	-3.4	3.9
dimethoxyetane	12.8	12.9	8.2	16.0	7.4	14.4	7.7	14.4
di(methoxyethyl)ether	27.3	23.4	16.9	40.4	14.9	28.1	15.2	28.2
BLYP								
dimethylether	192.1	187.7						
diethylether	8.1	9.7	4.0	17.9	0.3	10.2	-3.3	4.3
dimethoxyetane	12.8	16.3	8.2	16.9	7.4	14.6	8.5	15.6
di(methoxyethyl)ether	27.3	25.6	17.2	44.1	14.8	28.5	11.7	32.7

(a) The QM/MM calculations employ models B and C with link atom options L1-L3. The QM/MM division involves cutting CC bond(s) next to the protonated oxygen atom. (b) From ref. 23.

Table IV. Proton affinities (kcal/mol): Values relative to methanol[a]

Y(QM)	X(MM)	exp.[b]	QM	L1 model B	L1 model C	L2 model B	L2 model C	L3 model B	L3 model C
AM1									
HOCH$_2$	CH$_2$CH$_3$	8.9	7.1	3.7	6.2	3.4	9.4	1.8	6.2
HOCH$_2$CH$_2$	CH$_3$			8.2	8.9	6.9	9.9	1.9	2.9
HOCH$_2$	CH$_2$CH$_2$CH$_3$	9.2	7.5	3.6	6.6	2.9	9.7	1.2	6.3
HOCH$_2$CH$_2$	CH$_2$CH$_3$			8.1	9.0	9.4	12.8	5.0	7.0
HOCH$_2$CH$_2$CH$_2$	CH$_3$			7.9	8.2	7.1	8.4	4.4	4.7
RHF/6-31G*									
HOCH$_2$	CH$_2$CH$_3$	8.9	6.2	2.8	12.8	0.5	6.7	-2.1	3.3
HOCH$_2$CH$_2$	CH$_3$			6.1	10.5	5.0	8.1	2.8	4.8
HOCH$_2$	CH$_2$CH$_2$CH$_3$	9.2	7.4	2.6	13.9	3.2	10.9	0.5	6.6
HOCH$_2$CH$_2$	CH$_2$CH$_3$			6.3	12.1	6.3	9.9	4.7	7.1
HOCH$_2$CH$_2$CH$_2$	CH$_3$			7.3	10.2	6.5	8.1	5.3	6.3
BLYP/6-31G*									
HOCH$_2$	CH$_2$CH$_3$	8.9	8.4	3.0	12.9	1.3	8.4	-1.9	3.4
HOCH$_2$CH$_2$	CH$_3$			7.9	12.3	6.4	10.0	4.1	6.3
HOCH$_2$	CH$_2$CH$_2$CH$_3$	9.2	9.6	2.7	13.9	3.2	11.0	0.7	6.7
HOCH$_2$CH$_2$	CH$_2$CH$_3$			7.1	13.0	7.9	11.8	5.3	7.9
HOCH$_2$CH$_2$CH$_2$	CH$_3$			9.4	12.5	8.4	10.3	7.2	8.4

(a) The QM/MM calculations employ models B and C with link atom options L1-L3. QM/MM division as indicated. (b) From ref. 23.

Table V reports AM1 and RHF/6-31G* proton affinities for HO-CH$_2$-CH$_2$-CH$_2$-OH (1), HO-CH$_2$-CH$_2$-CH$_2$-NH$_2$ (2), and H$_2$N-CH$_2$-CH$_2$-CH$_2$-NH$_2$ (3), to check the influence of electronegative heteroatoms in the MM part on the protonation that occurs in the MM part. The different computational approaches generally predict the same order of proton affinities, i.e. **1**>**2** for O-protonation and **2**<**3** for N-protonation, but the QM/MM calculations tend to underestimate the corresponding differences for O-protonation when compared with the QM data. At the RHF/6-31G* level, the results from option L2 appear most reasonable, while the AM1 results are of similar quality for options L1 and L2. The QM/MM proton affinities in Table V refer to a QM/MM cut in the CC bond adjacent to the protonation site. When the QM/MM boundary is moved to the second CC bond (i.e. close to the MM heteroatom) the results remain reasonable for AM1, but become less stable for RHF/6-31G* (data not shown) which indicates the sensitivity of the ab initio wavefunction to high MM charges in the vicinity.

The coupling models considered do not allow for charge transfer between the QM and MM subsystems. The results in Tables I-V suggest that the trends in the calculated proton affinities are not adversely affected by this constraint. In general, such charge transfer effects should be minimized by a proper choice of the QM/MM boundary.

Conclusions

In QM/MM calculations with electronic embedding (models B and C), link atom options L1 and L2 both seem suitable at the semiempirical level. Option L2 is preferred at the ab initio and DFT levels where the combination of option L1 and model C yields unrealistic results. Option L2 should generally be more robust than L1 because it neglects any electrostatic interactions between Y-L (i.e. the QM region including link atoms) and the first neighboring charge group in the MM region X, and it may also be quite balanced since it includes the interactions between Y-L and the more distant parts of X evenly. Option L3 incorporates a maximum number of Y-L/X interactions, but cannot be recommended because the results are normally less consistent than those for L1 and L2.

Table V. Absolute proton affinities (kcal/mol)[a]

Y(QM)	X(MM)	exp.[b]	QM	L1		L2		L3	
				model B	model C	model B	model C	model B	model C
AM1									
HOCH$_2$	CH$_2$CH$_2$OH	-	190.1	184.5	191.6	181.5	189.7	184.3	191.9
HOCH$_3$	CH$_2$CH$_2$NH$_2$	-	177.2	176.2	179.4	176.0	182.8	174.3	179.4
H$_2$NCH$_3$	CH$_2$CH$_2$OH	228.6	220.2	217.3	219.7	215.7	219.8	216.0	219.6
H$_2$NCH$_3$	CH$_2$CH$_2$NH$_2$	234.1	224.4	220.6	225.2	219.8	226.1	219.0	224.3
*RHF/6-31G**									
HOCH$_3$	CH$_2$CH$_2$OH	-	202.0	200.5	220.3	190.7	198.7	191.3	199.9
HOCH$_3$	CH$_2$CH$_2$NH$_2$	-	189.0	185.0	197.8	185.9	192.7	183.2	189.5
H$_2$NCH$_3$	CH$_2$CH$_2$OH	228.6	225.2	231.6	244.8	222.0	226.0	222.5	225.3
H$_2$NCH$_3$	CH$_2$CH$_2$NH$_2$	234.1	236.8	232.6	247.6	229.1	236.6	228.3	235.4

(a) The QM/MM calculations employ models B and C with link atom options L1-L3. QM/MM division as indicated. The QM heteroatom is protonated.

(b) From ref. 23.

None of the link atom schemes is perfect. As expected theoretically, semiempirical QM methods are usually less sensitive than ab initio or DFT methods with regard to details of the link atom treatment. In general, it seems advisable to minimize link atom problems by moving the QM/MM boundary far away from the electronically important sites in the QM region. The numerical results for proton affinities suggest that there should be at least two "insulating" CC bonds between such sites and the link atoms, and that one should also avoid QM/MM cuts close to highly charged MM atoms.

Acknowledgment

This work was supported by the Schweizerischer Nationalfonds.

Literature Cited

1. Warshel, A.; Levitt, M. *J. Mol. Biol.* **1976**, *103*, 227.

2. Gao, J. In *Reviews in Computational Chemistry*; Lipkowitz, K.B.; Boyd, D.B., Eds.; VCH Publishers: New York, 1995, Vol.7; pp. 119-186.

3. Théry, V.; Rinaldi, D.; Rivail, J.-L.; Maigret, B.; Ferenczy, G. *J. Comp. Chem.* **1994**, *15*, 269.

4. Singh, U.C.; Kollman, P.A. *J. Comp. Chem.* **1986**, *7*, 718.

5. Field, M.J.; Bash, P.A.; Karplus, M. . *J. Comp. Chem.* **1990**, *11*, 700.

6. Bakowies, D.; Thiel, W. *J. Phys. Chem.* **1996**, *100*, 10580.

7. Bakowies, D. *Hybridmodelle zur Kopplung quantenchemischer und molekül-mechanischer Verfahren*; Ph.D. Thesis, University of Zürich; Hartung-Gorre Verlag: Konstanz, 1994.

8. Thole, B.T. *Chem. Phys.* **1981**, *59*, 341.

9. Thompson, M.A.; Schenter, G.K. *J. Phys. Chem.* **1995**, *99*, 6374.

10. Thompson, M.A. *J. Phys. Chem.* **1996**, *100*, 14492.

11. Gao, J.; *J. Comp. Chem.* **1997**, *18*, 1061.

12. Maseras, F.; Morokuma, K. *J. Comp. Chem.* **1995**, *16*, 1170.

13. Eichler, U.; Kölmel, C.M.; Sauer, J. *J. Comp. Chem.* **1997**, *18*, 463.

14. Smith, P.E., van Gunsteren, W.F. In *Computer Simulations of Biomolecular Systems*; van Gunsteren, W.F.; Weiner, P.K.; Wilkinson, A.J., Eds.; ESCOM: Leiden, 1993, Vol.2; pp. 182-212.

15. Bakowies, D.; Thiel, W. *J. Comp. Chem.* **1996**, *17*, 87.

16. CADPAC5.2 *The Cambridge Analytic Derivatives Package Issue 5.2*, Cambridge, 1992. A suite of quantum chemistry programs developed by Amos, R.D. with contributions from Alberts, I.L.; Colwell, S.M.; Handy, N.C.; Jayatilaka, D.; Knowles, P.J.; Kobayashi, R.; Koga, N.; Laidig, K.E.; Maslen, P.E.; Murray, C.W.; Rice, J.E.; Sanz, J.; Simandiras, E.D.; Stone, A.J.; Su, M.-D.

17. Thiel, W. *Program MNDO95*; University of Zürich, 1995.

18. Pearlman, D.A.; Case, D.A.; Caldwell, J.W.; Ross, W.S.; Cheatham III, T.E.; Ferguson, D.M.; Seibel, G.L.; Singh, U.C.; Weiner, P.K.; Kollman, P.A. *AMBER4.1* University of California: San Francisco, 1995.

19. Hehre, W; Radom, L.; Schleyer, P.v.R.; Pople, J.A. *Ab Initio Molecular Orbital Theory*; Wiley: New York, 1986.

20. Parr, R.G.; Yang, W. *Density Functional Theory of Atoms and Molecules*; Oxford University Press: Oxford, 1989.

21. Dewar, M.J.S.; Zoebisch, E.G., Healy, E.F.; Stewart, J.J.P. *J. Am. Chem. Soc.* **1985**, *107*, 3902.

22. Cornell, W.D.; Cieplak, P.; Bayly, C.I.; Gould, I.R.; Merz, K.M.; Ferguson, D.M.; Spellmeyer, D.C.; Fox, T.; Caldwell, J.W.; Kollman, P. *J. Am. Chem. Soc.* **1995**, *117*, 5179.

23. Lias, S.G.; Liebman, J.F.; Levin, R.D. *J. Phys. Chem. Ref. Data* **1984**, *13*, 695.

Chapter 5

A Method of Hybrid Quantum–Classical Calculations for Large Organometallic–Metallobiochemical Systems

Applications to Iron Picket-Fence Porphyrin and Vitamin B$_{12}$

Isaac B. Bersuker[1,2], Max K. Leong[1], James E. Boggs[1], and Robert S. Pearlman[2]

[1]Institute of Theoretical Chemistry, Department of Chemistry and [2]College of Pharmacy, University of Texas, Austin, TX 78712

For large molecular systems where neither quantum mechanical (QM) nor molecular mechanics (MM) calculations, applied separately, can solve the problem, we worked out a method of combined QM/MM calculations with an electronically transparent interface between the quantum and classical fragments. Three necessary conditions of (1) fragmentation, (2) interfragment self-consistency, and (3) QM-MM continuity are formulated and satisfied by (a) cutting the system on a $2s2p$ atom (border atom) that participates with its hybridized orbitals in both fragments and does not serve as a π bridge between them, (b) introducing an intermediate fragment which is treated by both QM (electronic structure) and MM (geometry optimization), and (c) using a special iterative procedure of double (intrafragment and interfragment) self-consistent (DSC) calculations which realizes the electronically transparent interface and charge transfers between the fragments.
The method is implemented in a package of computer programs based on semiempirical INDO/1 for QM calculations, improved IEH method (ICONC) for the DSC procedure, and SYBYL for the MM treatment, all packages modified accordingly. The calculations for two large molecular systems—iron picket-fence porphyrin and vitamin B$_{12}$—yielded results in good agreement with the experimental data available. For the first system separated in six fragments, the interfragment self-consistency was achieved after 8 iterations. The charge transfers between the fragments are significant, confirming the importance of electronic transparency of the QM/MM interface. The ~0.4 Å out-of-plane position of the iron atom with respect to the porphyrin ring is a quantum effect which cannot be reproduced by pure MM treatment. For vitamin B$_{12}$ (181 atoms), the system is separated in eight fragments. Interfragment self-consistency was reached after 7–10 iterations. The HOMO-LUMO separation for this system compares well with spectral data and explains the origin of its red color.

The importance of combined quantum mechanical (QM)/molecular mechanics (MM) treatment of molecular systems increases with the size of the system to be modeled in various chemical and biological problems. For relatively large systems, neither QM nor MM methods, applied separately, can solve the problem. In principle, *ab initio* or (semi *ab initio*) density functional QM methods with geometry optimization should be able to solve any molecular problem, but with increasing numbers of electrons the application of these methods is so far (and will be at least in the near future) impractical even for molecules of relatively moderate size in the biochemical sense.

On the other hand, semiempirical methods can handle much larger systems [1], but they too fail in geometry optimization when the number of atoms increases to some hundreds. Unlike *ab initio* calculations that include electronic correlation, semiempirical methods neglect the electron correlation effects in the interaction of non-bonded atoms in the Hamiltonian. Together with the accumulation of errors of semiempirical parameterization, this obviously limits the size of calculable systems. The failure to consider properly the electron correlations in steric (van der Waals) interactions is a factor of special concern when semiempirical methods are applied to large molecular (and moderate) systems where even small errors in bond angles and interatomic distances in each local nonbonded interaction may accumulate and increase by orders of magnitude. As an example of this failure in application to moderate systems, semiempirical method cannot explain the origin of attraction between two stacking aromatic rings.

The methods of MM are devoid of this failure since electron correlation effects are implicitly included in the atom-atom potentials [2]. However, these methods are classical by definition and do not describe the quantum effects which are essential in the overwhelming majority of problems, in fact in all the problems in which the electronic states should be considered explicitly. A major class of molecular systems that *a priori* cannot be considered by MM methods, in general, is provided by transition metal compounds. For these systems the *d* electron heterogeneity introduced by the transition metal atom makes the MM force field approach inapplicable (although there may be some restricted particular cases where the MM method for such systems is useful; see Ref.[3,4] and references therein).

Indeed, due to the three-dimensional *d* electron delocalization [3], there are no localized metal-ligand bonds which might be described by approximately constant parameters. In the coordination center ML_n with *n* equal or different ligands, the parameters of each bond (*e.g.*, $M–L_1$) are, in general, strongly dependent on all the other bonds (*e.g.*, $M–L_2$, $M–L_3$, *etc.*) formed by the same metal atom. Hence, these parameters can by no means be transferred *a priori* to another center that has the same $M–L_1$ bond but other (by number and type) ligands $L_{2'} \neq L_2$, and/or $L_{3'} \neq L_3$, *etc.* For instance, the Cu–O bond length varies between 1.8 Å and 3.0 Å depending on other bonds in the coordination center and the environment. This effect is well known in coordination chemistry as the mutual influence of ligands, the *cis-* and *trans*-effects, the Jahn-Teller and pseudo Jahn-Teller effects, *etc.* [3,5,6]. Even the multiplicity of the ground state may change under the influence of other ligands (*cf.* high-spin $\Leftrightarrow$ low-spin transitions induced by small changes in the ligand environment [3]).

It follows that the parameters to be used in the force field presentation of MM for a given metal center are nontransferable to other centers that may have the same metal-ligand bonds. This nontransferability is one of the main reasons for the failure of MM in application to transition metal systems. Among the other effects which make difficult the direct application of the MM method to transition metal systems, we mention here electronic excitation (chemical activation) of the ligands by coordination (via back-donation) and electron-conformational effects [3].

A question emerges whether the two methods, QM and MM, can be used in a combination in which the QM method is applied to the part of the molecular system in which the quantum effects are essential and should be taken into account explicitly, while MM is used to consider the remaining part of the system and the interaction between the QM and MM parts. There are several papers devoted to this problem [7–19] (see also the papers in this book). Warshel and Levitt [7] apparently were the first to use combined QM/MM calculations in the study of enzymic reactions. In this work a QM treatment was applied to a small part of the substrate where the bond cleavage and charge redistribution in the chemical reaction takes place, while the remaining enzyme system and its interaction with the reaction site were considered by means of MM. In the papers of Singh and Kollman [8], Field *et al.* [9] (see also [10,11]), Gao *et al.* [12] (see also [13–15]), Stanton *et al.* [16], Maseras and Morokuma [17], Bakowies and Thiel [18,19], several aspects of the QM/MM problem with different rules of QM/MM fragmentation and methods of calculation of the QM fragments and its interaction with the MM part are considered.

A common feature of the majority of these studies is that they ignore possible charge transfer between the QM and MM fragments. This is quite understandable in view of the systems considered in these works. In most cases the QM and MM parts are different molecular systems that interact by intermolecular forces (*e.g.*, reactants and solvent, both organic systems), and hence no significant charge transfers between them are anticipated. In other cases the QM and MM parts are fragments of a molecular system which can be considered electronically homogeneous in the sense that it contains similar *nsnp* atoms only (organic molecules without strong heterogeneous implications); the charge transfers between these fragments may be small. In the paper of Maseras and Morokuma [17] charge redistributions are allowed but the method is obviously aimed at relatively small systems.

If a large molecular system contains transition metals (or even smaller main group heterogeneity) and hence significant charge transfers between the molecular fragments are expected, the above methods of QM/MM calculations [7–19] are obviously inapplicable. Even in case of solute-solvent interactions there may be significant charge transfer if there is strong heterogeneity between them. In a recent paper, Stavrev and Zerner [20] show that significant charge transfer between a ruthenium complex and the water molecules of the solvent should be taken into account to explain the origin of the observed absorption spectra.

In the present paper we suggest a general method of fragmentary QM/MM description of large molecular systems including strongly heterogeneous compounds with significant charge transfer between the fragments. In the next section we formulate three necessary conditions which should be satisfied in any rigorous QM/MM method, and show how these conditions are satisfied in our method. Then follows the scheme of computer implementation and results of calculations performed for an illustrative example system—the iron picket-fence porphyrin. A part of the method description and results were published earlier [21,22].

Theoretical foundation: conditions of fragmentation, interfragment self-consistency, and QM-MM continuity

First we formulate, in general, the conditions which must be satisfied in any rigorous method of QM/MM calculations as follows:

1. *Condition of fragmentation*: in dividing the molecular system in fragments, there should be significant theoretical grounds to assume that the main electronic features are not lost in the proposed fragmentation, that is the fragmentary description may serve as a zero-order approximation;
2. *Condition of interfragment self-consistency*: if the fragments are calculated separately, there should be a possibility to converge (and a criterion of convergence of) the results to those of the system as a whole which includes charge redistribution between the fragments;
3. *Condition of QM-MM continuity*: if some of the fragments are calculated by QM methods while the others are considered by MM, there should be a QM/MM interface that allows for a smooth transition between the QM (quantum) and MM (classical) description of near-neighbor fragments.

Qualitatively, the condition of fragmentation requires that the interfragment interactions be small as compared with intrafragment interactions, but in such a general statement there is no indication of how to compare these interactions. An attempt to give a quantitative criterion of this condition was given earlier [23] based on some mathematical theorems concerning characteristic numbers (roots) of matrices [24,25].

Consider two fragments, I and II (the procedure is easily extended to multi-fragment systems). In transition metal systems one of the fragments contains the metal center with its minimum possible environment, the other includes the remaining organic ligands. Choose the border between the fragments (the cut line) to cross an *nsnp*-electron atom which is not a bridge for π-electron delocalization (*i.e.* the border atom should not form π bonds with at least one of the two fragments), and assume that some of its (*sp*-hybridized) orbitals belong to fragment I, while the others pertain to fragment II. In this way the border atom is included in both fragments with different orbitals in each of them. This partitioning method in which the border between the fragments cuts an atom (not a bond), differs from others suggested so far; it has the advantage of allowing direct charge transfers between the fragments which is most important when there is strong electron heterogeneity, *e.g.*, some of the fragments contain transition metals.

The separate calculation of the electronic structure of the two fragments in the MO-LCAO approximation means that the off-diagonal elements in the secular equation H_{ij} (resonance integrals) that correspond to the i and j orbitals from different fragments are neglected. This means that H_{ij} should be small as compared with, say, the Coulomb integrals H_{ii} and H_{jj}. A more rigorous appreciation of the error introduced by such a simplification of the secular equation of the MO-LCAO method can be made [23] based on Gershgorin's theorem [24] which proves that, using an orthogonal basis, the roots ε of the secular matrix lie within the so-called Gershgorin circles (on the conjugated plane):

$$|H_{ii} - \varepsilon| < \sum_{j \neq i} |H_{ij}| \qquad (1)$$

that is, the deviation of the MO energy level from the i-th diagonal matrix element H_{ii} lies within the interval equal to the sum of absolute values of all the off-diagonal elements formed by the i-th orbital. Equation 1 shows that neglecting some of the H_{ij} values by cutting the system in fragments reduces the interval between the MO energy levels. The smaller the neglected value $\left|H_{ij}\right|$, as compared with the remaining ones and H_{ii}, the closer the fragmentary MO energy levels come to those of the non-fragmented system.

For σ bonded fragments the i-th atomic orbital of the border atom can be taken hybridized and orthonormal to other AO's thus minimizing the off-diagonal elements of interfragment interactions. For instance, if the border atom is a saturated carbon atom, then the i-th AO is a hybridized sp^3 orbital, and hence $H_{ii} = (1/4)\left(H_{ss} + 3H_{pp}\right)$ and $H_{ij} = (1/4)\left(H_{ss} - H_{pp}\right)$, where H_{ss} and H_{pp} are the Coulomb integrals for the s and p electrons. Since the values of these two integrals are very close to each other (in sp atoms!), the neglect of H_{ij} as compared with H_{ii} is quite reasonable as a zero-order approximation. This cannot be done when the i-th orbital forms π bonds with both fragments because its intrafragment and interfragment H_{ij} values for the π-type interactions have the same order of magnitude.

After the MO-LCAO calculations of such roughly separated fragments, an improvement of the results is possible taking into account their interaction via the border atoms. A perturbation treatment with respect to the neglected elements H_{ij} [23] based on the mathematical theorem of Fan and Hoffman [25] shows that the error in the MO energies ε introduced by this neglect is:

$$\left|\varepsilon_\alpha - \varepsilon\right| \le v\left|C_{\alpha i} - H_{ij}\right| \qquad (2)$$

where

$$2v^2 = \sqrt{4m+1} - 1$$

m is the number of MO's involved in the interaction via H_{ij} from both fragments, and $C_{\alpha i}$ is the LCAO coefficient of the border-atom's i-th AO in the α-th MO. The more delocalized the MO, the smaller $C_{\alpha i}$. Hence delocalization of the electrons within the fragments lowers the interfragment interaction and the error in the above fragmentary calculations.

Thus to obey the condition of fragmentation, one may cut the system by a border line which crosses $nsnp$ (better $2s2p$) atoms only, and these border atoms do not serve as π-electron bridges between the fragments. In such fragmentation, the fragmentary calculations may serve as a zero-order approximation for which the interfragment interactions are small enough and there may be a procedure to converge the results to that of the non-fragmented system.

In the error estimates (1) and (2) only the MO energies are considered. Meanwhile, as stated above, the charge redistribution, especially the orbital charge transfers, play a key role in the specific effects of coordination, including excitation and chemical activation. To correct both the energy level positions and orbital charge redistribution due to the interfragment interaction and to converge the results to that of the non-fragmented system, the *double self-consistency* (DSC) *procedure* [23] can be employed. This procedure also allows one to satisfy the condition of interfragment self-consistency. In semiempirical MO-LCAO calculations of any type, for instance the iterative extended Hückel (IEH) type [1,26–28], the charges on atoms, and more specifically their electronic configurations, are usually calculated by means of a Mulliken population analysis. (Other population analyses may also be efficient). For instance, for the p-electron density on atom A we have:

$$q_p^A = \sum_{i,v} \sum_{p \in A} n_i C_{ip} C_{iv} S_{pv} \tag{3}$$

where C_{ip} and C_{iv} are the i-th MO-LCAO coefficients of the orbitals p and v, respectively, S_{pv} is their overlap integral, and n_i is the occupancy number of the i-th MO. These orbital densities determine the energy of ionization of the valence state I_o^{vs} and the matrix elements H_{ii} and H_{ij} for the next iteration in the self-consistent calculations. In other semiempirical methods (*e.g.*, INDO, CNDO) these matrix elements are determined by the density matrix $P_{\mu v}$:

$$P_{\mu v} = \sum_i n_i C_{\mu i} C_{vi}^* \tag{4}$$

For the border atoms, as for the others, the atomic orbital charges and the density matrix are determined by the LCAO coefficients of all the AO's of the atoms that participate in the appropriate MO. In other words, the charges on the border atoms that determine the ionization energy of the valence state or the density matrix (which, in turn, form the matrix elements of the secular equation) depend on the results of the previous calculation of both fragments. If, as a result of the first iteration with separate fragments and arbitrary distribution of electrons, there is no balance of charge between the fragments, *i.e.* the atomic charges (or density matrix) for the border atom were underestimated in one of the fragments and overestimated in the other one, the density matrix evaluation for the next iteration would increase the corresponding matrix elements in the first fragment and decrease them in the second fragment. Similar charge redistribution takes place when other semiempirical methods are employed in the double self-consistency procedure.

The border atoms thus serve as *channels of charge transfer from one fragment to another*, and the final solution of such fragmentary calculations yields MO energies and density matrices that are self-consistent with respect to both intrafragment and interfragment charge distribution (double self-consistency). In this way the electronically transparent interface between the fragments is realized; it takes into account all the effects of charge transfer, excitation by coordination, *etc.*, mentioned above. We calculate the fragments in parallel and independently, but in fact, they are strongly interdependent, since after each iteration the charges, or the density matrix, that determine the matrix elements of the Hamiltonian for each fragment are calculated based on the results of the previous iteration obtained from both fragments.

This condition of interfragment self-consistency implies that both neighboring fragments are calculated quantum-mechanically. If one of these fragments is to be optimized by MM as stipulated by the QM/MM method, both conditions of interfragment self-consistency and QM/MM quantum-to-classical continuity can be satisfied simultaneously by an additional procedure as follows.

Assume that I is the QM fragment (usually the smaller one, *e.g.*, the transition metal and its near-neighbor environment compatible with the rules of fragmentation, or the reaction site in an enzymic or solvent influenced reaction), while II is the MM fragment—a large predominantly organic system. After calculating the electronic structure of fragment I (with geometry optimization) and optimization of fragment II by MM (for details, see below), we can separate from the MM part a smaller fragment, adjacent to the border atom—the intermediate fragment—and calculate its electronic structure for fixed nuclei determined by MM. Since the intermediate fragment is assumed to be small (the smallest compatible with the rules of fragmentation), this additional semiempirical electronic structure calculation for fixed nuclei does not cause any problems. If the organic fragment is not very large, its whole structure can be taken as the intermediate fragment.

Thus in addition to the smaller part of the system which is given a full QM treatment and its larger part which is optimized by MM, we separate an intermediate fragment between them (a part of the MM site) which is treated by both QM (electronic structure) and MM (geometry optimization). This allows us to reach double self-consistency at the border atom using the electronic structure data from both fragments, and to realize a smooth transition from the quantum site to the classical MM description of the organic ligands.

Computer implementation

Based on the discussion given above the following computational scheme has been followed to implement the QM/MM method of calculation of transition metal systems (the scheme for other strongly heterogeneous systems would be quite similar):

1. Divide the transition metal system into fragments, one of which, the central fragment, contains the transition metal with a (reasonable) minimal number of ligand atoms compatible with the requirement that the border atoms (the *nsnp* atoms cut by the border line) are not π-electron bridges between the fragments. In case of multi-center transition metal systems (or, in general, several quantum sites), each transition metal (quantum site) may form its own central fragment, provided they are not connected by direct bonds or indirect π bonds.

2. Optimize the geometry of the central fragment(s) using semiempirical QM methods and dummy atomic groups to saturate the open valencies, as discussed below.

3. With the fixed central fragment geometry, optimize the whole system—the ligand fragments in the presence of the fixed central fragment—using MM. If existing software packages are used for this procedure, some modifications may be required to include non-bonding interactions of the ligands with the transition metal atom. However, since the interfragment distances are relatively large, these interactions are expected to be small.

4. *Second iteration.* With the geometry from step 3, separate the intermediate fragment adjacent to the border atom from the MM side and calculate semiempirically its electronic structure. If not too large, the whole ligand can be taken as the intermediate fragment. The separation of a fragment from the ligand should follow the same rules of (reasonable) minimal number of atoms (compatible with the requirement of non-π-bridged border atoms) and attached dummy atoms to imitate the cut valencies (see below).

5. Using equations 5–10 given below, change the matrix elements H_{ij} of the central fragment and reoptimize its geometry. At this point, the influence of the non-bonded atoms of the ligands can also be taken into account, if necessary, by including additional Coulomb and van der Waals terms in the Hamiltonian. (There is already such an option in some existing programs of semiempirical calculations, *e.g.*, in the ZINDO package [1]).

6. With the new fixed geometry of the central fragment, determine again the conformation of the rest of the system using MM, as indicated above. Repeat the whole procedure from step 2 until the new charges or energies coincide with those of the previous iteration within user-specified criteria.
7. There may be two or more conformations of the transition metal system with close energies; the computational procedure should be repeated for each of them.

The dummy atomic groups needed in step 2 and the formulas required in step 5 are as follows. In the semiempirical methods used in the electronic structure calculations, the matrix elements H_{ij} for the next iteration are functions of the density matrix $P_{\mu\nu}$ which, for the border atom, depend on the LCAO coefficients of the MO's from both fragments obtained from the intrafragment calculations of the previous iteration. Complications occur when one tries to employ existing programs of electronic structure calculation with geometry optimization to realize step 2 of this computational procedure (QM optimization of the metal-containing fragment) because of the "free valences" at the border atoms in the separate fragments. These artificial unsaturated valences produce completely unrealistic geometry. To avoid this difficulty the well known method of dummy atoms (usually hydrogens) which saturate the free valences is employed. In our case dummy atomic groups instead of dummy atoms seem to be more appropriate. The dummy atomic groups for a given fragment should be chosen and attached to the border atom in such a way as to substitute its real bonds with the near-neighbor atoms of the other fragment and to imitate the bonds of the latter atoms with the next atoms. The imitation is realized by substituting the latter by hydrogen atoms. For instance, in the example of iron picket-fence porphyrin considered below in more detail, the dummy groups for one of the border atoms are shown in Figure 1.

It is seen that the use of dummy groups instead of dummy atoms provides the most similar electronic environment for the border atom in the two fragments which, in turn, simplifies the procedure of electronically transparent interfacing and double self-consistency. If semiempirical methods of the type of INDO or CNDO are used in the calculations, the difference $\Delta P_{\mu\nu}^{A}$ between the density matrix elements for the border atom calculated in the central fragment $P_{\mu\nu}^{A}(\text{CF})$ and in the ligand fragment $P_{\mu\nu}^{A}(\text{L})$,

$$\Delta P_{\mu\nu}^{A} = P_{\mu\nu}^{A}(\text{CF}) - P_{\mu\nu}^{A}(\text{L}) \tag{5}$$

may serve as a measure of interfragment self-consistency. With $\Delta P_{\mu\nu}^{A}$ obtained from the previous iteration, the density matrix elements of the central and ligand fragments for the next iteration should be calculated as

$$P_{\mu\nu}^{'A}(\text{CF}) = P_{\mu\nu}^{A}(\text{CF}) - \kappa\Delta P_{\mu\nu}^{A}$$
$$P_{\mu\nu}^{'A}(\text{L}) = P_{\mu\nu}^{A}(\text{L}) + \kappa\Delta P_{\mu\nu}^{A} \tag{6}$$

where κ is an arbitrary coefficient chosen empirically to get faster convergence. For large fragments, a larger κ may provide a faster redistribution of the electronic charge over the system (note that $\Delta P_{\mu\nu}^{A}$ and Δq below characterize the

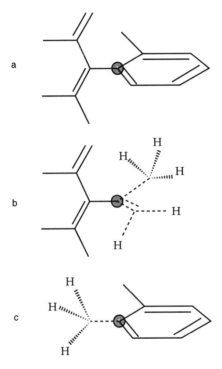

Figure 1.   The dummy atomic groups for one of the border atoms (carbon) in the fragmentation of iron picket-fence porphyrin (see below in Figure 2):(a) the real bonds formed by the border atom (the latter is shown in encircled dots); (b) the dummy groups (shown by dashed lines) in the calculations of the central fragment; (c) the dummy groups for the ligand fragment. (Reproduced with permission from ref. 22, Copyright 1997 Wiley.)

corresponding differences in, respectively, density matrix and charges for one atom, namely the border atom).

If the IEH method is employed, the difference between the atomic charges of the border atom calculated in the central fragment $q^A(\text{CF})$ and in the ligand fragment $q^A(\text{L})$,

$$\Delta q^A = q^A(\text{CF}) - q^A(\text{L}) \tag{7}$$

may serve as a measure of interfragment self-consistency instead of the $\Delta P^A_{\mu\nu}$ values. With Δq obtained from the previous iteration, the matrix elements of the central fragment and ligand for the next iteration should be calculated with the atomic charges $q'^A(\text{CF})$ and $q'^A(\text{L})$, respectively,

$$q'^A(\text{CF}) = q^A(\text{CF}) - \kappa\Delta q^A$$
$$q'^A(\text{L}) = q^A(\text{L}) + \kappa\Delta q^A \tag{8}$$

where κ has the same meaning as above.

Equations 5 and 6 (or 7 and 8) realize, in principle, the electronically transparent interface procedure, discussed above. However, if the INDO or CNDO type methods are used in the semiempirical calculations, their existing formulation and parameterization cannot be employed directly to carry out this procedure. Indeed, due to the charge transfers between the fragments, the fragmentary calculations inevitably yield *fractional charges* on the fragments when passing to the second and next iterations. Existing programs of the previously mentioned semiempirical methods aimed at electronic structure calculation and geometry optimization of whole molecules (not fragments) usually do not include the option of fractional molecular charges.

To modify and reprogram the semiempirical methods to include the option of fractional charges and other special features of the double self-consistency method, we impose a special condition on orbital population numbers n_i in the Roothaan SCF procedure, in particular in the charge density matrix of equation 4. To comply with the condition of fractional charges on the fragments, the n_i value (which is restricted to integer numbers 2, or 1, or 0 in the existing methods) should be fractional for the highest occupied MO, and the total number of electrons should be fractional, too. The MO occupancy thus should be

$$n'_i = \kappa\Delta q_i \tag{9}$$

for the LUMO of the fragment that acquires charge, and

$$n'_i = n_i - \kappa\Delta q_i \tag{10}$$

for the HOMO of the fragment that loses charge. If $\kappa\Delta q$ is larger than n_i (for HOMO) or larger than 2 (for LUMO), more than one orbital should be involved in the charge transfer. With the occupancies of equations 9 and 10 the open-shell versions of the semiempirical methods must be used in the above calculations. We reprogrammed the INDO/1 version of the ZINDO package [1] to include these options. Although the charges on separate fragments are fractional, the number of electrons on the whole systems remains integral and does not change. Therefore there are no problems of fractional charges in defining observables. The latter are defined for the system as a whole where charges are integral.

Iron picket-fence porphyrin

As a first example to verify the method as a whole, specify its implementation and reveal its efficiency for transition metal systems, we performed the calculations for a metallobiochemical model system, the iron picket-fence porphyrin [29], following the scheme of computations outlined above. Figure 2 shows the chemical formula of iron picket-fence porphyrin with indication (by a dashed line) of the fragmentation used in the calculations. The border atoms (four carbon atoms at four "fence" ligands and the nitrogen from the axial ligand) do not form π bonds with the central fragment because of steric restrictions (the corresponding two molecular groups from both sides of the border atom are not co-planar).

The choice of dummy atomic groups was discussed above (Figure 1). The central fragment was optimized using INDO/1 from the ZINDO package [1] reprogrammed to include the MO occupancy numbers following equations 9 and 10, while the MM treatment has been realized by using the Tripos Force Field [30] within the SYBYL molecular modeling package [31]. The calculations were carried out on an IBM RISC/6000 workstation.

The optimization procedure for the central fragment is rather fast (several minutes) for the initial (closed shell – low spin configuration) fragment and becomes much slower for the open-shell calculations after fractional charge transfers. The results reproduce the out-of-plane position of the iron atom with respect to the porphyrin ring at a distance $R = 0.42$ Å the experimental value is 0.40 Å [32] which is usual in the group of similar porphyrins [33]. This out-of-plane position of the central atom is an important feature of some iron (and other metal) porphyrins which is due to the vibronic mixing of the occupied $a_{2u}(\pi)$ MO of the porphyrin ring with the unoccupied $a_{1g}(d_{z^2})$ MO of (mainly) the iron atom by the out-of-plane displacement of the iron atom [34]. It is one of the effects which cannot be taken into account in a proper way without a QM treatment; MM does not involve electronically excited states.

The optimization of the system as a whole by MM with the fixed central fragment is more involved. Since SYBYL does not contain parameters for the non-bonded interaction of the ligand atoms with iron, the latter has been replaced by a dummy atom when optimizing the whole system with the fixed central fragment. In general, this procedure would be unacceptable. Fortunately, in the case under consideration, due to the large ligand-metal distances and delocalized charges, this procedure does not strongly influence the optimized configuration. For other cases, the use of other MM force field programs which include approximate nonbonded interactions with metals would be more appropriate.

The optimization procedure proved to be more convenient when starting from the central fragment and gradually (step-by-step) increasing its environment, the ligand size. As usual in conformation searches for large molecular systems where there may be more than one low-energy minima, the optimized conformation may be dependent on the starting geometry of the ligands. For instance, if the starting position of the ligand on the 4-fold axis of the porphyrin ring (the ligand with the nitrogen border atom, Figure 2) has an angle significantly smaller than 90°, it forms additional bonds with the atoms of the benzene rings of the other ligands (in the optimized geometry) which resemble the corresponding sterically encumbered derivatives of picket-fence porphyrin [29]. Starting with the 90° model, we get the proper configuration discussed below.

To compare different methods and find the one which is most appropriate to our problem, we performed electronic structure calculations with geometry optimization for a relatively large transition metal system $[Co(en)_2NH_3Cl]^{n+}$, where $n = 0, +1, +2, +3, +4$, and $en=NH_2CH_2CH_2NH_2$, using a variety of semiempirical methods, but without fragmentation [35]. The methods employed are: INDO/1 [1], IEH/ICONC-93 [27], PM3 (tm)/Spartan [36], DFT/BLYP/STO-3G** from Gaussian 94 [37], HF/STO-3G** from Gaussian 92 [38], HF/3-21G** from Gaussian 92 [38], and MP2/STO-3G** from Gaussian 92 [38]. Briefly, the results show that different methods yield significantly different charge distributions. The closest to the *ab initio* results on charges were obtained by INDO/1. PM3 (tm) yields unrealistic large negative charges on the metal. In some of the semiempirical versions (as well as *ab initio* without correlation) the atomic charges on some atoms (including the border atoms) decrease when the charge of the system is increased. This might be acceptable for small charge transfers, but in the above case the tendency remains also for large changes of the total charge n up to n=+4 and makes the convergence of the DSC procedure very difficult. The most consistent charge redistribution is given by ICONC [27]. This is quite understandable in view of the iterative charge and configuration self-consistency (described by nine parameters for the metal $nd(n+1)s(n+1)p$ atom and three parameters for each of the $nsnp$ atoms) employed by this method.

However ICONC does not contain direct geometry optimization. Therefore we chose to optimize the geometry by means of INDO/1, while the charge distribution and the DSC procedure were carried out by ICONC. Apparently, it is a general tendency that semiempirical Hamiltonians which perform geometry optimization well may be not equally good for charge distribution. To coordinate the two methods, we skipped the procedure of geometry reoptimization after each iteration in the example under consideration. This simplication can be avoided if a method which yields both reasonablecharge distribution and geometry is available.

Interfragment self-consistency was reached after 9–10 cycles. Figigure 3 illustrates the convergence process. The DSC charge redistribution between the fragments is illustrated by the data in Table I and in Figure 4. It is seen that each of the four "fence" fragments loses approximately 0.5–0.6 electron, while the axial ligand gains about 0.6 electron, transferred via the central fragment, the balance of the latter thus being approximately about -1.6 electrons. [Note that if we take into account that the oxidation state of the iron atom is two which means that we can start the count of charge transfers taking the central fragment charged +2, the charge transfer to the central fragment is about -3.6 electrons.] This substantial charge transfer between the fragments confirms that the non-charge-transfer models of QM/MM fragmentation, suggested earlier [7–19], may be invalid for transition metal systems. The size of these charge transfers seems to be correct in order of magnitude. A ligand-to-metal transfer of 0.5–1.0 electron is not unexpected and the large total transfer to the metal fragment may be reasonable in view of the high "charge capacitance" [3] of the porphyrin ring.

The optimized system appears as shown in Figure 5. It agrees quite well with the experimental data available, in particular with X-ray results [32], as shown in Table II for some major structural parameters. The remaining discrepancies (especially in the flexible part of the system) can be attributed to the differences between free and lattice-packed molecules and partly to the omitted geometry reoptimization after each iteration. The experimental measurements were carried out under the assumption of four-fold symmetry of the system. Owing to the disordered random orientations of the axial ligand in the crystal state, the atomic

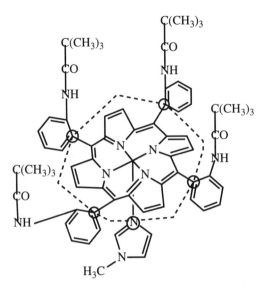

Figure 2. Structure of iron picket-fence porphyrin. (Reproduced with permission from ref. 22, Copyright 1997 Wiley.)

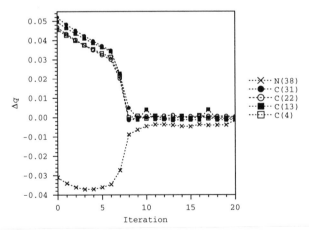

Figure 3. Atomic charge differences $\Delta q = q(\text{CF}) - q(\text{L})$ for the four carbon and one nitrogen border atoms versus iteration, demonstrating the convergence to interfragment self-consistency in the DSC procedure. (Reproduced with permission from ref. 22, Copyright 1997 Wiley.)

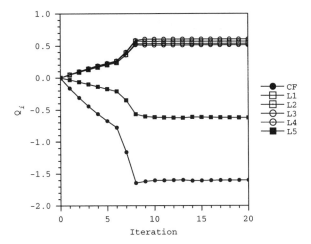

Figure 4.   Charge redistribution between the six fragments (CF is the central fragment and L_n is the n-th ligand fragment) in the DSC procedure versus iteration. It is seen that self-consistency is achieved after the 8th iteration.

Table I. Comparison of the geometry parameters of iron picket–fence prophyin obtained by our method with the experimental X-ray data. The distances R in Å, and the angles in degress. $R(Fe-N_i)$ are the iron–nitrogen bond distances; h is the out-of-plane displacement of the Fe atom; $R(N-N)$, $R(O-O)$, and $R(C-C)$ are the corrsponding inter-ligand interaomic distances between, respectively, the nitrogen, oxygen, and butyl–carbon atoms of the four "fence" ligands.

	This work	Experiment [32]
$R(Fe-N_1)$	2.088	2.075
$R(Fe-N_2)$	2.088	2.075
$R(Fe-N_3)$	2.088	2.068
$R(Fe-N_4)$	2.087	2.068
$R(Fe-N_5)$	2.126	
$\angle N_1-Fe-N_2$	156.54	158.02
$\angle N_3-Fe-N_4$	156.54	157.54
h	0.424	0.394
	0.424	0.404
$R(N-N)$	9.976	10.203
	9.400	9.602
	7.213	7.077
	6.915	7.077
	6.830	6.956
	6.454	6.956
$R(O-O)$	11.639	13.090
	9.985	12.340
	9.851	9.047
	7.341	9.047
	7. 021	8.965
	6. 565	8.965
$R(C-C)$	9.010	9.139
	7.201	8.079
	6.046	6.183
	5.865	6.183
	5.645	6.035
	5.542	6.035

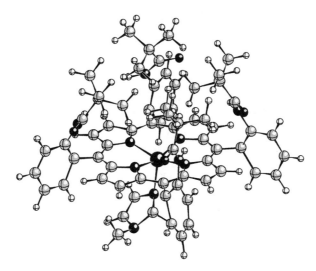

Figure 5.   The optimized structure of iron picket-fence porphyrin.
(Reproduced with permission from ref. 22, Copyright 1997 Wiley.)

Table II. Charges on the central and ligand fragments and convergence parameter κ vs. iteration cycle for iron picket-fence prophyrin.

iteration	Fragments						κ
	Center	1	2	3	4	5	
0	0.000	0.000	0.000	0.000	0.000	0.000	0.00
1	-0.163	0.046	0.050	0.047	0.052	-0.031	1.00
2	-0.310	0.088	0.096	0.090	0.100	-0.065	1.00
3	-0.443	0.128	0.140	0.131	0.146	-0.101	1.00
4	-0.564	0.166	0.181	0.168	0.188	-0.138	1.00
5	-0.676	0.201	0.219	0.203	0.228	-0.175	1.00
6	-0.779	0.234	0.256	0.235	0.265	-0.211	1.00
7	-1.163	0.360	0.393	0.356	0.404	-0.350	4.00
8	-1.643	0.534	0.575	0.517	0.585	-0.568	8.00
9	-1.618	0.534	0.571	0.511	0.604	-0.603	4.00
10	-1.603	0.535	0.569	0.511	0.603	-0.616	2.00
11	-1.602	0.535	0.573	0.512	0.603	-0.620	1.00
12	-1.599	0.535	0.572	0.512	0.604	-0.624	1.00
13	-1.597	0.535	0.572	0.512	0.604	-0.626	0.50
14	-1.608	0.535	0.571	0.513	0.603	-0.614	0.50
15	-1.606	0.535	0.570	0.513	0.604	-0.616	0.50
16	-1.603	0.535	0.570	0.513	0.604	-0.618	0.50
17	-1.601	0.535	0.569	0.514	0.603	-0.620	0.50
18	-1.601	0.535	0.570	0.514	0.603	-0.621	0.25
19	-1.600	0.535	0.570	0.514	0.603	-0.622	0.25
20	-1.598	0.535	0.569	0.514	0.603	-0.623	0.25

positions for this ligand could not be determined from the X-ray measurements. In this respect the modeling has advantageous possibilities; it reveals the real structure of individual entities also when the experimental methods fail to do it.

Vitamin B₁₂

ß-Cyano-(5',6'-dimethylbenzimidazolyl)-cobamide, or vitamin B_{12}, has 181 atoms. Even disregarding the failures of semiempirical QM methods in application to very large molecular systems, outlined in the Introduction, the attempt to model this system by any semiempirical method using a computer of the type IBM RISC/6000 with 256 Mb physical memory and 544 Mb virtual memory running AIX 3.2.5 failed because of the huge (minimal) basis set of 729 atomic orbitals and lack of usable symmetry.

To apply our combined QM/MM method, the whole system was divided into eight fragments as indicated by the dashed lines in Figure 6. Figure 7 shows the atom-numbering system. The central fragment has 116 atoms which is near to the maximum allowable in our scheme of calculations. Of the ligand fragments, all but the seventh (which is a *bichelate* ligand) have only one border atom. The whole calculation was carried out following the scheme of our QM/MM method outlined above.

Figure 8 illustrates the convergence process. Interfragment self-consistency was reached after 7–10 cycles. The DSC charge redistribution between the fragments is illustrated by the data in Table III and in Figure 9. It is seen that, when starting from charges +1 on the central fragment and -1 on the 7th fragment, the former loses additionally about one electron while the latter acquires about 0.7 electron, the changes of the charges on the other fragments being much smaller. The central fragment acts as a buffer zone—some fragments donate electrons to the central fragment and others, on the other hand, draw electrons from the central fragment. It is also obvious from Table III that faster convergence was achieved by manipulating the parameter κ. It would have taken a longer time if the parameters κ were not employed.

The optimized system appears as shown in Figure 10, and the major structural parameters for the flexible part of the system are listed in Table IV. They agree well with the experimental X-ray results [39]. Even the calculated interfragment bond distance between the Co and P atoms from the central fragment and the 7th fragment, respectively, coincides with the experimental data. The calculated out-of-plane distances h, measured from the Co atom to the planes of C(2)–C(3)–C(4) and C(2)–C(4)–C(5), and the torsion angle θ, defined as the angle between the normal vector of the benzimidazole plane and that of the cobalamide plane, are very close to those obtained from the experimental data.

Some of the absolute values of differences in the interatomic distances between atoms of different ligand fragments are relatively large, although they look less significant when taken as relative values. For example, in the largest deviation, the theoretical and experimental values of the interatomic distances between O(91) and O(175) are 6.50 Å and 7.02 Å, respectively, which yields about 7% discrepancy. On the other hand, some of these discrepancies, especially in the flexible part of the system, can be attributed to the differences between free and lattice-packed molecules.

Figure 11 demonstrates the energy levels of the HOMO, LUMO, and second HOMO (HOMO-1) of vitamin B_{12} calculated in this work. The energy difference between LUMO and HOMO-1 is 2.338 eV. An electronic transition between these two states HOMO-1→LUMO is equivalent to absorption of visible (green) light in

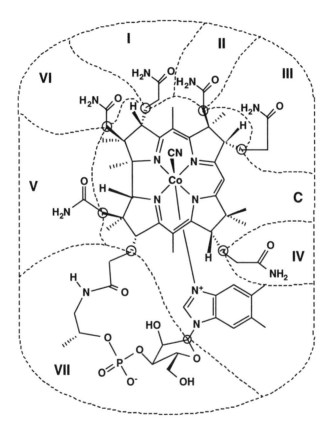

Figure 6.   Chemical structure of vitamin B_{12}. The circles indicate the border atoms and the dashed lines show the fragmentation of the whole molecule.

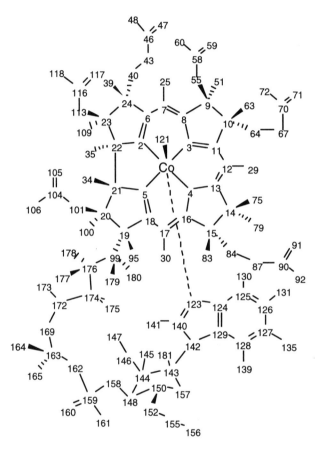

Figure 7. Atom-numbering system of vitamin B_{12} used in the QM/MM calculations (N_{122} from the $C_{121}N_{122}^-$ group is not shown).

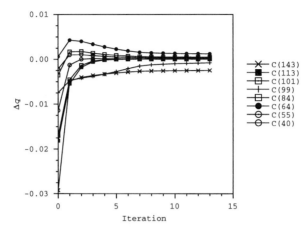

Figure 8. Atomic charge differences $\Delta q = q(\text{CF}) - q(\text{L})$ for the border atoms versus iteration, demonstrating the convergence to interfragment self-consistency in the DSC procedure for vitamin B_{12}.

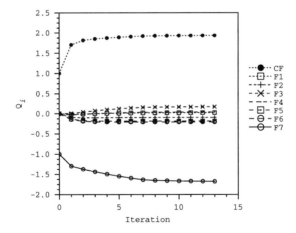

Figure 9. Charge redistribution between the six fragments in the DSC procedure versus iteration in vitamin B_{12}. CF is the central fragment and Fn is the n-th ligand fragment. It is seen that self-consistency is achieved after the 8th iteration.

Table III. Charges on the central and ligand fragments and convergence parameter κ vs. iteration cycle for vitamin B_{12}.

Iteration	Fragments								κ
	Center	1	2	3	4	5	6	7	
0	1.000	0.000	0.000	0.000	0.000	0.000	0.000	-1.000	0.00
1	1.705	-0.017	-0.091	0.005	-0.028	-0.136	-0.145	-1.293	8.00
2	1.816	-0.009	-0.101	0.039	-0.015	-0.172	-0.189	-1.370	8.00
3	1.852	-0.001	-0.101	0.071	-0.000	-0.183	-0.203	-1.436	8.00
4	1.873	0.006	-0.100	0.098	0.011	-0.186	-0.207	-1.495	8.00
5	1.890	0.011	-0.099	0.121	0.020	-0.187	-0.209	-1.548	8.00
6	1.906	0.015	-0.098	0.139	0.027	-0.187	-0.209	-1.594	8.00
7	1.920	0.019	-0.097	0.154	0.033	-0.186	-0.209	-1.634	4.00
8	1.926	0.020	-0.096	0.161	0.035	-0.186	-0.209	-1.651	2.00
9	1.928	0.021	-0.096	0.163	0.036	-0.186	-0.209	-1.658	2.00
10	1.931	0.021	-0.096	0.166	0.037	-0.186	-0.208	-1.666	1.00
11	1.932	0.022	-0.096	0.168	0.038	-0.186	-0.208	-1.669	1.00
12	1.933	0.022	-0.096	0.169	0.038	-0.186	-0.208	-1.673	1.00
13	1.935	0.022	-0.096	0.170	0.039	-0.186	-0.208	-1.676	1.00

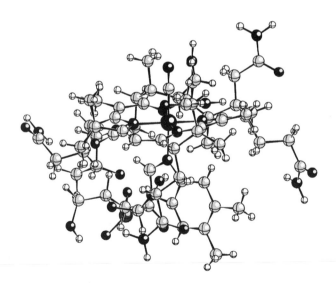

Figure 10. The optimized structure of vitamin B_{12}.

Table IV. Comparison of some geometry parameters of vitamin B_{12} obtained by this method with the experimental X-ray data in the flexible part of the system. The distances R are in Å, and the angles in degress; h is the out-of-plane displacement of the Co atom, and θ is the dihedral angle between two planes of dimethylbenz-imidazolyl and cobamind, respectively

	This work	Experiment [39]
$R(C_{121}-N_{122})$	1.156	1.153
$R(Co-N_{123})$	2.011	2.011
$R(Co-P_{159})$	8.976	8.976
$\angle Co-C_{121}-N_{122}$	178.92	178.920
h	0.122	0.122
	0.121	0.120
θ	93.85	93.84
$R(N_{48}-N_{60})$	10.661	10.820
$R(N_{60}-N_{72})$	8.484	8.610
$R(N_{72}-N_{92})$	10.046	9.730
$R(N_{92}-N_{172})$	6.437	6.924
$R(N_{172}-N_{106})$	5.463	5.846
$R(N_{106}-N_{118})$	6.594	6.287
$R(N_{118}-N_{48})$	7.994	7.847
$R(O_{47}-O_{59})$	10.513	10.238
$R(O_{59}-O_{71})$	7.014	6.660
$R(O_{71}-O_{91})$	10.901	10.873
$R(O_{91}-O_{175})$	6.500	7.015
$R(O_{175}-O_{105})$	5.683	5.397
$R(O_{105}-O_{117})$	7.171	7.403
$R(O_{117}-O_{47})$	5.856	5.770

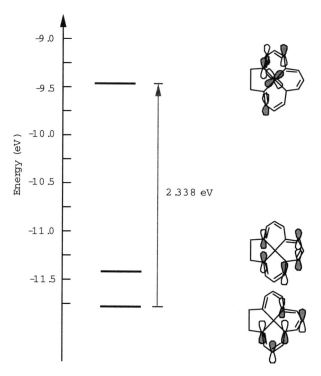

Figure 11. HOMO-1, HOMO, and LUMO energy levels of vitamin B_{12} (from bottom to top) with indication of the dominant five atomic orbitals in the corresponding MO-LCAO.

the region of 530 nm. This absorption explains why vitamin B_{12} has a red color [40], a complement of green light.

Conclusion

A method of combined QM/MM calculations for large molecular systems with strong electronic heterogeneity has been worked out and proved to be efficient. Distinguished from all the previous suggestions of QM/MM treatments, this method has no restriction in molecular size (in reasonable limits) and allows for charge transfers between the QM and MM fragments realized by means of a procedure of double (intrafragment and interfragment) charge self-consistency. In application to metallobiochemical model systems described in this paper, iron picket-fence porphyrin and vitamin B_{12}, the method proved to work well and yield results in reasonable agreement with existing experimental data. In continuation, the method is now in use to treat more complicated metallobiochemical systems including cobalamin and Vitamin B_{12} coenzyme. On the other hand we found out that existing methods of semiempirical calculations have some significant shortcoming [35] that call for improvements which would allow one to obtain both good geometries and charge distributions; this in turn will improve the convergence in the DSC procedure of our method.

Further developments will also include studies of the influence of charge transfer on spectra and reactivity that originate from the QM part (*e.g.*, the influence of solvent molecules on the spectra of transition metal compounds in solutions, as revealed in the work of Stavrev and Zerner [20] for a ruthenium complex in water). Strong effects of charge transfer are expected also in chemical activation of small molecules coordinated to the metal center of enzymes (*e.g.*, carbon monoxide and/or the oxygen molecules in hemoglobin, cytochrome P-450, peroxidase, *etc.*); it is strongly dependent on the charge distribution in the active site which in our method is influenced by the whole metallobiochemical system. Note that the interaction of the biological system with other molecular systems via electrostatic and van der Waals forces (ligand-receptor docking) is dependent on charge distribution on the peripheral atoms which can be evaluated by our method.

Acknowledgements

The authors are deeply thankful to Dr. Michael C. Zerner for his package of semiempirical calculations ZINDO, discussions, and cooperation, and to Dr. G. Calzaferri for his improved version of the program ICONC. This material is based in part upon work supported by the Texas Advanced Research Program under Grant No. 003658-345, and has been supported in part by the Welch Foundation.

References

1. Zerner, M. In: *Reviews in Computational Chemistry*; Lipkowitz, K. B.;. Boyd, D. B, Eds.; VCH, New York, 1991, Vol. 2, pp 313–365; *ZINDO, A General Semiempirical Program Package*; Dept. of Chemistry, The University of Florida, Gainsville, Florida 32611; in: *Metal-Ligand Interactions: from Atoms, to Clusters, to Surfaces*; Salahub, D. R.; Russo, N., Eds.; Kluwer, 1992, pp 101–123.
2. Burkert, U.; Allinger, N. L. *Molecular Mechanics*; ACS Monograph 177, Am. Chem. Soc., Washington, 1982; Bowen, J. P.; Allinger, N. L. In: *Reviews in Computational Chemistry*; Lipkowitz, K. B.; Boyd, D. B. Eds.; VCH, NY, 1991, Vol. 2, p. 81.
3. Bersuker, I. B. *Electronic Structure and Properties of Transition Metal Compounds. Introducion to the Theory*; Wiley, New York, 1996.

4. Comba, P.; Hawbly, T. W. *Molecular Modeling of Inorganic Compounds*; VCH, Weinheim, 1995.
5. Cotton, F. A.; Wilkinson, G. *Advanced Inorganic Chemistry: A Comprehensive Text*; 5th Edition; Wiley, New York, 1992.
6. Bersuker, I. B. *The Jahn-Teller Effect and Vibronic Interactions in Modern Chemistry*; Plenum, New York, 1984; Fischer, G. *Vibronic Coupling*; Academic Press, London, 1984; Levin, A. A. *Soviet Sc. Revs. B, Chemistry Revs.*; Vol'pin, M. E. Ed.; Harwood, 1987, Vol. 9, p. 279; *New J. Chem.* **1993**, *17*, 31; Bersuker, I. B.; Polinger, V. Z. *Vibronic Interactions in Molecules and Crystals*; Springer, New York, 1989; Burdett, J. K. *Molecular Shapes. Theoretical Models of Inorganic Stereochemistry*; Wiley, New York, 1980.
7. Warshel, A.; Levitt, M. J. Mol. Biol. **1976**, *103*, 227–249.
8. Singh, U. C.; Kollman, P. A. *J. Comput. Chem.* **1986**, *7*, 718-730.
9. Field, M. J.; Bash, P. A.; Karplus, M.; *J. Comput. Chem.* **1990**, *11*, 700–733.
10. Bash, P. A.; Field, M. J.; Karplus, M. *J. Am. Chem. Soc.* **1987**, *109*, 8092–94.
11. Bash, P. A.; Field, M. J.; Daveuport, R. C.; Petsko, G. A.; Ringe, D.; Karplus, M. *Biochemistry* **1991**, *30*, 5826–32.
12. Gao, J.; Xia, X. *Science* **1992**, *258*, 631–35.
13. Gao, J.; Chou, L. W.; Auerbach, A. *Biophys. J.* **1993**, *65*, 43–47.
14. Gao, J. *J. Comput. Chem.* **1997**, *18*, 1061–71.
15. Gao, J. *J. Phys. Chem.* **1997**, *101*, 657–663.
16. Stanton, R. V.; Hartsough, D. S.; Merz, K. M., Jr. *J. Phys. Chem.* **1993**, *97*, 11868–70.
17. Maseras, F.; Morokuma, K. *J. Comput. Chem.* **1995**, *16*, 1170.
18. Bakowies, D.; Thiel, W. *J. Comput. Chem.* **1996**, *17*, 87–108.
19. Bakowies, D.; Thiel, W. *J. Phys. Chem.* **1996**, *100*, 10580–94.
20. Stavrev, K. K.; Zerner, M. C. *J. Am. Chem. Soc.* **1995**, *117*, 8684–85.
21. Bersuker, I. B.; Pearlman, R. S. In: *Fifteenth Austin Symposium on Molecular Structure*; The University of Texas, 1994, P5, p. 17; Bersuker, I. B.; Leong, M. K.; Boggs, J. E.; Pearlman, R. S. *Proceedings of the First Electronic Computational Chemistry Conference*; Bachrach, S. M.; Boyd, D. B.; Gray, S. K.; Hase, W.; Rzepa, M. S., Eds.; Rzepa, ARInternet, Landover, MD, 1995, Paper 8.
22. Bersuker, I. B.; Leong, M. K.; Boggs, J. E.; Pearlman, R. S. *Internat. J. Quant..Chem.* **1997**, *63*, 1051–63.
23. Bersuker, I. B. *Teor. i. Eksp. Khim.* **1973**, *9*, 3–12; (English transl.: *Theoret. & Exp. Chem.* **1973**, *9*, 1–7).
24. Gershgorin, S. A. *Izvestia AN SSR, Ser. VII, OFM* 1931, N6, 749–752.
25. Fan, K.; Hoffman, A. *J. Nat. Bureau Standards, Appl. Math. Ser.* **1954**, *39*, 117.
26. Brandle, M.; Calzaferri, G. *Helv. Chim. Acta* **1993**, *76*, 924–951; Savary, F.; Weber, J.; Calzaferri, G. *Phys. Chem.* **1993**, *97*, 3722–27.
27. Calzaferri, G.; Brandle, M. *ICONC&INPUTC, QCPE* **1992**, *12*, 73.
28. Zerner, M.; Gouterman, M. *Theoret. Chim. Acta* **1966**, *4*, 44.
29. Collman, J. P.; Brauman, J. I.; Rose, E.; Suslick, K. S. *Proc. Natl. Acad. Sci. USA* **1978**, 75, 1052–55; Kim, Fettinger, K;. J.; Sessler, J. L.; Cyr, M.; Hugdahl, J.; Collman, J. P.; Ibers, J. A. *J. Am. Chem. Soc.* **1989**, *111*, 403–405; Collman, J. P.; Brauman, J. I.; Doxsee, K. M.; Halbert, T. R.; Bunnenberg, E.; Linder, R. IE.; LaMar, G. N.; Del Gaudio, J.; Lang, G.; Spartalian, K. *J. Am. Chem. Soc.* **1980**, *102*, 4182–92.
30. Clark, M.; Cramer, R. D., III; Van Opdenbosch, N. *J. Comput. Chem.* **1989**, *10*, 982–1012.

31. Distributed by Tripos, Inc., 1699 S. Hanley Road, Suite 303, St. Louis, MO 63144.
32. Jameson, G. B.; Molinaro, F. S.; Ibers, J. A.; Collman, J. P.; Brauman, J. I.; Rose, E.; Suslick, K. S. *J. Am. Chem. Soc.* **1980**, *102*, 3224.
33. Sheidt, W. R.; Gouterman, M. In: Iron Porphyrins; Lever, A. B. P.; Gray, H. B., Eds.; Part I; Addison-Wesley, London, 1983, p. 89.
34. Bersuker, I. B.; Stavrov, S. S. *Coord. Chem. Revs.* **1988**, *88*, 1–68; In: *Light in Biology and Medicine*; Douglas, R. H.; Moan, J.; Ronto, G., Eds.; Plenum, New York, 1991, Vol. 2; pp 357–365.
35. Bersuker, I. B.; Leong, M. K.; Boggs, J. E.; Pearlman, R. S. (to be published).
36. *Spartan Release 4.1*; Wavefunction, Inc., 18401 Von Karman Ave., Suite 370, Irvine, CA 92715.
37. Frisch, M. J.; Trucks, G. W.; Schlegel, H. B.; Gill, P. M. W.; Johnson, B. G.; Robb, M. A.; Cheeseman, J. R.; Keith, T.; Petersson, G. A.; Montgomery, J. A.; Raghavachari, K.; Al-Laham, M. A.; Zakrzewski, V. G.; Ortiz, J. V.; Foresman, J. B.; Cioslowski, J.; Stefanov, B. B.; Nanayakkara, A.; Challacombe, M.; Peng, C. Y.; Ayala, P. Y.; Chen, W.; Wong, M. W.; Andres, J. L.; Replogle, E. S.; Gomperts, R.; Martin, R. L.; Fox, D. J.; Binkley, J. S.; Defrees, D. J.; Baker, J.; Stewart, J. P.; Head-Gordon, M.; Gonzalez, C.; Pople, J. A. *Gaussian 94*, Revision C.3 Gaussian, Inc., Pittsburgh PA, 1995.
38. Frisch, M. J.; Trucks, G. W.; Head-Gordon, M.; Gill, P. M. W.; Wong, M. W.; Foresman, J. B.; Johnson, B. G.; Schlegel, H. B.; Robb, M. A.; Replogle, E. S.; Gomperts, R.; Andres, J. L.; Raghavachari, K.; Binkley, J. S.; Gonzalez, C.; Martin, R. L.; Fox, D. J.; Defrees, D. J.; Baker, J.; Stewart, J. J. P.; Pople, J. A. *Gaussian 92*, Revision C Gaussian, Inc., Pittsburgh PA, 1992.
39. Krautler, B.; Konrat, R.; Stupperich, E.; Farber, G.; Gruber, K.; Kratky, C. *Inorg. Chem.* **1994**, *33*, 4128.
40. Folders, K. In: B_{12}; Dolphin, D. Ed.; Wiley, New York, 1982. Vols. I and II.

Chapter 6

A Methodology for Quantum Molecular Modeling of Structure and Reactivity at Solid–Liquid Interfaces

Eugene V. Stefanovich and Thanh N. Truong

Department of Chemistry, University of Utah, Salt Lake City, UT 84112

A QM/MM methodology for modeling chemical reactions at solid-liquid interfaces is presented. This new method combines advances in dielectric continuum solvation models for describing polarization of the liquid with the embedded cluster approach for treating interactions in the solid. In addition, a new method for simple, yet accurate, incorporation of the Madelung potential effect in embedded cluster calculations is discussed. The advantages and accuracy of this method are demonstrated in a number of test calculations. Geometries and adsorption binding energies of H_2O at the NaCl(001)/water interface are calculated and compared with those at the NaCl(001)/vacuum interface.

In recent years the emphasis in quantum chemistry has been shifting from properties of gas-phase molecules and reactions toward challenging areas of condensed phase systems. In particular, molecular processes at solid-liquid interfaces have attracted much attention as they play important roles in environmental chemistry, biochemistry, electrochemistry, heterogeneous catalysis, and other fields. For example, sodium chloride crystals, apart from being a useful model system for theoretical developments, participate in a number of atmospheric processes. Sea salt aerosoles react with various gases, such as NO_2, in the earth's troposphere (1,2). Adsorbed water certainly plays a key role in these reactions. Despite much experimental progress, very little is known for certain regarding mechanisms of interfacial processes. Many experimental surface sensitive techniques such as thermodesorption, scanning tunneling, and photo-electron spectroscopies, require low coverage, ultra-high vacuum, or extreme temperature conditions, thus may be not directly relevant to the solid-liquid interfacial systems found in Nature. Theory can play a crucial role here.

Accurate theoretical modeling of functional groups and sorption complexes at the interface, however, is a difficult task because interfacial energetics are driven by a complicated balance between hydration forces and crystal-solute interactions. Any theory for realistic modeling of chemical reactions at solid-liquid interfaces should provide an accurate description of bond-forming and -breaking processes and interactions of adsorbates and surface defects with the crystal lattice and solvent. Below we discuss some advantages and weaknesses of three common theoretical approaches that can be employed to study interfacial chemical processes, namely the periodic quantum mechanical, classical molecular mechanics, and quantum embedded cluster methods.

The main advantage of *ab initio periodic supercell* calculations is that they permit an accurate description of interactions between the active surface site and the rest of the crystal. However, periodic models suffer from difficulties with representing the statistical nature of the solvent and cannot avoid spurious interaction between defects or active sites in adjacent unit cells (3). Often these methods are not practical for modeling low-symmetry defect sites and exploring potential energy surfaces. Furthermore, periodic boundary conditions are not particularly suitable for studying charged defects. Full *ab initio* quantum mechanical calculations of solid-liquid interfaces to date are limited to studies of monolayer or bilayer adsorption of water molecules using small surface unit cells (4-13) .

Classical *molecular mechanics*, in conjunction with Monte Carlo or molecular dynamics methods, is a powerful tool for analyzing structure and thermodynamic properties of condensed phase systems. In particular, valuable information has been obtained for solvent structure near the MgO-water (14) and NaCl-water (15-17) interfaces, acidities of hydroxyls at the Fe_2O_3-vacuum interface (18), and stabilization of Al_2O_3 surfaces by hydroxyls (19). However, current molecular force fields do not provide an accurate representation of reactive processes such as dissociative chemisorption. Performance of these approaches depends on advances in development of molecular mechanics force fields which is a difficult and time-consuming task (20).

The quantum *embedded cluster* approach (21-23) takes advantage of the localized nature of surface chemical processes, so that only a relatively small molecular system (active center + adsorbate, or *quantum cluster*) can be treated quantum mechanically by accurate *ab initio* or DFT methods using large basis sets. The rest of the crystal lattice and solvent (*environment*) are treated classically. Their action on electrons in the cluster is represented by adding an embedding potential $V_{embed}(\mathbf{r})$ to the quantum Hamiltonian of the cluster. Flexibility and comparatively low computational cost of the embedded cluster approach makes it especially suitable for studies of chemical interactions at solid-liquid interfaces.

In this paper, we discuss applications of our recently proposed quantum cluster methodology, called CECILIA (Combined Embedded Cluster at the Interface with LIquid Approach), for modeling chemical phenomena at non-metal solid-liquid interfaces (24). CECILIA combines advantages of the embedded cluster method discussed above with the dielectric continuum method for solvation. We focus on two important aspects that are critical for the accuracy of the CECILIA model: the solvent polarization field and Madelung potential from the crystalline lattice. These are two long-range components which often make dominant contributions to the total embedding potential $V_{embed}(\mathbf{r})$ for a cluster located at the solid-liquid interface. For solvent polarization, the dielectric continuum approach provides a cost effective methodology and is an active area of research in liquid phase solvation studies (see, for example Generalized COnductor-like Screening MOdel (GCOSMO) calculations in Refs. (25-31)), however, its application for solid-liquid interfaces is a new and unexplored area. For ionic and semi-ionic compounds, the Madelung potential from the crystal lattice makes an important contribution to $V_{embed}(\mathbf{r})$. For this contribution we recently proposed a new SCREEP method (Surface Charge Representation of the Electrostatic Embedding Potential; Stefanovich, E.V.; Truong, T.N. *J. Phys. Chem..*, submitted) that can accurately replace the Madelung potential active on a quantum cluster by a potential from a finite set of point charges located on a surface surrounding the cluster. In this chapter we will briefly describe both

GCOSMO and SCREEP methods focusing on their use of a common mathematical technique, i.e., solution of the electrostatic Poisson equation with the boundary condition for an ideal conductor. We also present calculation results for geometries and adsorption binding energies of H_2O at the NaCl(001)/water and NaCl(001)/vacuum interfaces. In the next section we briefly describe the physical model of the solid-liquid interface employed in this study.

A Physical Model of the Solid-Liquid Interface

In our CECILIA approach, the whole system (surface defect + crystal + solvent) is divided into three main regions (Figure 1) designed to maximize the chemical accuracy while keeping the problem tractable by modern computers.

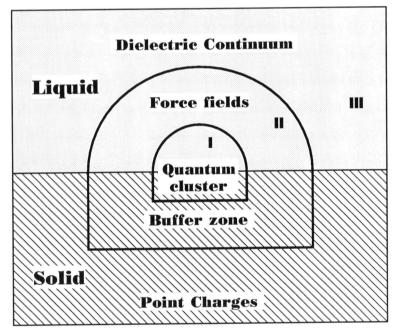

Figure 1. The CECILIA model. (Adapted from Ref. (24))

The innermost quantum mechanical region (I), or *quantum cluster* where chemistry occurs, is treated accurately by either *ab initio* molecular orbital or density functional methods. Normally, the quantum cluster may consist of several lattice atoms near the defect site, the adsorbate, and a few water molecules making strong hydrogen bonds with the surface complex. The *buffer zone* (II) may include several dozen atoms in the crystal lattice and several solvent molecules treated as classical particles surrounding the quantum cluster. This region is designed to describe short-range forces between nuclei and electrons in the quantum cluster and the surrounding medium. The *peripheral zone* (III) provides for a correct Madelung potential from the crystal lattice and the long-range solvent polarization potential in the quantum cluster and buffer regions. In this paper, we are primarily concerned with an accurate description of the peripheral zone.

In the CECILIA approach, a self-consistent treatment of the solvent polarization is achieved by using the GCOSMO dielectric continuum solvation model in which the liquid is modeled as a dielectric continuum separated from the solute (in our case crystal surface and adsorbate) by a sharp boundary. In homogeneous solvation studies, this boundary is constructed as a set of interlocking spheres centered on nuclei and having fixed radii fitted to experimental hydration free energies of ions and molecules (31). The solvent polarization field is then represented as charge density on the boundary determined self-consistently with the charge density distribution of the solute. We will demonstrate how this methodology can be applied for the solid-liquid interface situation.

To represent the Madelung potential from the crystal lattice acting inside the quantum cluster is not a trivial task even if we assume that the crystal in the peripheral zone is composed of point ions situated at lattice sites. Although in such a case one can easily evaluate the corresponding electrostatic potential $V_{el}(\mathbf{r})$ at any given point in the cluster, for instance by using Ewald summation formulas, the difficulty arises from the necessity to calculate matrix elements $\langle \mu | V_{el}(\mathbf{r}) | \nu \rangle$ over cluster basis functions. One possibility is to perform a direct calculation of matrix elements of the Ewald potential (32). However, implementation of this method (see DICAP (33) and EMBED (34) codes) requires significant modifications in standard molecular quantum chemistry programs. In addition, no analytical energy derivatives are currently available. A more common methodology is to substitute the infinite lattice by a finite number of point charges placed at ideal lattice sites $\mathbf{R}_j$ that have values corresponding to ionic charges in the crystal (35-40). This method has the advantage that matrix elements of the point-charge potential, as well as their first and second derivatives, are readily available in most quantum chemistry programs (as nuclear attraction integrals). However, construction of such finite lattice models becomes rather difficult for complex low-symmetry crystals. A well-known difficulty for such an approach is that results converge very slowly, if converge at all, when the size of the explicitly considered lattice is increased. Thus, there is no simple way for systematic improvement of results.

To illustrate this poor convergence, consider the electrostatic potential from an infinite 5-layer slab of point charges (±1) parallel to the (001) plane of the rocksalt (NaCl) lattice. Let us first demonstrate the performance of the traditional embedding scheme in which the infinite lattice is substituted by a neutral block with dimensions $n \times n \times m$ (n=6-14, m= 3-5), m layers deep. As an indication of the accuracy we calculated the RMS deviation of the model potential from the exact Ewald potential at 21 equidistant points having z-coordinates between 1.0Å and 5.0Å directly above the surface cation. These results are presented in Table I below. As expected, they show rather poor convergence when the size of the explicitly treated lattice is increased.

Table I. RMS deviations (in kcal/mol) from the exact Madelung potential above the NaCl(001) surface for finite lattice models with dimensions $n \times n \times m$.

$n \times n$	6×6	8×8	10×10	12×12	14×14
m					
3	3.64	1.71	0.97	0.60	0.39
4	2.42	0.83	0.35	0.17	0.09
5	3.10	1.40	0.80	0.51	0.34

In what follows we want to show that accurate representation of both the Madelung potential and the solvent polarization field can be achieved by using basically identical mathematical techniques, i.e., applying the boundary condition for a conductor in the external electrostatic field.

Boundary Condition for a Conductor in the External Electric Field

Consider a region of space C where the charge density is zero and the electrostatic potential is produced by the charge distribution $\rho(\mathbf{r})$ lying entirely outside C. A well-known theorem from electrostatics states that no matter what is the charge distribution $\rho(\mathbf{r})$ *outside* C, its electrostatic potential $V_{el}(\mathbf{r})$ *inside* C can be rigorously replaced by some surface charge density $\sigma(\mathbf{r})$ located on the *boundary* S of the volume C. The demonstration goes as follows. First assume that we have filled the volume C (interior of the surface S) with an ideal metallic conductor. The electrostatic potential inside S becomes exactly zero independent of the external potential $V_{el}(\mathbf{r})$. Physically, this condition is satisfied due to creation of the charge density $-\sigma(\mathbf{r})$ on the surface S whose potential $-\oint_S \dfrac{\sigma(\mathbf{r}')}{|\mathbf{r} - \mathbf{r}'|} d^2 r$ exactly compensates the external potential $V_{el}(\mathbf{r})$ for all points $\mathbf{r}$ on the surface S and in its interior.

$$V_{el}(\mathbf{r}) - \oint_S \frac{\sigma(\mathbf{r}')}{|\mathbf{r} - \mathbf{r}'|} d^2 r = 0 \tag{1}$$

This means that the electrostatic potential generated by the charge density $\sigma(\mathbf{r})$ on the surface S and in its interior is exactly equal to the original potential $V_{el}(\mathbf{r})$. (The potential generated by $\sigma(\mathbf{r})$ *outside* the surface S is generally different from $V_{el}(\mathbf{r})$).

For computational reasons, we resort to the boundary element method to represent the continuous surface charge density $\sigma(\mathbf{r})$ on the surface S by a set of M point charges q_j located at the centers $\mathbf{r}_j$ of surface elements with areas S_j.

$$q_j \approx \sigma(\mathbf{r}_j) S_j \tag{2}$$

This approximation is accurate when the number of surface points M is large enough, and the charge distribution $\sigma(\mathbf{r})$ is sufficiently smooth. Then Eq. (1) can be approximated by a matrix equation

$$V - \mathbf{Aq} = 0 \tag{3}$$

from which the vector of surface charges $\mathbf{q}$ can be determined by applying any common technique available for solving systems of linear equations. For example, one can use the matrix inversion method.

$$q = A^{-1}V \qquad (4)$$

In Eqs. (3) and (4), the vector V contains values of the external electrostatic potential at points r_j ($V_j = V_{el}(\mathbf{r}_j)$), and A is the $M \times M$ non-singular matrix with matrix elements.

$$A_{ij} = \frac{1}{|\mathbf{r}_j - \mathbf{r}_i|} \text{ for } j \neq i, \text{ and } A_{jj} = 1.07\sqrt{\frac{4\pi}{S_j}} \qquad (5)$$

Non-diagonal elements A_{ij} represent a generic Coulomb interaction between surface elements r_i and r_j. The diagonal elements A_{jj} describe the self-interaction of the surface element r_j. This self-interaction was discussed in detail by Klamt and Shüürmann (41), and the coefficient 1.07 was fitted by these authors for better numerical accuracy.

The CECILIA Model

The computational method described above was first applied by Klamt and Shüürmann (41) in their COSMO (COnductor-like Screening MOdel) dielectric continuum model for solvation in the bulk liquid. The virtue of using conductor boundary conditions for aqueous solvent comes from the fact that water has a rather high dielectric constant ($\varepsilon = 78.3$), thus screening properties of the solvent are similar to those of an ideal conductor. Therefore, in the first approximation of the CECILIA model, liquid can be represented as a conductor in the electrostatic field generated by the crystal and adsorbate. Two points should be made. First, one should correct conductor surface charges to account for the finite dielectric constant of water. Second, surface charges should be determined self-consistently with the electronic and atomic structure of the quantum cluster. Thus, instead of Eq. (4), the equation for surface charges is

$$q = -\frac{\varepsilon - 1}{\varepsilon} A^{-1}V \qquad (6)$$

where the vector V contains electrostatic potentials produced by the charge density of the crystal at points r_j on the dielectric cavity. This potential can be conveniently separated into contributions from classical particles (atomic nuclei in the quantum cluster and classical particles in the buffer and peripheral zones) with charges z_i and positions $\mathbf{R}_i$ and from the electron density $\rho(\mathbf{r})$ in the quantum cluster

$$V_j = V_j^{class} + V_j^{el} = \sum_i \frac{z_i}{|\mathbf{r}_j - \mathbf{R}_i|} - \int \frac{\rho(\mathbf{r})}{|\mathbf{r} - \mathbf{r}_j|} d^3r \qquad (7)$$

Correspondingly, surface charges in Eq. (6) can be separated into "classical" $\mathbf{q}^{class}$ and "electronic" $\mathbf{q}^{el}$ components. The classical surface charges are taken into account by adding the term

$$H_{\mu\nu}^s = \mathbf{q}^{class}\mathbf{L}_{\mu\nu} + W_{\mu\nu}^{buf} + \langle\mu|V_{el}(\mathbf{r})|\nu\rangle \qquad (8)$$

to the one-electron part of the Fock matrix for an isolated cluster ($H_{\mu\nu}^0$). Here $L_{\mu\nu}^j$ are matrix elements of the potential generated by a unit point charge at the point $\mathbf{r}_j$ on the dielectric cavity. In contrast to the GCOSMO Hamiltonian for a solute in the bulk liquid (30), this expression contains additional matrix elements of the short-range embedding potential $W_{\mu\nu}^{buf}$ from the buffer zone and Madelung potential $\langle\mu|V_{el}(\mathbf{r})|\nu\rangle$ from the buffer and peripheral zones. The former term may be represented by different embedding techniques including the pseudoatom (42), localized orbitals (43), density functional (44), or pseudopotential (33,45-48) methods. For such ionic compound as NaCl, the pseudopotential form of $W_{\mu\nu}^{buf}$ has been found to be rather accurate. The Madelung potential term will be discussed in more detail in the next Section. Another difference with the bulk solvation case is that the surface of the cavity is not closed. Only the quantum cluster and its nearest neighbors on the crystal surface need to be solvated to obtain relative energies of surface configurations (24). The contribution of electronic surface charges $\mathbf{q}^{el}$ to the two-electron part of the Fock matrix is given by

$$G_{\mu\nu}^s = \mathbf{q}^{el}\mathbf{L}_{\mu\nu} \qquad (9)$$

Then, the self-consistency between the electron density and solvent polarization field is achieved in a single SCF procedure by calculating $\mathbf{q}$ and $G_{\mu\nu}^s$ from Eqs. (6) and (9), respectively, at each iteration.

The total energy of the quantum cluster at the solid-liquid interface in HF and DFT methods is expressed as

$$E_{tot} = \sum_{\mu\nu}\left[P_{\mu\nu}\left(H_{\mu\nu}^0 + H_{\mu\nu}^s\right) + \tfrac{1}{2}P_{\mu\nu}\left(G_{\mu\nu}^0 + G_{\mu\nu}^s\right)\right]$$
$$+\mathbf{q}^{class}V^{class} + E_{nn} + E_{non-els} \qquad (10)$$

where $P_{\mu\nu}$ is the converged density matrix of the quantum cluster, E_{nn} is the energy of interaction between classical particles, i.e. nuclei in the QM cluster and ions in the buffer and point charge zones. Apart from Coulomb interactions, this term may include short-range potentials taken from various force-fields. $E_{non-els}$ contains non-electrostatic dispersion-repulsion and cavity formation contributions to the solvation free energy. First and second derivatives of the total energy (6) with respect to the coordinates of atoms in the cluster and buffer zone are available (25). This allows for efficient geometry optimization of adsorbate structures at solid-liquid interfaces.

The SCREEP Model

The conductor boundary conditions discussed above can also be used to conveniently replace the Madelung potential in the quantum cluster by a finite set of point charges. This allows for straightforward calculation of corresponding matrix elements $\langle \mu | V_{el}(\mathbf{r}) | \nu \rangle$. Let us first define a closed surface S such that all quantum cluster atoms and the GCOSMO dielectric cavity lie inside this surface (see Figure 2).

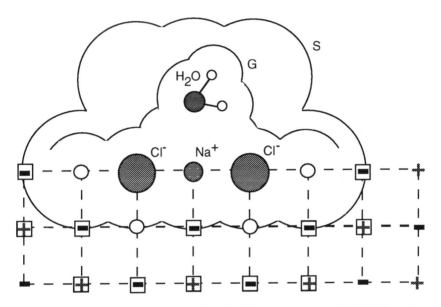

Figure 2. Side view of the $[Na_5Cl_4+H_2O]^+$ cluster on the NaCl(001) surface and sketch of the CECILIA model. Shaded circles are ions from the quantum cluster; empty circles are Na^+ ECPs from the buffer zone; "+" and "-" symbols inside squares denote point charges in the explicit zone; the potential from other lattice charges is modeled by the charge density on the SCREEP surface S; surface G is the dielectric continuum cavity.

As in continuum solvation methods, surface S can be constructed as a set of interlocking spheres of radius R_a centered on cluster nuclei and discretized into a finite number of surface elements. Then Madelung potentials on the surface elements can be calculated by using the Ewald summation technique (note that the contribution from lattice ions inside the surface S should be subtracted from the Ewald sum) and values of surface charges q can be obtained from Eq. (4). The potential generated by these charges can be used as a substitute for the Madelung potential inside the surface S, i.e., in the quantum cluster and on the GCOSMO dielectric cavity.

Our analysis of the SCREEP approximation revealed two important points. First, the radius R_a should be large enough so that a major part of the cluster wave function is contained inside the surface S. Second, for better accuracy of the

SCREEP model one should treat explicitly the potential from lattice ions lying close to the quantum cluster. This means that we directly include the potential of these ions in calculation of $\langle \mu | V_{el}(\mathbf{r}) | \nu \rangle$ and subtract this potential from the vector V used in Eq. (4). Moreover, the explicit zone (shown schematically by "+" and "-" symbols inside squares in Figure 2) should be selected in such a way that its total charge is zero or small.

Calculation Details

In the rest of this paper we consider applications of the SCREEP and CECILIA methods to study water molecule adsorption on the unrelaxed rock-salt NaCl(001) surface and NaCl(001)-water interface. The quantum cluster $[Na_5Cl_4+H_2O]^+$ selected for these studies contained nine surface atoms that form a square 3×3 substrate and adsorbed water molecule (Figure 2). As NaCl is a rather ionic crystal, we do not expect that slight deviation from stoichiometry and assignment of the integer charge (+1) to this cluster will affect results in any significant way. Geometries for water molecules in different environments (gas phase, liquid phase, adsorbed on the clean surface, and adsorbed at the interface) were fully optimized at the pseudopotential Hartree-Fock level. The oxygen atom was described by the SBK effective core pseudopotentials (ECP) and CEP-31+G* basis set (49). The 311++G** basis set was used for hydrogens. Na and Cl ions were described by the Hay-Wadt ECP and standard valence double-zeta basis sets (50). Electron correlation was included at the MP2 level as single point calculations at HF optimized geometries. Dissociation and adsorption energies were calculated as total energy differences between the compound system and its separated fragments.

The short-range embedding potential $W_{\mu\nu}^{buf}$ (exchange, Pauli repulsion) was represented by Hay-Wadt ECP for Na^+ cations that are nearest neighbors to the quantum cluster. More distant Na^+ ions can be rather accurately treated as point charges (q=+1). Actually, when they were treated as pseudopotentials, cluster energies changed by less than 0.1 kcal/mol. We are not aware of any accurate whole-ion pseudopotential representation of lattice Cl^-. However, as indicated in Refs. (33,45), the ground-state electron density in the quantum cluster only slightly penetrates surrounding anions, therefore they can be described rather accurately in the point charge (q=-1) approximation.

Cavities for SCREEP and CECILIA calculations were constructed using the gepol93 algorithm (51) as a set of interlocking spheres centered on atoms. The dielectric cavity adjusted automatically when atoms in the quantum region moved during the geometry optimization, while the SCREEP surface remained fixed. Each complete atomic sphere contained 60 surface charges. Atomic radii for the CECILIA cavity were taken from our previous work (24,31): 1.172 Å for H, 1.576 Å for O, 1.61 Å for Na, and 1.75 Å for Cl. The cavity boundary was truncated so that only atoms in the cluster and their nearest neighbors on the surface were solvated. The effect of such truncation was found to be insignificant. For example, when the boundary was truncated so that only cluster atoms were solvated, the adsorption energies changed by less than 0.1 kcal/mol.

The SCREEP surface S was constructed by making spheres of radius R_a= 2.5 Å around each atom in the cluster and around 9 additional centers having z-coordinates of 4.0 Å just above each atom in the cluster (Figure 2). Introduction of these additional centers comes from the necessity to provide enough space for placing adsorbate molecules inside the SCREEP surface above the crystal. Using the same accuracy criterion as in Table I, we found that the SCREEP method gives a much more accurate (RMS error of less than 0.01 kcal/mol) Madelung potential than any finite lattice $n \times n \times m$ representation shown in Table I.

The term $E_{non-els}$ in eq. (10) was calculated using methods described in Refs. (52,53) with the optimized solvent water radius of 1.29 Å (31). We used the OPLS force field parameters (54) for calculating dispersion-repulsion interactions with the solvent in this work. All calculations were performed by using our locally modified version of the Gaussian92/DFT computer code (55).

Results and Discussion

Let us first compare three embedding schemes for H_2O adsorption on the NaCl(001)-vacuum interface: no embedding at all, i.e., adsorption on the bare $[Na_5Cl_4]^+$ cluster; embedding in the finite $8 \times 8 \times 4$ lattice of point charges (±1) (247 explicit lattice ions outside cluster); and the SCREEP embedding scheme (121 lattice ions in the explicit zone plus 422 point charges on the surface S). These results are presented in the first three columns in Table II.

Table II. Geometry (distances in Å, angles in degrees)[a] and adsorption energy (kcal/mol) for H_2O molecule adsorbed on the NaCl(001) surface calculated using the $[Na_5Cl_4+H_2O]^+$ quantum cluster and different embedding schemes. The most accurate results are shown in bold.

	Solid-vacuum interface				Solid-liquid interface	
Madelung potential	no	$8 \times 8 \times 4$	SCREEP	$8 \times 8 \times 4$	$8 \times 8 \times 4$	SCREEP
Buffer zone	no	no	no	ECP	ECP	ECP
Solvent effects	no	no	no	no	yes	yes
Oz	2.390 [2.43][c]	2.384	2.380	**2.401**	2.599	**2.597**
Ox=Oy	0.294	0.374	0.375	**0.409**	0.057	**0.047**
OH	0.948	0.949	0.949	**0.949**	0.952	**0.952**
HOH angle	106.4	106.2	106.2	**98.3**	104.5	**104.4**
tilt angle	19.6 [30.5][c]	5.4	5.9	**0.3**	11.9	**12.8**
Adsorption energy	7.5 [7.8][c]	8.3	8.3	6.7 (8.1)[b]	3.6 (5.5)[b]	3.6 (5.5)[b]

[a] The origin was placed at the central Na^+ ion in the cluster; x and y axes were directed toward nearest surface Cl^- ions; the z-axis pointed outside the crystal.
[b] In parentheses results of MP2 single-point calculations.
[c] In square brackets theoretical results from Ref. (56).

H_2O adsorbs on the NaCl(001) surface with oxygen above the Na^+ site and hydrogen atoms pointing symmetrically toward nearest anions and slightly away from the surface (24). As seen from Table II, the neglect of the Madelung field in the cluster leads to some increase of the oxygen-surface distance and molecular tilt angle (the angle between the molecular axis and the surface). This comparison may explain why recent *ab initio* calculations of H_2O/NaCl(001) adsorption using the Na_9Cl_9 cluster without any embedding potential (56) overestimate these parameters. The adsorption energy calculated by these authors (7.8 kcal/mol) is also close to our result for the bare cluster (7.5 kcal/mol). Results for $8 \times 8 \times 4$ and SCREEP embedding schemes agree with each other but differ significantly from those for the bare cluster. Comparison of the $8 \times 8 \times 4$ and SCREEP embedding schemes serves two purposes. First, it confirms our correct implementation of the SCREEP method. Second, it indicates that traditional finite lattice embedding (the $8 \times 8 \times 4$ lattice in our case) may be quite successful for simple crystal lattices if proper care is taken in selecting the finite lattice size *(24,44,57,58)*. Thus, real benefits of using SCREEP embedding can be revealed in studies of complex crystal lattices, such as zeolites, where traditional embedding models are not easy to construct.

The importance of the non-electrostatic short-range embedding potential ($W_{\mu\nu}^{buf}$ in eq. (8)) can be seen by comparing results from the second column in Table II with calculations in which cations nearest to the quantum cluster were represented by Na^+ effective core pseudopotentials (4th column in Table II). The latter more accurate treatment leads to a decrease of the adsorption energy by 1.6 kcal/mol, decrease of the molecular tilt angle by 5.1 degree and increase of the adsorption distance by 0.017 Å. Taking into account correlation effects at the MP2 level, the adsorption energy of 8.1 kcal/mol is in reasonable agreement with experimental data (8.5-13.1 kcal/mol, see Ref. *(24)*). Note that the adsorption energy is likely to be overestimated in experiments due to the presence of surface defects.

Results for H_2O adsorption at the NaCl-water interface are shown in the 5th ($8 \times 8 \times 4$ embedding) and 6th (SCREEP embedding) columns in Table II. In both cases, Na^+ pseudopotentials were used in the buffer zone. As expected, both embedding models give similar results. The screening effect of the solvent reduces attraction of the H_2O molecule to surface ions (compare with results in the 4th column). The distance of the oxygen atom from the surface increases by 0.2 Å. H atoms no longer feel a strong attraction to Cl^- lattice anions, therefore, the tilt angle between the molecular axis and surface plane increases from 0.3 to 12-13 degrees, the lateral shift of the molecule along the <110> axis decreases substantially, and the HOH angle increases to the value of 104.4 characteristic for the hydrated water molecule. In general, the internal structure of the H_2O molecule adsorbed at the interface is much closer to the geometry of hydrated H_2O than to the geometry of water adsorbed at the clean NaCl(001) surface. In agreement with these results, dielectric screening by the solvent reduces the interaction energy between H_2O and the solid surface by about 2.6 kcal/mol. However, there is still a noticeable attraction (5.5 kcal/mol) which immobilizes the H_2O molecule near the NaCl surface in agreement with previous molecular dynamics simulations*(15,17)* and a helium atom scattering study *(16)*. About half of this adsorption energy (2.6 kcal/mol) is due to non-electrostatic solvation effects *(24)*.

Conclusion

We presented a general methodology for *ab initio* embedded cluster calculations of reactivity at the solid-liquid interface (CECILIA model). A new important component of this model is the SCREEP method which allows to accurately represent the Madelung potential in embedded cluster calculations of solids. With appropriate choice of computational parameters, the SCREEP method can easily provide a RMS deviation of less than 0.1 kcal/mol from the exact lattice potential. It is important that this method requires only a negligible increase (1-2%) of the computational time in our quantum embedded cluster calculations as compared to the bare molecule calculation, and the accuracy level can be improved in a simple and systematic way. Moreover, no modifications to existing molecular quantum chemistry programs were required. Positions and values of surface charges need to be determined only once from a separate simple calculation. In addition to efficient geometry optimization, our current implementation of the CECILIA and SCREEP approaches allows for calculation of vibrational frequencies and excited states. Extensive studies of adsorption and reactions at MgO-water, TiO_2-water, and Al_2O_3-water interfaces are currently underway in our laboratory.

Acknowledgment

This work was supported in part by the National Science Foundation through a Young Investigator Award to T.N.T.

Literature Cited

(1)   Vogt, R.; Finlayson-Pitts, B. J. *J. Phys. Chem.* **1995**, *99*, pp 17269-17272.
(2)   Langer, S.; Pemberton, R. S.; Finlayson-Pitts, B. J. *J. Phys. Chem. A* **1997**, *101*, pp 1277-1286.
(3)   Anchell, J. L.; Hess, A. C. *J. Phys. Chem.* **1996**, *100*, pp 18317-18321.
(4)   Goniakowski, J.; Gillan, M. J. *Surf. Sci.* **1996**, *350*, pp 145-158.
(5)   Lindan, P. J. D.; Harrison, N. M.; Holender, J. M.; Gillan, M. J. *Chem. Phys. Lett.* **1996**, *261*, pp 246-252.
(6)   Fahmi, A.; Minot, C. *Surf. Sci.* **1994**, *304*, pp 343-359.
(7)   Podloucky, R.; Steinemann, S. G.; Freeman, A. J. *New J. Chem.* **1992**, *16*, pp 1139-1143.
(8)   Scamehorn, C. A.; Hess, A. C.; McCarthy, M. I. *J. Chem. Phys.* **1993**, *99*, pp 2786-2795.
(9)   Scamehorn, C. A.; Harrison, N. M.; McCarthy, M. I. *J. Chem. Phys.* **1994**, *101*, pp 1547-1554.
(10)   Taylor, D. P.; Hess, W. P.; McCarthy, M. I. *J. Phys. Chem. B* **1997**, *101*, pp 7455-7463.
(11)   Orlando, R.; Pisani, C.; Ruiz, E.; Sautet, P. *Surf. Sci.* **1992**, *275*, pp 482-492.
(12)   Ursenbach, C. P.; Voth, G. A. *J. Chem. Phys.* **1995**, *103*, pp 7569-7575.
(13)   Langel, W.; Parrinello, M. *J. Chem. Phys.* **1995**, *103*, pp 3240-3252.
(14)   McCarthy, M. I.; Schenter, G. K.; Scamehorn, C. A.; Nicholas, J. B. *J. Phys. Chem.* **1996**, *100*, pp 16989-16995.
(15)   Anastasiou, N.; Fincham, D.; Singer, K. *J. Chem. Soc. Faraday Trans. II* **1983**, *79*, pp 1639-1651.
(16)   Bruch, L. W.; Glebov, A.; Toennies, J. P.; Weiss, H. *J. Chem. Phys.* **1995**, *103*, pp 5109-5120.
(17)   Wassermann, B.; Mirbt, S.; Reif, J.; Zink, J. C.; Matthias, E. *J. Chem. Phys.* **1993**, *98*, pp 10049-10060.
(18)   Rustad, J. R.; Felmy, A. R.; Hay, B. P. *Geochim. Cosmochim. Acta* **1996**, *60*, pp 1563-1576.
(19)   Nygren, M. A.; Gay, D. H.; Catlow, C. R. A. *Surf. Sci.* **1997**, *380*, pp 113-123.
(20)   Chacon-Taylor, M. R.; McCarthy, M. I. *J. Phys. Chem.* **1996**, *100*, pp 7610-7616.
(21)   *Electronic properties of solids using cluster methods*; Kaplan, T. A.; Mahanti, S. D., Eds.; Plenum: New York, 1995.
(22)   Pacchioni, G.; Bagus, P. S. In *Cluster Models for Surface and Bulk Phenomena*; G. Pacchioni, Ed.; Plenum Press: New York, 1992.
(23)   *Quantum mechanical cluster calculations in solid state studies*; Grimes, R. W.; Catlow, C. R. A.; Shluger, A. L., Eds.; World Scientific: Singapore, 1992.
(24)   Stefanovich, E. V.; Truong, T. N. *J. Chem. Phys.* **1997**, *106*, pp 7700-7705.
(25)   Stefanovich, E. V.; Truong, T. N. *J. Chem. Phys.* **1996**, *105*, pp 2961-2971.

104

(26) Truong, T. N.; Stefanovich, E. V. *J. Phys. Chem.* **1995**, *99*, pp 14700-14706.
(27) Truong, T. N.; Stefanovich, E. V. *J. Chem. Phys.* **1995**, *109*, pp 3709-3716.
(28) Truong, T. N.; Nguyen, U. N.; Stefanovich, E. V. *Int. J. Quant. Chem.: Quant. Chem. Symp.* **1996**, *30*, pp 403-410.
(29) Truong, T. N.; Truong, T.-T. T.; Stefanovich, E. V. *J. Chem. Phys.* **1997**, *107*, pp 1881-1889.
(30) Truong, T. N.; Stefanovich, E. V. *Chem. Phys. Lett.* **1995**, *240*, pp 253-260.
(31) Stefanovich, E. V.; Truong, T. N. *Chem. Phys. Lett.* **1995**, *244*, pp 65-74.
(32) Saunders, V. R.; Freyria-Fava, C.; Dovesi, R.; Salasco, L.; Roetti, C. *Mol. Phys.* **1992**, *77*, pp 629-665.
(33) Puchin, V. E.; Shluger, A. L.; Tanimura, K.; Itoh, N. *Phys. Rev. B* **1993**, *47*, pp 6226-6240.
(34) Pisani, C.; Orlando, R.; Nada, R. In *Cluster Models for Surface and Bulk Phenomena*; G. Pacchioni; P. S. Bagus and F. Parmigiani, Ed.; Plenum: New York, 1992; pp 515-531.
(35) Harding, H. H.; Harker, A. H.; Keegstra, P. B.; Pandey, R.; Vail, J. M.; Woodward, C. *Physica B & C* **1985**, *131*, pp 151-156.
(36) Allouche, A. *J. Phys. Chem.* **1996**, *100*, pp 1820-1826.
(37) Allouche, A. *J. Phys. Chem.* **1996**, *100*, pp 17915-17992.
(38) Ferrari, A. M.; Pacchioni, G. *J. Phys. Chem.* **1996**, *100*, pp 9032-9037.
(39) Ferro, Y.; Allouche, A.; Corà, F.; Pisani, C.; Girardet, C. *Surf. Sci.* **1995**, *325*, pp 139-150.
(40) Pacchioni, G.; Ferrari, A. M.; Márquez, A. M.; Illas, F. *J. Comp. Chem.* **1997**, *18*, pp 617-628.
(41) Klamt, A.; Schüürmann, G. *J. Chem. Soc., Perkin Trans. II* **1993**, pp 799.
(42) Sauer, J. *Chem. Rev.* **1989**, *89*, pp 199-255.
(43) Gorb, L. G.; Rivail, J.-L.; Thery, V.; Rinaldi, D. *Int. J. Quant. Chem.: Quant. Chem. Symp.* **1996**, *30*, pp 1525-1536.
(44) Stefanovich, E. V.; Truong, T. N. *J. Chem. Phys.* **1996**, *104*, pp 2946-2955.
(45) Winter, N. W.; Pitzer, R. M.; Temple, D. K. *J. Chem. Phys.* **1987**, *87*, pp 2945-2953.
(46) Kunz, A. B.; Vail, J. M. *Phys. Rev. B* **1988**, *38*, pp 1058-1063.
(47) Kunz, A. B.; Klein, D. L. *Phys. Rev. B* **1978**, *17*, pp 4614-4619.
(48) Shidlovskaya, E. K. *Latvian J. Phys. Techn. Sci.* **1996**, *4*, pp 57-82.
(49) Stevens, W.; Basch, H.; Krauss, J. *J. Chem. Phys.* **1984**, *81*, pp 6026-6033.
(50) Wadt, W. R.; Hay, P. J. *J. Chem. Phys.* **1985**, *82*, pp 284-298.
(51) Pascual-Ahuir, J. L.; Silla, E.; Tuñon, I. *J. Comp. Chem.* **1994**, *15*, pp 1127-1138.
(52) Floris, F. M.; Tomasi, J.; Pascual-Ahuir, J. L. *J. Comp. Chem.* **1991**, *12*, pp 784-791.
(53) Pierotti, R. A. *J. Phys. Chem.* **1963**, *67*, pp 1840-1845.
(54) Jorgensen, W. L.; Tirado-Rives, J. *J. Am. Chem. Soc.* **1988**, *110*, pp 1657-1666.
(55) *GAUSSIAN 92/DFT*; Frisch, M. J.; Trucks, G. W.; Schlegel, H. B.; Gill, P. M. W.; Johnson, B. G.; Wong, M. W.; Foresman, J. B.; Robb, M. A.; Head-Gordon, M.; Replogle, E. S.; Gomperts, R.; Andres, J. L.; Raghavachari, K.; Binkley, J. S.; Gonzalez, C.; Martin, R. L.; Fox, D. J.; Defrees, D. J.; Baker, J.; Stewart, J. J. P.; Pople, J. A., Gaussian, Inc.: Pittsburgh, PA, 1993.
(56) Jug, K.; Geudtner, G. *Surf. Sci.* **1997**, *371*, pp 95-99.
(57) Stefanovich, E. V.; Truong, T. N. *J. Chem. Phys.* **1995**, *102*, pp 5071-5076.
(58) Johnson, M. A.; Stefanovich, E. V.; Truong, T. N. *J. Phys. Chem. B* **1997**, *101*, pp 3196-3201.

AB INITIO DYNAMICS

Chapter 7

Dual-Level Methods for Electronic Structure Calculations of Potential Energy Functions That Use Quantum Mechanics as the Lower Level

José C. Corchado[1,2] and Donald G. Truhlar[1]

[1]Department of Chemistry and Supercomputer Institute, University of Minnesota, Minneapolis, MN 55455–0431
[2]Departamento de Quimica Física, Universidad de Extremadura, Badajoz 06071, Spain

This chapter overviews the status of three dual-level approaches for potential surface calculations that use quantum mechanics for both the upper and lower level. Three types of approach are singled out for discussion: SEC and SAC calculations, IMOMO calculations with harmonically capped correlated small systems, and dual-level direct dynamics. The scaling external correlation (SEC) and scaling all correlation (SAC) methods are semi-*ab initio* approaches to the calculation of bond energies and barrier heights for chemical reactions. The IMOMO calculations are very similar in spirit to QM/MM methods, but the lower level is quantum mechanical. Dual-level direct dynamics is a general technique for combining levels in dynamics calculations that include quantum mechanical tunneling contributions.

Dual-level methods have become very popular in modern quantum chemistry. The recent surge of interest in methods that combine quantum mechanical solutions for a small part of a system with molecular mechanics for the rest of the system (*1–6*) prompts one to place these methods in the perspective of a larger set of dual-level and multi-level approaches. These approaches have certain elements in common, but additionally they have interesting differences. The purpose of the present overview is to discuss some dual-level methods from this viewpoint.

The diversity of dual-level methods is such that even classifying them creates a stimulating challenge (and, like all review tasks, is doomed to incompleteness). Nevertheless, here is an attempt to enumerate some of the main varieties:

1. Double slash (//) methods for energies of stationary points. This is perhaps the most ubiquitous dual-level approach. A common example would be an MP2/6-311G(d,p)//HF/6-31G(d) calculation in which a stationary-point geometry is optimized at the HF/6-31G(d) level, and then a more accurate energy is calculated by the MP2/6-311//G(d,p) method (*7*). This method can be used for both minima and saddle points. One can denote such a theory L2//L1 where L2 and L1 denote the higher and lower "levels." A variant is L3//L2[L1] where the geometry is optimized at level L1 and vibrational frequencies are also calculated at this level; then the geometry is optimized at a higher level L2, but without a vibrational frequency calculation. The L2 geometry is then used for a single-point calculation at level L3.

2. Gaussian-2 theory. The Gaussian-2 (often abbreviated G2) approach to estimating thermodynamic quantities (*8*) has been extremely successful, and has spawned several variants. G2 theory and its variants are actually carefully layered multi-level theories. The theory assumes that various effects of improving the lower level electronic structure calculation (i.e. by using a larger one-electron basis set or a more accurate treatment of the electron correlation) are additive. Thus, instead of carrying out a full higher level calculation, a sequence of "intermediate" level calculations directed to estimating the effects of several enhancements on the lower level is carried out.

3. CBS. The truncation of the one-electron basis sets used in *ab initio* calculations is frequently the major source of error in the results. The increase in the basis set size involves an increase in the required computational resources that makes impossible to converge a calculation with respect to the size of the basis set even in most small systems. When the interest is mainly focused on differences in the calculated energy for a series of systems or a series of points on a potential energy surface, one can expect some compensation between the errors introduced in the various energy calculations, but even for relative energies, the incompleteness of the one-electron basis set may be severe. In order to minimize the effects of the basis set truncation, an extrapolatory scheme derived by means of perturbation theory and based on the asymptotic basis-set dependence of the pair natural orbital energies has been proposed (*9*). The goal is to obtain an energy comparable to the one that would be obtained with a complete basis set (CBS). Several variants have been proposed (*10*), accounting for a variety of choices of basis sets and methods for including correlation energy.

4. SEC and SAC. The calculation of bond energies is a critical step in the calculation of heats and free energies of reaction, and the calculation of barrier heights is a first step in the calculation of enthalpies and free energies of activation. Unfortunately, although there has been great progress in *ab initio* electronic structure methods, including the G2 and CBS approaches mentioned above, the calculations can be very expensive. Thus semiempirical dual-level approaches can also prove useful, especially for calculating barrier heights, and in this paper we discuss two such methods, namely the scaling-external-correlation (SEC) (*11*) and scaling-all-correlation (SAC) (*12–15*) methods. These methods are designed to circumvent the difficulty that the differential electron correlation energies between species with made and broken bonds or partial broken bonds (as at transition states) are very slowly convergent with respect to the mixing of electronic configurations and the completeness of the one-electron basis sets for electronic structure calculations.

Section 2 of the present chapter provides a review of the SEC and SAC methods.

5. QM/MM methods. The most rapidly growing area of activity in dual-level methods is in the development of algorithms that combine quantum mechanics (QM) with molecular mechanics (MM), where "molecular mechanics" refers to any "classical" force field. Since this subject is covered extensively by the other chapters in the present volume, we will not review it here, but we simply mention that there are two main approaches. The first approach, which is more common (*1–5*), involves applying QM to the primary subsystem and MM to the rest. The second approach, called IMOMM (*6*), involves applying QM to the primary subsystem and applying MM to both the primary subsystems and the entire system—then combining all three calculations. Both methods involve technological challenges in how to treat the boundary, where one locates so called "link atoms" and/or "capping atoms."

6. IMOMO methods. An appealing feature of the IMOMM method mentioned above is that it is very general as to the kinds of levels that are combined, and in particular one can equally well choose a QM level as an MM level for treating the entire system. This is called

IMOMO (*16*). For moderate-size systems, the QM level can be as high as one can afford (*17*), although for very large systems, economic considerations may limit one to semiempirical molecular orbital theory. An alternative approach for very large systems is to use a triple-level scheme, and IMOMO is easily generalized to accommodate this (*18*).

Section 3 of the present chapter reviews two aspects of IMOMO. The first is its ability to include electronic substituent effects ("inductive effects") of the rest of the system on the primary subsystem (*17,19,20*). Second is a new approach we have developed, called a harmonic cap, for optimizing geometries with IMOMO methods (*21*). The approach is actually equally applicable to QM/MM methods, and we believe that it eliminates the "link atom problem," and so it is liable to be particularly interesting to readers of this volume.

7. SCRF/SASA models. Another dual-level approach, albeit of quite a different nature, has been employed for the problem of modeling solvation effects. In this approach, one combines a quantum mechanical method for electrostatics with a classical mechanical method for first-solvation shell effects. In particular one uses the self-consistent reaction field (SCRF) molecular orbital method (*22*) for the former with a treatment based on solvent-accessible surface area (SASA) (*23,24*) and atomic surface tensions (*25,26*) for the latter. The most completely developed set of models of this type is the SM*x* series (*27*) of "solvation models." These have culminated in a suite of models called SM5 models, and these are overviewed in another chapter in this volume (*28*).

8. Dual-level direct dynamics. Direct dynamics is the calculation of dynamical quantities using electronic structure calculations for all required energies, forces, and Hessians, without the intermediacy of an analytic potential energy function (*29–31*). The treatment of atomic motion in direct dynamics calculations can be either classical (*29,32*) or quantal, although quantal effects on nuclear dynamics are included most efficiently by semiclassical (*30,31,33*) methods. (The distinction should not be confused with the use of quantum mechanics or classical molecular mechanics for the potential energy function, which is an entirely separate issue.) Dual-level direct dynamics has been developed so far for calculations based on quasiclassical variational transition state with semiclassical multidimensional tunneling contributions, usually abbreviated VTST/MT for variational transition state theory with multidimensional tunneling (*30,31,33-38*). In practice we use up to four levels, e.g., for OH + NH$_3$ (*37*) we used QCISD(T)//MP2/aug-cc-pVTZ[MP2/aug-cc-pVDZ]///MP2/6-31G* and ...///PM3-SRP, where L3//L2[L1] is explained above and /// separates the method for high-level calculations at stationary points (before the triple slash) and the method used (prior to interpolation corrections) for everywhere else (after the triple slash). The notation PM3-SRP denotes the use of specific reaction parameters (*30*) starting from the PM3 general parameters (*39*) at the neglect of diatomic difference overlap (NDDO) level of semiempirical molecular orbital theory.

Section 4 of this chapter presents a brief review of dual-level direct dynamics and SRP methods.

SEC and SAC

SEC. Correlation energy is defined as the difference between the Hartree-Fock energy and the true energy (*40*). This is usually calculated by optimizing the orbitals of a restricted form of wave function, called the reference state, and adding electron correlation by mixing in additional configurations, either by a variational configuration interaction (CI) method or by perturbation theory or a coupled cluster approach. For some molecules, primarily open-shell systems, transition states, systems with partially broken bonds, and systems with low-lying excited states, the Hartree-Fock self-consistent-field (HF-SCF) method based on a single configuration does not provide a good zero-order description. In such cases, we need a multi-configuration SCF (MCSCF) reference state for a qualitatively correct description (*41*), and when correlation energy is added to such a state, the resulting calculation is called a multi-reference one. The correlation energy may be decomposed into two parts: internal correlation, also called static correlation, which is the part included in a minimal MCSCF wave function, and external correlation (also called dynamical correlation), which is the remainder even after a good zero-order description has been achieved. The remainder would typically be calculated by a multi-reference CI calculation (MRCI) (42). One could equally well include such configuration mixing effects by multi-reference perturbation theory (MRPT), e.g., CASPT2 (43). Computationally we have found (11) that the fraction of external correlation energy is more a function of the level of theory and the one-electron basis set than of the geometry. By taking advantage of this for a given level, one can scale the external correction energy by a factor that is independent of geometry. In particular, if one assumes that one recovers a fraction F of the external correlation energy, independent of the geometry of the system, one can write (11)

$$E(SEC) = E(MCSCF) + \frac{E(MRCI) - E(MCSCF)}{F}. \tag{1}$$

The SEC method has proven particularly useful for the calculation of barrier heights for chemical reactions. In this case one determines the value of F such that the reactant and product bond energies are correct or the overall exoergicity is correct. Then with F determined on the basis of reactants and products, it can be used at transition state geometries to calculate barrier heights. Some examples (44–47) of SEC calculations of barrier heights are provided in Table 1. In all cases, the barrier heights calculated by the SEC method or potential energy surfaces (47–50) adjusted to have a barrier height equal to the SEC barrier heights are consistent with experiment.

SAC. Scaling the entire correlation energy, even the internal part, is expected to be less reliable than scaling the external correlation energy, but it is computationally much more tractable since MRCI and MRPT calculations are much more difficult than single-reference ones based on a single-configuration reference state. (A single-reference correlated calculation is one in which there is only a single configuration in the zero-order wave function that serves a starting point for estimating the effect of configuration mixing.) Thus scaling all correlation (SAC) can be very useful. In this approach we write

$$E(SAC) = E(HF) + \frac{E(SRPT) - E(HF)}{F} \tag{2}$$

Table 1. Computation of barrier heights by MRCI-SEC calculations

Reaction	Barrier (kcal)		Ref.
	MRCI	SEC	
F + H$_2$ collinear	3.7	1.6	(44)
bent	...	1.3	(45)
H + DF → HCl + D	43.2	38.4	(11)
H + DCl → HCl + D	20.0	18.1	(46)
Br + H$_2$	2.7	1.9	(47,48)
H + DBr → HBr + D	13.4	11.9	(47,48)

where *HF* denotes Hartree-Fock and *SRPT* denotes single-reference perturbation theory. One would probably not want to use single-reference CI (SRCI) in eq. (2) because it has severe size-consistency failures, whereas SRPT and single-reference coupled-cluster theory are size consistent. (We use "size consistency" and "size extensivity" as synonymous here. Size consistency (*51*) is not strictly required for using eq. (2), but a method that is not at least approximately size extensive would not be expected to have a geometry-independent F value. MRCI is not strictly size extensive, but it is better than SRCI.)

In early work with the method (*12–14*) we developed general scale factors for various types of bonds with both Møller-Plesset second-order (MP2) and Møller-Plesset fourth-order (MP4) perturbation theory, which are single-reference methods. These general scale factors were determined for several common basis sets. In later work (*15*), we considered other single-reference correlated methods, in particular the coupled-clusters method with double excitations (CCD) (*52,53*) and the quadratic configuration interaction method with single and double excitations (QCISD) (*54*). (The latter method, although named as a CI method, is actually size extensive.)

The basic conclusions of the early systematic studies of the SAC method (*12–14*) are that F is indeed reasonably consistent across systems (for a given basis set and order of perturbation theory), and furthermore there does not appear to be much advantage in MP4-SAC as compared to MP2-SAC. It is instructive to summarize some of the recent results (*15*) we obtain with the SAC method for bond energies. In order to test the SAC method with general scale factors, we considered 13 molecules with experimentally known atomization energies. This test suite consists of H_2, CH_4, NH_3, H_2CO, H_2O, HF, C_2H_2, HCN, F_2, CO, CO_2, N_2, and N_2O, and we will consider the results for the cc-pVDZ (*55*) basis set. For the MP2-SAC calculations, we found that we could not get particularly accurate results with a constant value of F (constant across systems), so we chose $F = 0.918 - 0.143x - 0.176x$, where x is the fraction of atoms that are H. (Note: that F is still independent of geometry, e.g., we use the same value for $CH_3 + H$ as for CH_4). We found a mean |error| per bond of 1.3 kcal per mol. For QCISD-SAC, we used a completely constant value of $F = 0.729$, and we found an average |error| per bond of 1.5 kcal. For CCD-SAC we used $F = 0.729$, and found an average |error| per bond of 1.4 kcal. Note that the average |error|s per bond without scaling the correlation energy are respectively 8.4, 14.5, and 15.6 kcal/mol. The results of the scaling are amazingly good considering none of the three correlation levels considered involves explicit consideration of triple excitations.

Recently Siegbahn *et al.* (*56,57*) proposed a closely related method, which they call PCI-X. In actual applications, PCI-X is basically the same as SAC except that these authors prefer to use a single universal F value.

Correlation Balance. Clearly the success of the SEC and SAC approaches is dependent on the quality of the basis set and in particular on its ability to treat the correlation energy of various geometries and bonds in a consistent way. The idea that a basis set should be electrostatically balanced so that it yields accurate dipole moments and bond polarities is a familiar one, but correlation balance has been less widely studied. Once one attempts to design extrapolation methods (*8–15*) for electronic structure calculations, the question of correlation-balanced basis sets becomes paramount, not only for basis sets but also for approaches to configuration mixing. For example, it is clear from the necessity (see above) to use a variable F in MP2-SAC that coupled-cluster theory gives a more uniform treatment of correlation in bonds to hydrogen and bonds between heavy atoms than MP2 theory does. Some attempts to study the question of correlation balance have been presented (*13,58–61*) but much more work needs to be done.

SAC calculations for kinetics. The area where the SAC method has proved most useful so far (*12,14,58–70*) is for kinetics calculations. This subsection reviews this body of work.

The first application of a SAC method to a reaction was an application of the MP2-SAC scheme for the calculation of the properties of the stationary points along the

$CH_4 + Cl \rightarrow CH_3 + HCl$ reaction-path (58). A 6-311G($2d,d,p$) basis set (71) was found to provide a balanced treatment of the effects of electronic correlation on the making and breaking bonds. Therefore, the average of the F values obtained for each bond using the MP2/6-311G($2d,d,p$) was used for the scaling. The MP2-SAC/6-311G($2d,d,p$) level was chosen for the optimization of the geometries and frequency calculations at the reactants, products, and saddle point for this reaction. The properties obtained in this work for the saddle point were employed as reference data for the calibration of an analytical potential energy surface (72), which was used in a VTST/MT calculation of the rate constants and kinetic isotope effects for this reaction.

The first application of MP2-SAC to the calculation of a rate constant was the study of the hydrogen abstraction from CH_4 by the OH radical (62). In this study, the basis set employed for SAC calculations was the 6-311G($2d,p$) basis set, since this is the basis set that provided the most balanced treatment of the correlation energy for the O–H and C–H bonds. In particular, the values of the factor F estimated from the H–OH and H–CH_3 bond dissociation energies were 0.875 and 0.864 respectively. The properties of reactants, products, and saddle point were therefore evaluated at the MP2-SAC/6-311G($2d,p$) level and used in a calculation of the rate constant by means of the zero-order interpolated variational transition-state theory (IVTST–0) (73), giving rate constants in good agreement with the experimental results.

The availability of a set of average values for the F factors (12) allows us to apply the SAC corrections in a more automatic way, without a check of the balance in the basis set for the treatment of the correlation energy. Thus, the MP2-SAC method was also used for calculating single-point energies in a study on the effects of hydration and dimerization of the formamidine rearrangement (63). In this study, where the MP2-SAC/6-31G(d,p) method was used for calculating only single-point energies, the value of the F factor utilized was the average between two previously determined values of the F factor, one for the N–H bond and one for the O–H bond. In a study on hydrogen abstraction reactions of ammoniacal compounds (64) involving systems for which experimental dissociation energies are not available, the MP4-SAC method was used with F equal to the average of the factors proposed in the original SAC paper for the basis sets and bonds involved in the calculations.

The same procedure was used in the calculation of the barrier height for the $NH_3 +$ H $\rightarrow NH_2 + H_2$ reaction (65). Table 2 shows a summary of the results obtained for this reaction, in which an MP4-SAC/6-311+G(d,p) single point calculation is compared to other schemes for extrapolating the Møller-Plesset series (74, 75). The MP4-SAC results are close to the more expensive QCISD(T)/6-311+G(d,p) results, and the use of an average F factor leads to a reasonable reaction endoergicity. We note that since the endoergicity is the difference between two bond energies, and since F is calculated using experimental bond energies, the use of an F factor obtained from one bond for calculations involving a different bond will give energies for this second bond closer to the experimental results when the F values for both bonds are more similar. Therefore, the accuracy in the endoergicity is a reflection of the adequacy of an average F factor for different reactions. The rate constants obtained using the MP4-SAC results (65, 66) are in good agreement with the experimental results. The same procedure was used for a VTST/MT study of the reaction OH + $NH_3 \rightarrow NH_2 + H_2O$ (67, 68). The MP4-SAC method with a standard F factor (14) has also recently been used for calibrating an analytical potential energy surface for the Cl + $H_2 \rightarrow$ ClH + H reaction (69).

A series of papers have been addressed to understanding the kinetics of the reactions between hydrocarbons and the OH radical. These studies involve the reactions of OH with methane (59, 70), ethane (60), and propane (61). In all the cases the MP2-SAC method was employed. A modified triple-ζ basis set was obtained in order to get a balanced

Table 2. Barrier height and endoergicity for the $NH_3 + H$ reaction calculated at different levels of calculation[a]

Method	ΔE[b]	$\Delta E^{\neq c}$
UHF	-0.2	23.7
PUHF	-2.0	18.6
UMP2	10.8	23.4
PUMP2	9.7	20.0
UMP4	5.3	18.2
PUMP4	4.6	16.2
MP4-SAC	5.7	15.7
PMP∞[d]	4.4	16.1
Feenberg[e]	4.1	15.5
QCISD(T)	5.3	16.8
Exp[f]	6.9	

[a]Results taken from Ref. (65). [b]ΔE is the Born-Oppenheimer classical energy of reaction, i. e., it excludes zero point energy. [c]$\Delta E^{\neq}$ is the Born-Oppenheimer classical energy of activation. [d]Extrapolation of the Moller-Plesset series by the methods in Ref. (74). [e]Extrapolation of the Moller-Plesset series by the methods in Ref. (75). [f]Ref. (76).

treatment of the electron correlation for the C–H and O–H dissociation energies, followed by a MP2-SAC single-point energy calculation, except in the propane reaction, for which a geometry optimization was also carried out at the MP2-SAC level. This was part of the input necessary for carrying out calculations of the rate constant and kinetic isotope effects. In the methane and ethane reactions, this information was completed with 2 or 1 extra points on the reaction path, respectively. Then a dynamical calculation was carried out by IVTST methods. The reaction OH + propane was simulated using a dual-level direct dynamics technique (see Section 4) that allows a more reliable calculation of the tunneling contribution.

In conclusion we can say that the use of SAC methods with economical dynamical methods has been a powerful tool for the accurate study of chemical reactions. Nevertheless the SAC methods, although being a good approximation to more accurate, much more expensive calculations, still require a large amount of computer resources for systems with many atoms. Thus, while the treatment of the correlation energies can be limited to a very affordable MP2 calculation, a high quality basis set is still required. The advances in computer hardware and software are opening doors for the study of medium-size systems, such as propane, but we are still far from being able to perform this kind of calculations in the large organic or biological systems, as well as condensed phase systems. In this sense, only the IMOMO (or related) methods seem to be able to propel us ahead.

IMOMO

CCSS. A special case of IMOMO is the use of a correlated capped small system (CCSS). Consider, for example, a calculation of the bond energy for the C-H bond in $H-CH_2CH_2NH_2$. One can take the small system as $H-CH_2$ and the capped small system as $H-CH_3$. One could calculate the entire system ($H-CH_2CH_2NH_2$) only at the lower level, e.g., AM1, HF/3-21G, HF/MIDI!, HF/6-31G*, or B3PW91/6-31G* (or—in the case of IMOMM—the lower level would be MM). One could calculate the capped small system at both lower and higher levels, where the higher (correlated) level might be, e.g., CCSD(T)/6-31G(d,p), QCISD(T)/cc-pVTZ, MP2/6-31G(d,p), or HF/cc-pVTZ. [The methods and basis sets are explained elsewhere: Ref. (*7*) for MP2, HF, 3-21G, 6-31G*, and 6-31G(d,p), Ref. (*77*) for AM1, Ref. (*78,79*) for B3PW91, Ref. (*80*) for CCSD(T), Ref. (*54*) for QCISD(T), Ref. (*81*) for MIDI!, and Ref. (*55*) for cc-pVTZ.] Then the dual-level energy, called the integrated (I) energy, is

$$E_{entire}^{I} = E_{entire}^{low} + (E_{small}^{high} - E_{small}^{low}) \tag{3}$$

$$= E_{small}^{high} + (E_{entire}^{low} - E_{small}^{low})$$

Table 3 gives examples of how well this approach works for the molecule given as an example above. Table 4 gives surveys of the performance for several small systems of this type (*17,20*). In these tables, the entire molecule is of the form $H-CHXCH_2Y$, and the capped small system is $H-CH_2X$, where X and Y are H, CH_3, NH_2, OH, and F. In both Table 3 and Table 4, standard geometries were used for all calculations—the goal was to test how well electronic effects are included when one uses the same geometry for the integrated calculation as for a more accurate calculation. Notice that in these examples the system is capped very close (geminal) to the bond that is broken, thus providing a challenging test. The goal is to combine a high-level calculation on the capped small system with a low-level calculation on the entire system such that the integrated result is more accurate than either single level. Accuracy is measured by deviation from a full high-level calculation on the entire system (which is feasible for these test cases but which would presumably be unaffordable for the entire system in real applications to large systems). The table shows that the integration of levels is successful, in fact quite dramatically successful. For example, in Table 3 we see that a dual-level calculation is much better than either a higher-level calculation on the capped subsystem or a lower-level calculation on the entire system.

We note specifically the especially good results obtained with the MIDI! basis set in Tables 3 and 4. This is very encouraging because the MIDI! basis set is the first example of a new breed—a basis set optimized specifically to serve as the lower level in dual-level methods (*81*). In particular the MIDI! basis is optimized to give good geometries so it can serve as L2 in L1//L2 calculations and to give balanced charge distributions so it can serve as a starting point for multi-level calculations in the condensed phase, where electrostatics is often of crucial importance.

IMOHC. The second topic we consider in IMOMO theory is the integrated molecular orbital harmonic cap (IMOHC) strategy (*21*) for geometry optimization in IMOMO and IMOMM calculations. One of the most challenging problems that one faces in most QM/MM applications is the treatment of the link atom. (Note: The "link atom" is the atom that becomes the capping atom in the capped small system.) In the case of a geometry optimization the problem of the link atom is very delicate, since usually the link atom has a different nature in the entire system and the capped small system. If the optimization is done in such a way that this atom is constrained to have identical geometrical parameters (e.g., bond lengths) in both systems, the optimization may be unphysical, or it may fail.

Table 3. Bond energy for H-CH$_2$CH$_2$NH$_2$ by IMOMO[a]

System	Method	D_e (kcal/mol)	error (kcal/mol)
entire system	QCISD(T)/cc-pVTZ	106.4[b]	0.0
capped small system[c]	QCISD(T)/cc-pVTZ	111.3	4.9
entire system	AM1	84.4	22.0
integrated	QCISD(T)/cc-pVTZ: AM1	105.6	0.8
entire system	HF/MIDI!	79.7	26.7
integrated	QCISD(T)/cc-pVTZ:HF/MIDI!	106.6	0.2

[a]From Ref. (17) [b]Presumed accurate. All other calculations are compared to this value using standard geometries. [c]H–CH$_3$

Table 4. Mean unsigned errors (kcal/mol) in bond energies for H – CHXCH$_2$Y $\rightarrow$ H + CHXCH$_2$Y

higher level	lower level	< |error| >		
		higher level capped small system	lower level entire system	integrated
test set 1[a] (nine cases; X, Y = H, CH$_3$, NH$_3$, OH, F)				
QCISD(T)/cc-pVTZ	HF/6-31G*	2.1	23.9	0.4
"	HF/MIDI!	2.1	26.1	0.3
"	HF/3-21G	2.1	23.3	0.6
"	AM1	2.1	23.4	1.3
test set 2[b] (ten cases; X, Y + H, NH$_2$, OH, F, Cl)				
MP2/6-31G(d,p)	HF/3-21G	1.6	21.4	0.9
CCSD(T)/6-31G(d,p)	HF/3-21G	1.9	21.3	0.9
"	HF/6-31G*	1.9	22.1	0.3

[a]From Ref. (17). [b]From Ref. (20)

The IMOHC strategy is to circumvent this difficulty by allowing an independent optimization of the link atom in the entire and capped small systems. Thus, if the entire system is formed by a number of atoms N_{entire}, the geometry optimization will involve $3(N_{entire} + 1)$ coordinates, instead of the usual $3N_{entire}$. The extra coordinates occur because the link atom occurs both as the link atom itself (denoted L) in the entire system and also, with different coordinates, as the capping atom (denoted M) in the capped small system. All the other atoms in the capped small system have the same coordinates as in the entire system.

The geometry optimization is finished when the $3(N_{entire} + 1)$ components of the gradient are zero. By applying the distributive law of differentiation, we can write an equation similar to the first half of the equation 3 for the gradients:

$$\nabla E_{entire}^{I} = \nabla E_{entire}^{low} + \left(\nabla E_{small}^{high} - \nabla E_{small}^{low} \right). \tag{4}$$

An examination of this equation leads us to the conclusion that it will not generally succeed in giving a physical location for the link atom. The behavior we should expect from the optimization algorithm is to find the geometrical conformation that gives nonzero values that cancel out each other for each component of the three terms on the right-hand side of equation 4, thereby giving a zero value for the gradient. However, in practice, the optimization may lead to a geometry with the link atom infinitely separated from the rest of the system, which gives zero components of the gradient for the individual terms for the gradient components related to the link atom. In practice, equation 4 can also lead to unphysically distorted geometries for which $E_{small}^{high} - E_{small}^{low}$ is negative although E_{small}^{high} and E_{small}^{low} are both high and positive. This is obviously an undesirable situation, since the link atom of the capped small system (i.e., the capping atom) needs to be in a physical location in order to carry out its capping role in the electronic structure calculations on the small system.

The solution we devised for this problem is to include additive harmonic terms that keep the capping atom at a physically meaningful location in the small system. Thus, the first harmonic cap term we introduce is a correction that prevents the optimization algorithm from locating the capping atom unphysically far from the atom, A, to which it is bonded in the small system. This is given by

$$T_R = \tfrac{1}{2} k_R (R_{A-M} - R_{eq})^2. \tag{5}$$

The parameters k_R and R_{eq} are calculated by means of three single-point energy calculations for the capped small system, so that the computational cost is kept to a minimum and accurate results are obtained (21).

Similarly, in order to avoid unphysical values for the angles in the capped small system, a second harmonic correction can be introduced,

$$T_\theta = \tfrac{1}{2} k_\theta (\Delta\theta)^2. \tag{6}$$

We define the angle $\Delta\theta$ by

$$\Delta\theta = \arccos(\hat{R}_{A-M} \cdot \hat{R}_{A-L}), \tag{7}$$

so that the link atom L and capping atom M have the same orientation. The calculation of k_θ is carried out by means of two or three single-point energy calculations for the capped small system.

Finally, a third term can be defined in order to restrict the possible values of torsion angles in the capped system,

$$T_\phi = \tfrac{1}{2} k_\phi (\Delta \phi)^2, \tag{8}$$

where ϕ indicates the deviation of an L–A–B–C dihedral angle in the entire system from the M–A–B–C dihedral angle in the capped subsystem, where B is an atom bonded to A, and C is an atom bonded to B. Once again, k_ϕ will be estimated by means of a few single-point energy calculations. (Neither of the systems considered in this paper has an L–A–B–C dihedral angle because the capped small system is so small, but we mention the torsion for completeness.)

The total energy for the system will therefore be given by:

$$E_{entire}^{I} = E_{entire}^{low} + (E_{small}^{high} - E_{small}^{low}) + \\ + \tfrac{1}{2} k_R (R_{A-M} - R_{eq})^2 + \tfrac{1}{2} k_\theta (\Delta\theta)^2 + \tfrac{1}{2} k_\phi (\Delta\phi)^2 \tag{9}$$

The IMOHC method has been tested for the optimization of the geometry of ethane and also for calculating its vibrational frequencies and the C–H bond energy. Some results are shown in Tables 5 and 6. For the results in these tables, k_R and R_{eq} were calculated from three higher-level single-point calculations on methane, with one at the lower-level minimum-energy geometry and the other two having slightly shorter and larger C–H bonds. For this system we found that the only capping term needed is the bond-length controlling term; the parameters k_θ and k_ϕ were therefore set to zero.

Since the most one can expect of any integrated method is to reproduce the results obtained from a complete higher-level calculation, the IMOHC results are compared to those obtained by means of the higher level used in the calculation, MP2/6-31G, as well as to those for the small system (methane and methyl radical) at the higher level of calculation and those for the entire system as the lower level (HF/3-21G). The optimized geometry is much closer to the higher-level optimized geometry than to either the higher-level capped small-system or the lower-level entire-system calculation. The results for the C–H bond energy (Table 6) are also excellent, the error being less than 0.1 kcal/mol, undoubtedly preferable to the 2.6 and 13.7 kcal/mol errors of the single-level calculations.

In a second test, the geometry of the ethylamine molecule (the same molecule as studied in Table 3) was optimized, using the same levels as in the ethane test just described. Once again, the capped system was the methane molecule, with the capping atom being a hydrogen atom and the link atom being a carbon atom. The k_R and R_{eq} parameters were taken from the previous example. In this case, it was necessary to also include the harmonic term for the bend. The bending force constant k_θ was calculated by means of a single-point MP2/6-31G//HF-3-21G calculation at the HF/3-21G optimum geometry for methane, and an MP2/6-31G calculation on a geometry for which one of the C–H bonds in methane deviated by 0.5 degrees from its optimal position at the lower level. (In a less symmetric system, we could need three points.) We obtained $k_\theta = 2.033 \times 10^{-2}$ kcal mol^{-1} deg^{-2}. This calculation is somewhat arbitrary, since the resulting force constant is dependent on the direction in which the distortion took place. However, we expect that the final results will generally show little dependence on the precise value of k_θ.

Table 5. C–H bond distance (in Å) as predicted by single-level and IMOHC calculations[a]

Level	H–CH$_3$	H–CH$_2$	H–CH$_2$CH$_3$	H–CHCH$_3$
HF/3-21G	1.0830	1.0717	1.0841	1.0734
MP2/6-31G	1.0959	1.0830	1.0988	1.0867
MP2/6-31G:HF/3-21G		1.0973	1.0854	

[a]All the results in this table are taken from Ref. (21).

Table 6. Energy of the C–H bond (kcal/mol) as predicted by single-level and IMOHC calculations[a]

Molecule	Level	ΔE[b]	ΔH_0^0
CH$_4$	MP2/6-31G	99.66	89.90
C$_2$H$_6$	HF/3-21G	84.15	73.58
C$_2$H$_6$	MP2/6-31G	97.02	87.25
C$_2$H$_6$	MP2/6-31G:HF/3-21G	97.09	87.27

[a]All the results in this table are taken from Ref. (21). [b]ΔE is the Born-Oppenheimer classical energy of dissociation, i. e., it excludes zero point energy; ΔH_0^0 is the standard-state enthalpy of dissociation at 0 K, including zero point energy.

Table 7. Average unsigned error in the three C$_\beta$–H bond distances in ethylamine (in Å) as predicted by single-level and IMOHC calculations. The reference value is the MP2/6-31G average C$_\beta$–H bond length in ethylamine

Level	H–CH$_3$	H–CH$_2$CH$_3$NH$_2$
HF/3-21G	0.0157	0.0149
MP2/6-31G	0.0028	0.0000[a]
MP2/6-31G:HF/3-21G		0.0018

[a]Zero by definition

Table 7 shows the average unsigned error in the C–H bond distance for the β-carbon (where we label the molecule $H_2NC_\alpha H_2C_\beta H_3$), and Table 8 gives the average error in three $H-C_\beta-H$ bond angles of the small system computed at each single level and also by the dual-level calculations. The "errors" are actually deviations from the optimized MP2/6-31G geometry. Once again, the results are encouraging. The dual-level calculation gives bond distances and bonds angles in the small system much closer to the high-level entire-system calculation than are obtained from either a low-level calculation on the entire system or a high-level calculation on the capped small system.

Table 8. Average unsigned error in bond angles in ethylamine (in degrees) as predicted by single-level and IMOHC calculations. The reference value is the MP2/6-31G average $C_\alpha-C_\beta-H$ bond angle in ethylamine[a]

Level	$H-CH_3$	$H-CH_2CH_3NH_2$
HF/3-21G	0.985	0.187
MP2/6-31G	0.985	0.000[b]
MP2/6-31G:HF/3-21G		0.078

[a]If the hydrogens are labelled D–J as follows: $DENC_\alpha FGC_\beta HIJ$, then the quantity tabulated is the average of the unsigned errors in these three bond angles: $HC_\beta I$, $HC_\beta J$, and $IC_\beta J$. [b]Zero by definition

Dual-level direct dynamics

The dual-level direct dynamics approach (36-38) is an analog for reaction dynamics of the // approximation in electronic structure calculations on bound-state properties. In the dual-level approach a reaction path is constructed and a complete variational transition state theory calculation including tunneling is carried out at a lower level; the results are then improved by carrying out a reduced number of electronic structure calculations at a higher level without recalculating the reaction path. If HL and LL denote the higher and lower levels, then HL///LL denotes the result of using them together in a dynamics calculation; this is a direct generalization of the popular // notation. Thus, HL///LL indicates a dual-level dynamics calculation based on a potential energy surface calculated at the LL, improved by a small number of HL calculations. One critical distinction between HL//LL and HL///LL, however, is that HL//LL does not involve any geometry optimizations at the higher-level, but HL///LL allows for the possibility that the saddle point and reagent geometries be re-optimized at the higher-level since it is usually dangerous not to do so.

The information from the higher-level is introduced into the lower-level surface by means of interpolated corrections (IC) (36, 38). The method in its simplest form consists of defining an error function that is calculated by means of the HL and LL values for a property and interpolating that function all along the reaction path. For example, let $\omega_m^{HL}(s)$ be the value of the vibrational frequency of the mode m at a distance along the reaction path given by the reaction coordinate, s, as calculated at the higher level. (In the formulation used so far, the only points along the reaction path calculated at the higher level are stationary points, i.e., reactants, products, saddle point, and/or wells on the reaction path.) Let $\omega_m^{LL}(s)$ be the value of the vibrational frequency of the same mode at the same value of s as calculated at the lower level. We define a function of s that measures how far the lower-level frequency for that mode is from the higher-level one; for example,

$$f(s) = \ln\left[\omega_m^{HL}(s)/\omega_m^{LL}(s)\right]. \tag{10}$$

The value of this function will be available only for those values of s for which the HL calculation has been performed; nevertheless, by using the appropriate functions (36) we can interpolate the values of $f(s)$ for any point on the reaction path. Thus, $\omega_m^{DL}(s)$, the dual-level estimated frequency for the mode m at s, will be given by

$$\omega_m^{DL}(s) = \omega_m^{LL}(s)\exp\left[f^{ICL}(s)\right], \tag{11}$$

where

$$f^{ICL}(s) = \text{interpolant}\left\{\ln\left[\omega_m^{HL}(s)/\omega_m^{LL}(s)\right]\right\}, \tag{12}$$

and where ICL indicates "interpolated corrections based on the logarithm" (38). Following this idea, the method corrects the energy, frequencies, moments of inertia, and reduced moments of inertia for hindered rotors along the reaction path, although geometries and eigenvectors cannot be corrected by this method. For a detailed discussion on the interpolatory functions and the different ways of defining the corrections the user is referred to the original IC methodology papers (36, 38).

As the higher level the best general options are to take values obtained from *ab initio* calculations at a level as high as possible or from SAC calculations, although some other options are in principle applicable, for example experimentally deduced properties of the saddle point.

As a lower level, the range of options is even broader, since the accuracy in the energies is not as important as in the higher-level calculation. Thus, an analytical potential energy surface for the reaction between atomic oxygen and methane has been used as lower level in a recent application of the dual-level methods (82), although this is not expected to be the most widely used choice, since the construction of an analytical potential energy surface can be an extremely time consuming (and perhaps tedious) task. For $CH_4 + O$ the analytical potential energy surface employed was the J1 surface (83) for the $CH_4 + H$ reaction slightly modified in order to reproduce the theoretical and experimental results for the $CH_4 + O$ reaction. But the lower-level in most applications to date has been a direct dynamics calculation. In a direct dynamics calculation, whenever an energy, gradient, or Hessian is required in the dynamical calculation, it is computed "on the fly" by means of an electronic structure calculation.

One possibility for the electronic structure level for the lower level in a direct dynamics calculation is to use a density functional theory (DFT) calculation (84). This kind of calculation is much more affordable for large systems than *ab initio* calculations of comparable accuracy, and for some types of reactions it is more accurate than semiempirical calculations. DFT is an especially appropriate choice for many systems containing metal atoms, where semiempirical theory appears to be less reliable than for purely organic systems without metals. Thus, we modeled the C–H bond activation involving rhodium complexes by means of DFT. In particular, for the rearrangement of *trans*-$Rh(PH_3)_2Cl(\eta^2\text{-}CH_4)$ to $Rh(PH_3)_2\text{-}Cl(H)(CH_3)$ (33), we used the B3LYP DFT level (85), with the LANL2DZ basis set (86) as the lower level. A dual-level technique was employed for the correction of only the energy along the reaction path, by using energies for reactants, products, and saddle point computed at the B3LYP level, with a larger basis set.

Some studies have been carried out using as a lower level an *ab initio* calculation using small basis sets and/or inaccurate treatment of the electron correlation. Thus, the first application of the IC method with an *ab initio* lower level was the study of the rate constant of the OH + NH_3 hydrogen abstraction reaction (*37*) using MP2/6-31G(*d,p*) as the lower level. Further comment on the results for this reaction will be provided below.

In calculations carried out so far, the most widely used choice for the lower-level direct dynamics calculation has been the semiempirical neglect-of-diatomic differential-overlap (NDDO) (*39,77,87,88*) method. The main advantage of this way of calculating the lower level is its economy; semiempirical methods are fast and affordable even for large systems, allowing us to calculate larger regions of the potential energy surface. Nevertheless, the lack of quantitative accuracy of these methods is a concern. Any of the usual general parametrizations that make use of NDDO approximations, e.g., MNDO (*88*), AM1 (*77*), and PM3 (*39*), are based on a set of parameters fitted in order to minimize in an average way the discrepancies between calculated and observed properties for a broad training set of test molecules, usually organic molecules. Thus, although these methods behave qualitatively correctly for most reactions, predictive quantitative accuracy is almost never achieved, especially for saddle points (as well as any other nonclassical structure, since the parametrization of the methods did not include such structures). Nevertheless the general parametrizations are sometimes useful with no change in parameters. For example the kinetic isotope effect for the [1,5] sigmatropic rearrangement of *cis*-1,3-pentadiene (*89*) was calculated by direct dynamics based on the MINDO/3 (*90*), AM1, and PM3 semiempirical surfaces. The results are in excellent agreement with the experimental results (within 13%), reflecting the suitability of these semiempirical methods in the study this reaction. Furthermore, since the IC method corrects the deficiencies of the lower level surface, with results usually showing weak dependence on the lower level of calculation, the standard parametrizations of the semiempirical methods are even more useful as the lower-level of a dual-level calculation. However, one can generally do better, as discussed next.

Since the parameters of semiempirical calculations are, in some sense, arbitrary parameters fitted in order to minimize an average error, a solution to the lack of accuracy in the description of a particular reaction is the reparametrization of the semiempirical Hamiltonian in order to improve the description of a particular reaction. Thus, the NDDO methods with specific reaction parameters (SRP) (*30*) can provide us with a economic but accurate description of the potential energy surface suitable for direct dynamics calculations.

The reparametrization of the semiempirical method might seem to be a task as time consuming as the construction and parametrization of an analytical expression for the potential energy surface; but this is not the case. Several aspects of the task make it more affordable:

- The problem of finding a physically meaningful functional form for the mathematical expression of the energy is eliminated. All reactions, independent of the number of atoms involved and the type of reaction can be treated using the same mathematical tools, which have a clear physical meaning.
- The original parametrization constitutes a good starting point for the reparametrization. In fact, we have usually tried to modify the original parameters by no more than a certain percentage (usually 10% or 20%), and one need not change a large number of parameters.
- The parameters to be fitted have a more clear physical meaning than the parameters usually involved in analytical expressions of the potential energy. The

selection of the parameters to be optimized and their fitting can, in principle, be guided by chemical intuition.

- A set of parameters optimized for a reaction can be expected to be at least partially transferable to other similar reactions. For this reason we sometimes call SRP parameters "specific range parameters" when we try to make them useful for some range of systems.

- When the NDDO-SRP surfaces are corrected by means of dual-level techniques (which is the usual procedure), the lower-level description of the reaction does not need to be energetically accurate. Obviously, the final results are expected to be more reliable when the lower level reproduces well the high-quality *ab initio* or experimental data, but when the NDDO-SRP surface is reasonable, the dual-level methods depend only weakly on the lower-level results.

- Automatic or semiautomatic optimization procedures can be used for the fitting of the SRP parameters. This is accomplished by defining an error function to be minimized, where the error function may contain the average deviation of the NDDO-SRP predictions from a few accurately known energies, stationary-point geometries, and/or frequencies. Taking into account the nature of the problem (multidimensional minimization of an error function dependent on a large number of variables, without analytic derivatives and the possibility that the error surface presents a rugged landscape), a particularly useful way of finding the optimum parameters is by using genetic algorithms (*91-93*). Thus, we can define an error function dependent on the set of parameters that we want to fit (*91*), and we can use genetic algorithms to look for the global minimum of that function. The main decisions to be made are the construction of the error function (weighting the errors in those properties that we want to describe more accurately), the choice of starting parameters (usually AM1 or PM3), the choice of which parameters to vary, and the limits (e.g., ±5%, ±10%) over which are allow those parameters to vary from their standard values.

The first applications of NDDO-SRP methods were in single-level calculations. The first work using this approach was the study of the rate constants and primary and secondary kinetic isotope effects for the microsolvated S_N2 reaction $Cl^-(H_2O)_n + CH_3Cl$, with $n = 0, 1, 2$ (*30*). The parameters were fitted to reproduce the experimental value of the electron affinity on Cl (a very important parameter for describing solvation effects) and the classical barrier height inferred (*94*) from a modeling of the $n = 0$ case. The results agree very well with those obtained from an analytic surface (*94, 95*), which had been much harder to construct. Thus, the same set of parameters was used for a detailed description of the Cl^- solvation (*96*). In later work single-level NDDO calculations with SRP parameters fitted in order to describe experimental properties have been applied to modeling several other kinetic isotope effects (*97-100*).

Nevertheless, a more promising way of using NDDO-SRP methods for VTST/MT calculations is with IC methods. In particular we favor a bootstrap approach in which the higher-level information is used not only in the IC corrections but also for calibrating the SRP parameters (*31, 36-38, 61, 82, 95b, 101–103*). An important advantage that we have pointed out previously is the relative independence of the final results on the lower level calculation. As an example, in Table 9 we examine the HL///LL rate constants for the N_2H_2 + H $\rightarrow$ N_2H + H_2 reaction calculated using 4 different low-level surfaces, and we compare the results to a full HL calculation (*38*). At low temperatures the effects of a different lower level are more noticeable, but they all are within a reasonable range of the full HL results.

An important advantage of the IVTST-IC approach, especially for reactions with high curvature of the reaction path, is the possibility of including tunneling contributions from regions of the potential energy surface that are far from the reaction-path (the broad region of configuration space traversed by significant tunneling paths is called the reaction

Table 9. Rate constants (10^{-11} cm^3 molecule^{-1} s^{-1}) for the H + N$_2$H$_2$ reaction calculated using single-level and dual-level methods[a]

Method	T (K)			
	300	600	1500	3000
HL///MNDO	0.065	0.39	2.7	18
HL///AM1	0.088	0.35	3.5	17
HL///MNDO-SRP	0.120	0.41	3.5	17
HL///AM1-SRP	0.050	0.32	3.8	18
HL	0.068	0.35	3.4	20

[a]In this table, HL denotes MRCI55/cc-pVTZ//CASSCF/cc-pVDZ. the results in this table are all taken from Ref. (*38*).

swath). The portion of the reaction swath that is far removed from the minimum energy path on its concave side is very important for reactions with large reaction-path curvature in the region of the barrier, since most of the tunneling may take place in this region of the surface. The reliable calculation of tunneling effects in reactions with large curvature therefore requires information not only about the reaction path and the harmonic valley surrounding it but about energies in the wider reaction swath. The high cost of the kind of *ab initio* calculations we want to use makes this information unavailable in most cases, but when using semiempirical NDDO or NDDO-SRP methods the use of large-curvature approximations for tunneling is possible (*98*), and the IC methods can also correct the energy in this region of the surface (*36*) on the basis of limited stationary-point data at a higher level. As an example, we note that the reaction-path for the reaction OH + NH$_3$ reaction (*37*) has an important curvature. As a consequence, tunneling methods that don't include the tunneling probability through the farther out regions of the reaction swath give rate constants that are too low at low temperatures, where tunneling is more important. The use of a semiempirical lower-level surface allows an inexpensive calculation of tunneling along the reaction path and through the reaction swath, leading to low-temperature rate constants in agreement with experimental results (*37*).

Summary

Dual-level methods in which the lower level is quantum mechanical can be very useful for bond energies, barrier heights, reaction-path dynamics, and electronic substituent effects on a subsystem of a larger system. The methods may be *ab initio* or partly semiempirical. Basis sets (MIDI!) and semiempirical parameters (SAC, NDDO-SRP) may be optimized specifically for use in dual-level calculations. The following methods were illustrated:

> *applicable primarily to small systems*
> • Scaling external correlation (SEC)
> *applicable to medium-sized systems*
> • Scaling all correlation (SAC)

applicable to all-sized systems
- NDDO-SRP as lower level
- Variational transition state theory with interpolated corrections
- IMOMO and IMOHC for electronic substituent effects on energies and geometries with DFT, HF, or semiempirical molecular orbital theory as lower level.

Acknowledgments

We acknowledgment many contributors to the methods discussed here (see the references) and would especially like to single out the following contributions: SEC: Frank Brown and David Schwenke; SAC: Mark Gordon and Ivan Rossi; IMOMO: E. Laura Coitiño and Molli Noland; dual-level direct dynamics: John Chuang, E. Laura Coitiño, Joaquin Espinosa-García, Patton Fast, Wei-Ping Hu, Yi-Ping Liu, Kiet Nguyen, and Orlando Roberto-Neto. Our work on electronic structure theory is supported in part by the National Science Foundation, and our work on gas-phase variational transition state theory is supported in part by the U.S. Department of Energy, Office of Basic Energy Sciences. JCC aknowledges a Fulbright Scholarship.

Literature Cited

(1) (a) Kao, J. *J. Amer. Chem. Soc.* **1987**, *109*, 3817. (b) Luzhkov, V.; Warshel, A. *J. Comput. Chem.* **1992**, *13*, 199.

(2) (a) Weiner, S. J.; Singh, U. C.; Kollman, P. A. *J. Amer. Chem. Soc.* **1985**, *107*, 2219. (b) Singh, U. C.; Kollman, P. A. *J. Comp. Chem.* **1986**, *7*, 718.

(3) (a) Bash, P. A.; Field, M. J.; Karplus, M. *J. Amer. Chem. Soc.* **1987**, *109*, 8092. (b) Field, M. J.; Bash, P. A.; Karplus, M. *J. Comput. Chem.* **1990**, *11*, 700. (c) Field, M. J. In *Computer Simulation of Biomolecular Systems: Theoretical and Experimental Applications*, Vol. 2; van Gunsteren, W. F., Weiner, P. K., Wilkinson, A. J., Eds.; ESCOM: Leiden, 1993; p. 102.

(4) (a) Gao, J. *J. Phys. Chem.* **1992**, *96*, 537. (b) Gao, J. *Rev. Comp. Chem.* **1996**, *7*, 119.

(5) Cummins, P. L.; Gready, J. E. *J. Comput. Chem.* **1997**, *18*, 1496.

(6) (a) Maseras, F.; Morokuma, K. *J. Comput. Chem.* **1995**, *16*, 1170. (b) Matsubara, T.; Maseras, F.; Koga, N.; Morokuma, K. *J. Phys. Chem.* **1996**, *100*, 2573. (c) Deng, L.; Woo, T. K.; Cavallo, L.; Margl, P. M.; Ziegler, T. *J. Amer. Chem. Soc.* **1997**, *119*, 6177.

(7) Hehre, W. J.; Radom, L.; Schleyer, P. v. R.; Pople, J. A. *Ab Initio Molecular Orbital Theory*; Wiley: New York, 1986.

(8) Curtiss, L. A.; Raghavachari, K.; Trucks, G. W.; Pople, J. A. *J. Chem. Phys.* **1991**, *94*, 7221.

(9) Nyden, M. R.; Petersson, G. A. *J. Chem. Phys.* **1981**, *75*, 1843.

(10) (a) Petersson, G. A.; Tensfeldt, T. G.; Montgomery, J. A. *J. Chem. Phys.* **1991**, *94*, 6091. (b) Ochterski, J. W.; Petersson, G. A.; Montgomery, J. A. Jr. *J. Chem. Phys.* **1996**, *104*, 2598.

(11) Brown, F. B.; Truhlar, D. G. *Chem. Phys. Lett.* **1985**, *117*, 307.

(12) Gordon, M. S.; Truhlar, D. G. *J. Amer. Chem. Soc.* **1986**, *108*, 5412.

(13) Gordon, M. S.; Truhlar, D. G. *Int. J. Quantum Chem.* **1987**, *31*, 81

(14) Gordon, M. S.; Nguyen, K. A.; Truhlar, D. G. *J. Phys. Chem.* **1989**, *93*, 7356.

(15) Rossi, I.; Truhlar, D. G. *Chem. Phys. Lett.* **1995**, *234*, 64.

(16) (a) Humbel, S.; Sieber, S.; Morokuma, K. *J. Chem. Phys.* **1996**, *105*, 1959. (b) Svensson, M.; Humbel, S.; Morokuma, K. *J. Chem. Phys.* **1996**, *105*, 3654.

(17) Noland, M.; Coitiño, E. L.; Truhlar, D. G. *J. Phys. Chem. A* **1997**, *101*, 1193.

(18) Svensson, M.; Humbel, S.; Froese, R. D. J.; Matsubara, T.; Sieber S.; Morokuma, K. *J. Phys. Chem.* **1996**, *100*, 19375.

(19)   Coitiño, E. L.; Truhlar, D. G.; Morokuma, K. *Chem. Phys. Lett.* **1996**, *259*, 159.

(20)   Coitiño, E. L.; Truhlar, D. G. *J. Phys. Chem. A* **1997**, *101*, 4641.

(21)   Corchado, J. C.; Truhlar, D. G. submitted for publication

(22)   Tapia, O. In *Quantum Theory of Chemical Reactions*; Daudel, R., Pullman, A., Salem, L., Veillard, A., Eds.; Reidel: Dordrecht, 1980; Vol. 2, p. 25.

(23)   Lee, B.; Richards, F. M. *J. Mol. Biol.* **1971**, *55*, 379.

(24)   Hermann, R. B. *J. Phys. Chem.* **1972**, *76*, 2754.

(25)   Eisenberg, D.; McLachlan, A. D. *Nature* **1986**, *319*, 199.

(26)   Cramer, C. J.; Truhlar, D. G. *J. Amer. Chem. Soc.* **1991**, *113*, 8305. Erratum: **1991**, *113*, 9901.

(27)   Storer, J. W.; Giesen, D. J.; Hawkins, G. D.; Lynch, G. C.; Cramer, C. J.; Truhlar, D. G.; Liotard, D. A. *ACS Symp. Ser.* **1994**, *568*. 24.

(28)   Hawkins, G. D.; Zhu, T.; Li, J.; Chambers, C. C.; Giesen, D. J.; Cramer, C. J.; Truhlar, D. G., chapter elsewhere in this volume.

(29)   Parrinello, M. In *MOTEC: Modern Techniques in Computational Chemistry*; Clementi, E., Ed.; ESCOM: Leiden, 1991; p. 833.

(30)   Gonzàlez-Lafont, A.; Truong, T. N.; Truhlar, D. G. *J. Phys. Chem.* **1991**, *95*, 4618.

(31)   Truhlar, D. G. In *The Reaction Path in Chemistry;* Heidrich, D., Ed.; Kluwer: Dordrecht, 1995; p. 229.

(32)   (a) Wang, I.; Karplus, M. *J. Amer. Chem. Soc.* **1973**, *95*, 8160. (b) Warshel, A.; Karplus, M. *Chem. Phys. Lett.* **1975**, *32*, 11. (c) Malcome-Lawes, D. J. *J. Chem. Soc. Faraday Trans. 2* **1975**, *71*, 1183. (d) Leforestier, C. *J. Chem. Phys.* **1978**, *68*, 4406. (e) Truhlar, D. G.; Duff, J. W.; Blais, N. C.; Tully, J. C.; Garrett, B. C. *J. Chem. Phys.* **1982**, *77*, 764. (f) Margl, P.; Ziegler, T.; Blöchl, P. E. *J. Am. Chem. Soc.* **1995**, *117*, 12625.

(33)   Espinosa-García, J.; Corchado, J. C.; Truhlar, D. G. *J. Amer. Chem. Soc.* **1997**, *119*, 9891.

(34)   Truhlar, D. G.; Garrett, B. C. *J. Chim. Phys.* **1987**, *84*, 365.

(35)   Truong, T. N.; Lu, D.-h.; Lynch, G. C.; Liu, Y.-P.; Melissas, V. S.; Stewart, J. J. P.; Steckler, R.; Garrett, B. C.; Isaacson, A. D.; Gonzàlez-Lafont, A.; Rai, S. N.; Hancock, G. C.; Joseph, T.; Truhlar, D. G. *Comput. Phys. Commun.* **1993**, *75*, 143.

(36)   Hu, W.-P.; Liu, Y.-P.; Truhlar, D. G. *J. Chem. Soc. Faraday Trans.* **1994**, *90*, 1715.

(37)   Corchado, J. C.; Espinosa-García, J.; Hu, W.-P.; Rossi, I.; Truhlar, D. G. *J. Phys. Chem.* **1995**, *99*, 687.

(38)   Chuang, Y.-Y.; Truhlar, D. G. *J. Phys. Chem. A* **1997**, *101*, 3808.

(39)   Stewart, J. J. P. *J. Comp. Chem.* **1989**, *10*, 221.

(40)   Löwdin, P.-O. *Adv. Chem. Phys.* **1959**, *2*, 207.

(41)   (a) Werner, H.-J. *Adv. Chem. Phys.* **1987**, *69*, 1. (b) Shepard, R. *Adv. Chem. Phys.* **1987**, *69*, 63. (b) Roos, B. O. *Adv. Chem. Phys.* **1987**, *69*, 399.

(42)   (a) Shavitt, I. In *Advanced Theories and Computational Approaches to the Electronic Structure of Molecules*; Dykstra, C. E., Ed.; Reidel: Dordrecht, 1984; p. 185. (b) Werner, H.-J. In *Domain-Based Parallelism and Problem Decomposition Methods in Computational Science and Engineering*; Keyes, D. E., Saad, Y., Truhlar, D. G., Eds.; SIAM: Philadelphia, 1995; p. 239.

(43)   (a) Andersson, K.; Malmqvist, P.-A.; Roos, B. O.; Sadlej, A. J.; Wolinski, K. *J. Phys. Chem.* **1990**, *94*, 5483. (b) Andersson, K.; Malmqvist, P.-A.; Roos, B. O. *J. Chem. Phys.* **1992**, *96*, 1218.

(44)   Steckler, R.; Schwenke, D. W.; Brown, F. B.; Truhlar, D. G. *Chem. Phys. Lett.* **1985**, *121*, 475.

(45)   Schwenke, D. W.; Steckler, R.; Brown, F. B.; Truhlar, D. G. *J. Chem. Phys.* **1986**, *84*, 5706.
(46)   Schwenke, D. W.; Tucker, S. C.; Steckler, R.; Brown, F. B.; Lynch, G. C.; Truhlar, D. G.; Garrett, B. C. *J. Chem. Phys.* **1989**, *90*, 3110.
(47)   Lynch, G. C.; Truhlar, D. G.; Brown, F. B.; Zhao, J.-g. *Hyperfine Interactions* **1994**, *87*, 885.
(48)   Lynch, G. C.; Truhlar, D. G.; Brown, F. B.; Zhao, J.-g. *J. Phys. Chem.* **1995**, *99*, 207.
(49)   Lynch, G. C.; Steckler, R.; Schwenke, D. W.; Varandas, A. J. C.; Truhlar, D. G.; Garrett, B. C. *J. Chem. Phys.* **1991**, *94*, 7136.
(50)   Mielke, S. L.; Lynch, G. C.; Truhlar, D. G.; Schwenke, D. W. *Chem. Phys. Lett.* **1993**, *213*, 10. Erratum: **1994**, *217*, 173.
(51)   Pople, J. A.; Binkley, J. S.; Seeger, R. *Int. J. Quantum Chem. Symp.* **1976**, *10*, 1.
(52)   Pople, J. A.; Krishnan, R.; Schlegel, H. B.; Binkley, J. S. *Int. J. Quantum. Chem.* **1978**, *14*, 545.
(53)   Bartlett, R. J. *Annu. Rev. Phys. Chem.* **1981**, *32*, 359.
(54)   Pople, J. A.; Head-Gordon, M.; Raghavachari, K. *J. Chem. Phys.* **1987**, *87*, 5968.
(55)   Dunning, T. H. Jr. *J. Chem. Phys.* **1989**, *90*, 1007.
(56)   (a) Siegbahn, P. E. M.; Blomberg, M. R. A.; Svensson, M. *Chem. Phys. Letters* **1994**, *223*, 35. (b) Siegbahn, P. E. M.; Svensson, M.; Boussard, P. J. E. *J. Chem. Phys.* **1995**, *102*, 5377.
(57)   Siegbahn, P. E. M.; Crabtree, R. H. *J. Amer. Chem. Soc.* **1996**, *118*, 4442.
(58)   Truong, T. N.; Truhlar, D. G.; Baldridge, K. K.; Gordon, M. S.; Steckler, R. *J. Chem. Phys.* **1989**, *90*, 7137.
(59)   Melissas, V. S.; Truhlar, D. G. *J. Chem. Phys.* **1993**, *99*, 1013.
(60)   Melissas, V. S.; Truhlar, D. G. *J. Phys. Chem.* **1994**, *98*, 875.
(61)   Hu, W.-P.; Rossi, I.; Corchado, J. C.; Truhlar, D. G. *J. Phys. Chem. A*, **1997**, *101*, 6911.
(62)   Truong, T. N.; Truhlar, D. G. *J. Chem. Phys.* **1990**, *93*, 1761. Erratum: **1992**, 97, 8820.
(63)   Nguyen, K. A.; Gordon, M. S.; Truhlar, D. G. *J. Amer. Chem. Soc.* **1991**, *113*, 1596.
(64)   Espinosa-García, J.; Corchado, J. C.; Sana, M. *J. Chim. Phys.* **1993**, *90*, 1181.
(65)   Espinosa-García, J.; Tolosa, S.; Corchado, J. C. *J. Phys. Chem.* **1994**, *98*, 2337.
(66)   Espinosa-García, J.; Corchado, J. C. *J. Chem. Phys.*, **1994**, *101*, 1333.
(67)   Corchado, J. C.; Olivares del Valle, F. J.; Espinosa-García, J. *J. Phys. Chem.* **1993**, *97*, 9129. Erratum: **1994**, *98*, 5796.
(68)   Espinosa-García, J.; Corchado, J. C. *J. Chem. Phys.*, **1994**, *101*, 8700.
(69)   Allison, T.; Lynch, G. C.; Truhlar, D. G.; Gordon, M. S. *J. Phys. Chem.* **1996**, *100*, 13575.
(70)   Melissas, V. S.; Truhlar, D. G. *J. Chem. Phys.* **1993**, *99*, 3542.
(71)   (a) Krishnan, R.; Binkley, J. S.; Seeger, R.; Pople, J. A. *J. Chem. Phys.* **1980**, *72*, 650. (b) McLean, A. D.; Chandler, G. S. *J. Chem. Phys.* **1980**, *72*, 5639.
(72)   Espinosa-García, J.; Corchado, J. C. *J. Chem. Phys.*, **1996**, *105*, 3517.
(73)   Gonzàlez-Lafont, A.; Truong, T. N.; Truhlar, D. G. *J. Chem. Phys.* **1991**, *95*, 8875.
(74)   (a) Pople, J. A.; Frisch, M. J.; Luke, B. T.; Binkley, J. S. *Int. J. Quantum Chem. Symp.* **1983**, *17*, 307. (b) Handy, N. C.; Knowles, P. J.; Somasundram, K. *Theor. Chim. Acta* **1985**, *68*, 87.
(75)   (a) Feenberg, E. *Ann. Phys. (N.Y.)* **1958**, *3*, 292. (b) Wilson, S. *Int. J. Quantum Chem.* **1980**, *18*, 905.

(76) Chase, M. W. Jr.; Davies, C. A.; Downey, J. R. Jr.; Frurip, D. J.; McDonald, R. A.; Syverud, A. N. "JANAF Thermochemical Tables." *J. Phys. Chem. Ref. Data, Suppl.* **1985**, 14.

(77) (a) Dewar, M. J. S.; Zoebisch, E. G.; Healy, E. F. *J. Amer. Chem. Soc.*, **1985**, *107*, 3902. (b) Dewar, M. J. S.; Zoebisch, E. G. *J. Mol. Struct. (Theochem)*, **1988**, *180*, 1.

(78) (a) Becke, A. D. *J. Chem. Phys.* **1993**, *98*, 5648. (b) Perdew, J. P. In *Electronic Structure of Solids '91*; Ziesche, P., Eschrig, H., Eds.; Akademie Verlag: Berlin, 1991; p 11. (c) Perdew, J. P.; Chevary, J. A.; Vosko, S. H.; Jackson, K. A.; Pederson, M. R.; Singh, D. J.; Foilhaus, C. *Phys. Rev. B* **1992**, *46*, 6671. (d) Perdew, J. P.; Burke, K.; Ernzerhof, M. *ACS Symp. Ser.* **1996**, *629*, 453. (e) Perdew, J. P.; Burke, K.; Wang, Y. *Phys. Rev. B* **1996**, *54*, 16533.

(79) Stephens, P. J.; Devlin, F. J.; Chabalowski, C. F.; Frisch, M. J. *J. Phys. Chem.* **1994**, *98*, 11623.

(80) (a) Purvis, G. D.; Bartlett, R. J. *J. Chem. Phys.* **1982**, *76*, 1910. (b) Bartlett, R. J.; Watts, J. D.; Kucharski, S. A.; Noga, J. *Chem. Phys. Letters* **1990**, *165*, 513.

(81) Easton, R. E.; Giesen, D. J.; Welch, A.; Cramer, C. J.; Truhlar, D. G. *Theor. Chim. Acta* **1996**, *93*, 28.

(82) Corchado, J. C.; Espinosa-García, J.; Roberto-Neto, O.; Chuang, Y. Y.; Truhlar, D. G. Unpublished results.

(83) Joseph, T.; Steckler, R.; Truhlar, D. G. *J. Chem. Phys.* **1987**, *87*, 7036.

(84) Laird, B. B., Ross, R. B., Ziegler, T., Eds. *Density Functional Theory*; American Chemical Society: Washington, 1996.

(85) Stephens, P. J.; Devlin, F. J.; Ashuar, C. S.; Bak, K. L.; Taylor, P. R.; Frisch, M. J. In *Density Functional Theory*; Laird, B. B., Ross, R. B., Ziegler, T., Eds. Chemical Society: Washington, 1996; p. 105.

(86) (a) Dunning, T. H.; Hay, P. J. In *Modern Theoretical Chemistry*; Schaefer, H. F., Ed.; Plenum: New York, 1977; Vol. 2, pp 1-28. (b) Hay, P. J.; Wadt, W. R. *J. Chem. Phys.* **1985**, *82*, 270. (c) Wadt, W. R.; Hay, P. J. *J. Chem. Phys.* **1985**, *82*, 284.

(87) (a) Pople, J. A.; Santry, D.; Segal, G. *J. Chem. Phys* **1965**, *43*, S129. (b) Pople, J. A.; Beveridge, D. J. *Approximate Molecular Orbital Theory*; McGraw-Hill: New York, 1970.

(88) Dewar, M. J. S.; Thiel, W. *J. Amer. Chem. Soc.* **1977**, *99*, 4899, 4907.

(89) Liu, Y.-P.; Lynch, G. C.; Truong, T. N.; Lu, D.-h.; Truhlar, D. G; Garrett, B. C. *J. Amer. Chem. Soc.* **1993**, *115*, 2408.

(90) Bingham, R. C.; Dewar, M. J. S.; Lo, D. H. *J. Amer. Chem. Soc.* **1975**, *97*, 1285.

(91) Rossi, I.; Truhlar, D. G. *Chem. Phys. Letters* **1995**, *233*, 231.

(92) Goldberg, D. E. *Genetic Algorithms in Search, Optimization and Machine Learning*; Addison-Wesley: Reading, 1989.

(93) Bash, P. A.; Ho, L.L.; Mackerell, A.D.; Levine, D.; Hallstrom, P. *Proc. Natl. Acad. Sci. USA* **1996**, *93*, 3698.

(94) Tucker, S. C.; Truhlar, D. G. *J. Amer. Chem. Soc.* **1990**, *112*, 3338.

(95) (a) Tucker, S. C.; Truhlar, D. G. *J. Amer. Chem. Soc.* **1990**, *112*, 3347. (b) Truhlar, D. G.; Lu, D.-h.; Tucker, S. C.; Zhao, X. G.; Gonzàlez-Lafont, A.; Truong, T. N.; Maurice, D.; Liu, Y.-P.; Lynch, G. C. In *Isotope Effects in Chemical Reactions and Photodissociation Processes*; Kaye, J. A., Ed.; American Chemical Society Symposium Series 502: Washington, DC, 1992; pp. 16-36.

(96) Zhao, X. G.; Gonzàlez-Lafont, A.; Truhlar, D. G. *J. Chem. Phys.* **1991**, *94*, 5544.

(97) Viggiano, A. A.; Paschkewitz, J. S.; Morris, R. A.; Paulson, J. F.; Gonzàlez-Lafont, A.; Truhlar, D. G. *J. Amer. Chem. Soc.* **1991**, *113*, 9404.

(98) Liu, Y.-P.; Lu, D.-h.; Gonzàlez-Lafont, A.; Truhlar, D. G.; Garrett, B. C. *J. Amer. Chem. Soc.* **1993**, *115*, 7806.
(99) Corchado, J. C.; Espinosa-García, J. *J. Chem. Phys.* **1996**, *105*, 3160.
(100) Gonzàlez-Lafont, A.; Truhlar, D. G. In *Chemical Reactions in Clusters*; Bernstein, E. R., Ed.; Oxford University Press: New York, 1996; pp. 3-39.
(101) Robert-Neto, O.; Coitiño, E. L.; Truhlar, D. G. Unpublished results.
(102) Hu, W.-P.; Truhlar, D. G. *J. Amer. Chem. Soc.* **1995**, *117*, 10726.
(103) Hu, W.-P.; Truhlar, D. G. *J. Amer. Chem. Soc.* **1996**, *118*, 860.

Chapter 8

A Combined Car-Parrinello Quantum Mechanical–Molecular Mechanical Implementation for Ab Initio Molecular Dynamics Simulations of Extended Systems

Tom K. Woo, Peter M. Margl, Liqun Deng, and Tom Ziegler[1]

Department of Chemistry, University of Calgary, 2500 University Drive, N.W., Calgary, Alberta, T2N 1N4, Canada

We describe our implementation of a combined Car-Parrinello and molecular mechanics method for the *ab initio* molecular dynamics simulations of extended systems. We also introduce a combined QM/MM multiple time step method which allows for the differential sampling of the QM/MM regions within the molecular dynamics framework. We also provide a brief overview the Car-Parrinello method, with a particular focus on aspects of the method important to the combined QM/MM methodology and important in simulating reactions involving non-periodic molecular systems. Finally, several applications of the combined Car-Parrinello and molecular mechanics method are demonstrated. Namely, the chain termination reaction and monomer capture process in the Brookhart Ni-diimine olefin polymerization catalyst are simulated.

There have been many approaches introduced, such as multiple time scale techniques(1) and linear scaling methods,(2,3) to increase the efficiency of Car-Parrinello(4) based *ab initio* molecular dynamics simulations to treat extended molecular systems. Taking a different route, we have implemented the combined quantum mechanics and molecular mechanics (QM/MM) method(5) of Singh(6) and Field(7) into the Car-Parrinello *ab initio* molecular dynamics framework. The combined QM/MM method has recently gained significant attention(6-12) because of its potential to simulate large systems in a detailed, yet efficient manner. Our implementation includes a multiple-time step scheme(1) such that the molecular mechanics region is sampled at a faster rate than the quantum mechanics region, thereby providing better ensemble averaging during the calculation of the free energy barriers. In this article it is our intent to provide some details of our combined QM/MM *ab initio* molecular dynamics implementation and demonstrate the utility of the method on practical applications of the method. Additionally, we will briefly review some of the novel techniques we utilize, including the Car-Parrinello (CP) methodology, the PAW-CP implementation and the slow growth method for determining reaction free energy barriers.

[1]Corresponding author.

Car-Parrinello *Ab Initio* Molecular Dynamics

The *ab-initio* molecular dynamics (AIMD) method(4,13) is maturing into a powerful predictive tool for chemical processes, as it is one of the select few first-principles methods that are able to sample large regions of configuration space within a practical amount of time. Recent applications in chemical catalysis show the potential power of the method, allowing glimpses of real-time reaction simulations,(14) studies of sub-picosecond fluxionality,(15-17) reaction path scans,(17-19) and free energy calculations.(20-22)

Conventional *ab initio* molecular dynamics involves moving the nuclei according to Newton's equations of motion with the forces calculated from an electronic structure calculation. The electronic structure is normally described by a set of orthonormal molecular orbitals, φ_i, which are expanded in terms of a basis set, χ_k such that $\varphi_i = \sum_k c_{ik} \chi_k$. The optimal coefficients are solved variationally with the constraint that the molecular orbitals remain orthogonal. Generally this is done in a self consistent manner by the diagonalization of the Hamiltonian matrix or equivalent. An alternative method for determining the optimal coefficients draws analogy to nuclear dynamics which can be used to optimize nuclear geometries by applying friction to the motion. By assigning fictitious masses to the coefficients, fictitious dynamics can be performed on the coefficients which then move through electronic configuration space with forces given by the negative gradient of the electronic energy. The equivalent equations of motion are:

$$\mu \ddot{c}_{i,k} = -\frac{\partial E^{el}}{\partial c_{i,k}} - \sum_j \lambda_{i,j} c_{i,k} \langle \Psi_i | \Psi_k \rangle \qquad (1)$$

where μ is a fictitious mass, $\ddot{c}_{i,k}$ is the coefficient acceleration, and the last term corresponds to the constraint force imposed to maintain orthogonality. The coefficients move through electronic configuration space with a fictitious kinetic energy and by applying friction the coefficients, they can be steadily brought to settle into an optimal Born-Oppenheimer electronic configuration.

In 1985 Car and Parrinello(4) developed a scheme by which to perform the nuclear dynamics and the electronic coefficient dynamics in parallel as to improve the efficiency of the AIMD method. In this way the electronic MD and nuclear MD equations are coupled:

$$\mu \ddot{c}_{i,k} = -\frac{\partial E}{\partial c_{i,k}} - \sum_j \lambda_{i,j} c_{i,k} \qquad m_I \ddot{x}_I = -\nabla E(x_I, c_{i,k})$$

Formally, the nuclear and electronic degrees of freedom are cast into a single, combined Lagrangian:

$$\mathcal{L} = \sum \mu_i \langle \dot{\Psi}_i | \dot{\Psi}_i \rangle + \frac{1}{2} \sum M_I \dot{R}_I^2 - E_{DFT}(|\Psi\rangle, R) + \sum_{i,j} \Lambda_{ij} (\langle \Psi_i | \Psi_j \rangle - \delta_{ij}) \qquad (2)$$

In equation 2, the first two terms represent the kinetic energy of the wave function and nuclei, respectively, the third term is the potential energy and the last term accounts for the orthogonality constraint of the orbitals. The combined Car-Parrinello Lagrangian ensures that the propagated electronic configuration corresponds to the propagated nuclear positions. Although, the generated electronic configuration does not always correspond to the proper Born-Oppenheimer wavefunction, over time it generates a electronic structure that oscillates around it giving rise to stable molecular dynamics.(13) The coupled Car-Parrinello dynamics, therefore, results in a speed up over conventional AIMD since the electronic wave function does not have to be converged at every time step, instead it only has to be propagated. The primary disadvantage of the Car-Parrinello MD scheme is that the electronic configuration oscillates about the Born-Oppenheimer wavefunction at a high frequency. Therefore, in order to generate stable molecular dynamics a very small time step must be used, usually an order of magnitude smaller than in conventional *ab initio* molecular dynamics.

Although other "first-principles" methods can be used,(23,24) the Car-Parrinello coupled dynamics approach has mostly been implemented within the density functional framework with plane wave basis sets (as opposed to atom centered basis sets). Therefore, Car-Parrinello *ab initio* molecular dynamics generally refers only to this type of implementation. Applications of the Car-Parrinello AIMD method are concentrated in the area of condensed phase molecular physics. Applications in quantum chemistry have been limited until very recently because of certain deficiencies of the method which are only now being overcome.

The source of the limitations stem from the use of plane wave basis sets. Thes plane wave basis sets are advantageous in that the computational effort for the required integrations becomes minute on a per function basis. On the other hand, an enormous number of plane waves is required even when pseudo potentials are utilized to approximate the core potential. The number of plane waves required to accurately treat transition metals and first row elements becomes prohibitively large. This is clearly a problem for applications in chemistry since so much chemistry involves systems containing first row elements. Plane wave basis sets also introduce another problem since periodic images are created automatically and therefore the simulation actually describes a periodic crystal. If non-periodic molecular systems are simulated, the cell size of the periodic systems must be sufficiently large that the wave functions of the images no longer overlap. This requires a vacuum region of approximately 5 Å between images. This is an issue because the computational effort increases with the cell size. If charged systems or systems with large dipole moments are simulated, the long range electrostatic interactions between periodic images will lead to artificial effects.

Projector Augmented Wave (PAW) Car-Parrinello AIMD. The Projector Augmented Wave (PAW) Car-Parrinello AIMD program developed by Peter Blöchl overcomes the aforementioned problems such that AIMD simulations of molecular systems containing first row elements and transition metal complexes has become practical. PAW utilizes a full all-electron wavefunction with the frozen core approximation which allows both accurate and efficient treatment of all elements including first row and transition metal elements. In the PAW method, the plane waves are augmented with atomic based functions near the nuclei such that the rapidly oscillating nodal structure of the valence orbitals near the nuclei are properly represented. One can think of it as smoothly stitching in an atomic-like function into the plane waves such that the plane waves describe regions where the orbitals are smooth, allowing for rapid convergence of the plane waves. The plane wave expressions of PAW and those of the pseudopotential methods are similar enough that the numerical techniques for the most computationally demanding operations are related and equally efficient. Therefore, PAW combines the computational advantages of using plane waves with the accuracy of all-electron schemes. The details of the implementation are described elsewhere.(25,26) To deal with charged systems, PAW has a charge isolation scheme(27) to eliminate the spurious electrostatic interactions between periodic images. The charge isolation scheme involves fitting atomic point charges such that the electrostatic potential outside the molecule is reproduced (ESP fit). The ESP charges, which are recalculated at each time step, are then used to determine the electrostatic interaction between periodic images via Ewald sums. The spurious interactions between images is then subtracted. These features of the PAW package allows for the practical application of *ab initio* molecular dynamics to chemical systems.

Reaction Free energy barriers with AIMD. Reaction free energy barriers are routinely calculated from conventional static electronic structure calculations. Here, the excess free energy of the reactants and transition state can be determined by constructing a partition function from a frequency calculation and using a harmonic (normal mode) approximation. In most cases where the interactions are strong, the approximation is good. However, for processes where weak intermolecular forces dominate, the harmonic or quasi-harmonic approximation breaks down.(28) Alternatively, *ab initio* molecular dynamics simulations can be utilized to determine reaction free energy barriers. An MD simulation samples the available configuration space of the system as to produce a Boltzmann ensemble from which a partition function can be constructed and used to determine the free energy. However, finite MD simulations can only sample a restricted part of the total configuration space, namely the low energy region. Since estimates of the absolute free energy of a system requires a global sampling of the configuration space, only relative free energies can be calculated.

A number of special methodologies have been developed to calculate relative free energies. Since we are interested in reaction free energy barriers, the method we use in our research is derived from the method of thermodynamic integration.(29,30) Assuming we are sampling a canonical NVT ensemble the free energy difference, ΔA, between an initial state with $\lambda=0$ and a final state with $\lambda=1$, is given by Eqn 7.

$$\Delta A_{(0 \rightarrow 1)} = \int_0^1 \frac{\partial A(\lambda)}{\partial \lambda} d\lambda \tag{3}$$

Here the continuous parameter λ is such that the potential $E(\lambda)$ passes smoothly from initial to final states as λ is varied from 0 to 1. Since the free energy function can be expanded in terms of the partition function:

$$A(\lambda) = -kT \ln \left[\int \cdots \int e^{-\frac{E(X^N, \lambda)}{kT}} dX^N \right] \tag{4}$$

the relative free energy ΔA can be rewritten as:

$$\Delta A_{(0 \rightarrow 1)} = \int_0^1 \left(\int \cdots \int \frac{\partial E(X^N, \lambda)}{\partial \lambda} dX^N \bigg|_\lambda \right) \tag{5}$$

or

$$\Delta A_{(0 \to 1)} = \int_0^1 \left\langle \frac{\partial E(X^N, \lambda)}{\partial \lambda} \right\rangle_\lambda d\lambda \tag{6}$$

where the subscript λ represents an ensemble average at fixed λ. Since the free energy is a state function λ can represent any pathway, even non-physical pathways. However, if we choose λ to be a reaction coordinate as to represent a physical reaction path, this provides us with a means of determining an upper bound for a reaction free energy barrier by means of thermodynamic integration. The choice in reaction coordinate is important since a poorly chosen reaction coordinate will result in an unfavorable reaction path and potentially a significant over estimate of the barrier. The more the reaction coordinate resembles the intrinsic reaction coordinate (IRC)(31,32) the potentially better the estimate. The reaction coordinate can be sampled with discrete values of λ on the interval from 0 to 1 or carried out in a continuous manner in what is termed a "slow growth" simulation(20,30,33) by

$$\Delta A = \sum_{i=1}^{N_{steps}} \langle f_\lambda \rangle_i \Delta \lambda_i \tag{7}$$

where i indexes the step number. Here the free energy difference becomes the integrated force on the reaction coordinate and can be thought of as the work necessary to change the system from the initial to final state. The discrete sampling resembles a linear transit calculation such that a series of simulations is set up corresponding to successive values of the reaction coordinate from the initial to final state. For each sample point, the dynamics must be run long enough to achieve an adequate ensemble average force on the fixed reaction coordinate. In a slow growth simulation the reaction coordinate is continuously varied throughout the dynamics from the initial to the final state. Thus, in each time step the reaction coordinate is incrementally changed from that in the previous time step. Formally speaking the system is never properly equilibrated unless the change in the RC is infinitesimally small (reversible change). However, the smaller the rate of change the better the approximation. Since the RC is changed at each time step, the force on the reaction coordinate is biased depending on the direction in which the RC is varied. Therefore a forward and reverse scan of the RC is likely to give different results as depicted in Figure 1. This hysterisis as it is called is a direct consequence of the improper equilibration. Thus it is generally a good idea to perform both forward and reverse scans to reduce this error and to determine whether the rate of change of the reaction coordinate is appropriate. In other words, a slow growth simulation with virtually no hysteresis has its RC changed adequately slow, whereas a simulation with large hysteresis has its RC sampled too quickly. The advantage of the slow growth simulation is that the dynamics is not disrupted when the reaction coordinate is changed and hence the system only has to be thermally equilibrated once. On the other hand the method has the disadvantage that both the forward and reverse scans should be performed.

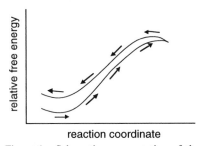

relative free energy

reaction coordinate

Figure 1. Schematic representation of the hysteresis in a slow growth free energy plot. The arrows designate the direction of the scan in terms of the reaction coordinate.

It should be noted that although the forces at each time step are determined from a full quantum mechanical electronic structure calculation, the dynamics itself is still classical. Therefore, quantum dynamical effects such as the tunneling are not included in the estimates of the reaction free energy barriers. Since the classical vibrational energy levels are continuous, $H_{vib} = \Sigma RT/2$ for all states, the zero point energy correction and ΔH_{vib} are also not included in the free energy barriers derived from the AIMD simulation.

Mass Rescaling. Since the configurational averages in classical molecular dynamics do no depend on the masses of the nuclei,[34] a common technique to increase the sampling rate involves replacing the true masses with more convenient ones. Since nuclear velocities scale with $m^{-1/2}$, smaller masses move faster and therefore in principle sample configuration space faster. As a result, the masses of the heavy atoms can be scaled down in order to increase sampling. For example, we commonly rescale the masses of C, N and O in our simulations from 12, 14 and 16 amu, respectively, to 2 amu. There is a limit to the mass reduction, however, because at some point the nuclei move so fast that the time step has to be reduced. At this point there is no gain in reducing the masses further because if the time step has to be shortened, we have to perform more time steps to achieve the same amount of sampling. It is for this reason, we generally scale our hydrogen masses up from 1 amu to 1.5 amu or higher in order to use a larger time step. The features of the mass rescaling technique will be utilized in our multiple time step scheme, which is described later.

Combined QM/MM Car-Parrinello *Ab Initio* Molecular Dynamics

The combined QM/MM involves determining the part of the potential energy surface of a system by an electronic structure calculation and the rest of the molecular potential by a molecular mechanics force field calculation. If the partition is made between separate molecules, then the combined QM/MM method is rather straightforward since the two regions interact only through non-bonded terms. However, if the partition occurs within the same molecule as illustrated in Figure 2, the coupling of the two methods is substantially more difficult and dubious. The difficulty lies in the fact that at least one covalent bond will involve an atom from the QM region and one from the MM region. The immediate problem that arises is how to truncate the electronic system of the QM region across the QM/MM bond. If the MM subsystems is simply removed, the unfilled valences left to the QM region would pose significant problems in a standard electronic structure calculation ultimately resulting in an unphysical picture. The simplest approach and the approach we have adopted to overcoming this truncation problem is to cap the electronic system with "dummy" hydrogen atoms as first introduced by Singh and Kollman(6) (Figure 2b). The capped system forms what is termed the *model QM system*(11) for which a standard electronic structure calculation is performed. The QM contributions to the hybrid forces and total energy are derived from the calculation of the model system. It is assumed that the capped model QM system with the molecular mechanics region acting as a perturbation represents an adequate model of the electronic structure of the active site within the real, extended system.

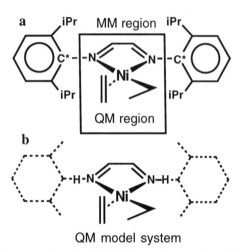

Figure 2. a) An example of a QM/MM partitioning in a Ni-diimine olefin polymerization catalyst. This structure represents the "real" system. b) The corresponding model QM system for which the electronic structure calculation is performed. The hydrogen atoms cap the electronic system and are termed "dummy" atoms. The dummy atoms correspond to "link" atoms in the real system that are labeled C^* in a).

This combined QM/MM approach with the capped QM model system can be easily embedded within the Car-Parrinello framework allowing for *ab initio* molecular dynamics simulations of extended systems to be performed efficiently. We have implemented the combined QM/MM methodology within the PAW code(25) by extending the Car-Parrinello Lagrangian of Eqn. 2 to include the MM subsystem:

$$\mathcal{L} = \sum \mu_i \langle \dot{\Psi}_i | \dot{\Psi}_i \rangle + \tfrac{1}{2} \sum M_{I,QM} \dot{R}_{I,QM}^2 - E_{DFT}(|\Psi\rangle, R_{QM}) + \sum_{i,j} \Lambda_{ij}(\langle \Psi_i | \Psi_j \rangle - \delta_{ij})$$

$$+ \tfrac{1}{2} \sum M_{I,MM} \dot{R}_{I,MM}^2 - E_{MM}(R_{QM}, R_{MM})$$

(8)

The last two terms in equation 8 are the kinetic energy of the MM nuclei and the potential energy derived from the MM force field. The first three terms of equation 8 refer to the "capped" QM model system and are equivalent to those in equation 2. Equation 8 essentially describes the coupled equations of motion of three subsystems: the QM nuclei, the QM wave function and the MM nuclei. A separate Nosé thermostat can be applied to each of the three subsystems in order to properly generate an NVT ensemble during a simulation.(35,36)

The energies and forces on the nuclei are combined by the prescription first introduced by Singh and Kollman(6). In this way all molecular mechanics potential terms that include at least one MM atom are included in the hybrid potential. The QM/MM scheme of Maseras and Morokuma(11) has also been implemented in a slightly modified fashion. Compared to the Singh and Kollman prescription, the Maseras and Morokuma QM/MM scheme has different set of rules for which MM potentials are to be included in the combined potential. Furthermore, with the Maseras and Morokuma scheme, the "link" atom of the MM region (C^* atom in Figure 2a) is always constrained to lie along the bond vector of the dummy bond. The forces on the link atom are then transferred to the dummy atom rigorously in an internal coordinate system. In our molecular dynamics scheme the use of the internal coordinate system is problematic, so we have used an alternative approach whereby a strong harmonic theta potential is applied to maintain the linearity of the QM-Dummy-Link atoms. (A SHAKE(37) constraint may also have been used but this has not been implemented.) The potential acts both to place the link atom along the vector of the dummy bond and to transfer the forces from the dummy atom to link atom in the spirit of the Maseras and Morokuma QM/MM scheme. It also does so in a fashion that allows for energy conserving dynamics.

Periodicity of the Car-Parrinello plane wave method is not problematic to the QM/MM approach. The cell size of the Car-Parrinello calculation is not dependent on the size of the molecular mechanics region. Even with electrostatic coupling of the QM and MM regions, the periodicity of the Car-Parrinello methodology can be overcome. The molecular mechanics region can itself have periodic boundary conditions with a cell size that is independent of the cell size of the QM calculation. The addition of periodic boundary conditions in the MM region is necessary for QM/MM solvent simulations.

The molecular mechanics code has been completely written in FORTRAN90 and the core components of the code are shared with our QM/MM implementation(38) within the ADF(39,40) program. To date, the molecular mechanics potentials that have been implemented are the AMBER95(41) force field of Kollman and coworkers and the Universal Force Field (UFF)(42) of Rappé and coworkers. Electrostatic coupling of the MM and QM systems is currently being implemented in our laboratory to allow for simulations of reactions in solvent to be carried out.(9)

Multiple Times Step QM/MM Method. A QM/MM simulation will most often involve a small QM region embedded within a much larger MM domain, such as with the simulation of the active site chemistry within an enzyme. Associated with the larger size of the MM region is a potential energy surface that is more complex and one which is likely to posses a higher degree of configurational variability. This necessitates an increased degree of sampling in the MM region in order to obtain meaningful ensemble averages from a simulation. Unfortunately, when the QM and MM regions are propagated simultaneously, there will be an unnecessary over sampling the QM region. This is undesirable because the computational expense of the QM calculation generally overshadows that of the MM calculation. The multiple time step method(1,43,44) in combination with mass rescaling can be applied to the QM/MM methodology as to differentially sample the QM and MM regions, thereby increasing the sampling of the MM domain without increasing the computational expenditure in the QM region.

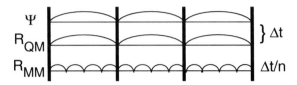

Figure 3. Representation of the multiple time step scheme. The wave function and the QM subsystems are propagated a time step Δt, while the MM subsystem is over sampled with a smaller time step $\Delta t/n$.

In a classical molecular dynamics simulation, the configurational averages do not depend on the masses of the nuclei,(34) and therefore the true nuclear masses can be replaced with more convenient values. Rescaling of nuclear masses to smaller values, which allows the particles to move faster, increases the rate of configurational sampling. Therefore, by decreasing the masses of the MM nuclei relative to those in the QM region, we can differentially sample the configuration space of the two regions. However, if the same time step is used for both regions a limit to the practical over sampling is quickly reached because as we rescale the masses and increase the speed of the MM nuclei, a smaller time step is required to maintain conservative dynamics. Unnecessarily small time steps in the QM region is very costly. Therefore, to permit significant over sampling of the MM region, a multiple time step scheme is required such as that represented in Figure 3. In the Car-Parrinello framework the wave function and the QM nuclei are propagated with a large time step, Δt, while the less computationally demanding MM region is n times over sampled with a time step of $\Delta t/n$. On a more detailed level, we have implemented the reversible multiple time step algorithm of Tuckerman et al.(43) which allows for numerically stable, energy conservative molecular dynamics to be performed. We have modified the original formalism which was based on the velocity Verlet propagation algorithm to accommodate the standard Verlet algorithm used in the PAW program.

To illustrate the stability of the multiple time step QM/MM method, the QM and MM kinetic energies and the total energy are plotted in Figure 4 for a dynamics simulation of a 4-ethyl-nonane. The calculation involved ethane as the model QM system for which the electronic structure was calculated at the gradient-corrected BP86 DFT level.(45-47) A time step of 7.3×10^{-17} s was used for the QM system whereas the MM subsystem was over sampled 20 times, such that the time step was 3.6×10^{-18} s. The simulation involved 10000 QM time steps and 20000 MM time steps. Masses of 12.0 amu and 1.0 amu were used for the C and H atoms, respectively, in the QM system whereas the masses in the MM region were rescaled 400 fold to enhance the sampling rate by approximately a factor of 20. (The carbon atom masses used in the simulation are displayed in the structure in Figure 4.) Masses of the link C atoms denoted with asterisks in Figure 4 where not rescaled down but kept at the true mass of 12.0 amu. This is done to minimize the effects of the strong coupling of the high frequency oscillations in the MM system with the QM system which might destroy the energy conserving dynamics in the QM subsystem. Shown in the lower part of Figure 4 is the total energy of the system including the fictitious kinetic energy of the electronic system. The plot demonstrates the stability of the multiple time step method which, for this simulation, exhibits no drift in the total energy over the period of the whole simulation.

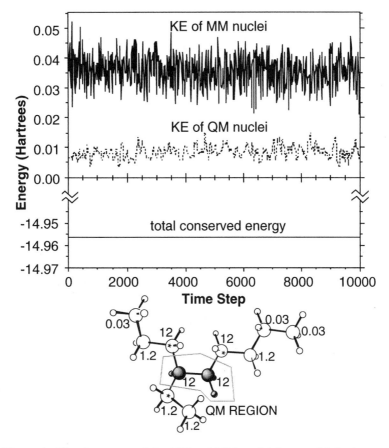

Figure 4. Kinetic energy of the QM and MM nuclei for a multiple time step QM/MM dynamics simulation of 4-ethyl-nonane where ethane is the model QM system. To demonstrate energy conservation of the dynamics, the total energy of the system is shown at the same scale as the kinetic energies. The MM system was over sampled 20 times with the multiple time step scheme. Shown below the graph is the structure with the QM/MM boundary shown. The masses (amu) used in the simulation label the carbon atoms.

Application of the Combined QM/MM AIMD Method

Our research interests lie in the modeling of transition metal based homogenous catalysis and recently, we have applied a "static" QM/MM approach to study a Ni(II) diimine based olefin polymerization catalyst.(10) We found that the QM/MM approach provided meaningful insight into the chemistry of the catalyst. Furthermore, the predicted reaction barriers were in good agreement with experiment. Since the static calculations showed that the QM/MM approximation is appropriate for studying the Brookhart catalyst system, our first practical applications of the combined QM/MM Car-Parrinello method have been on this system.

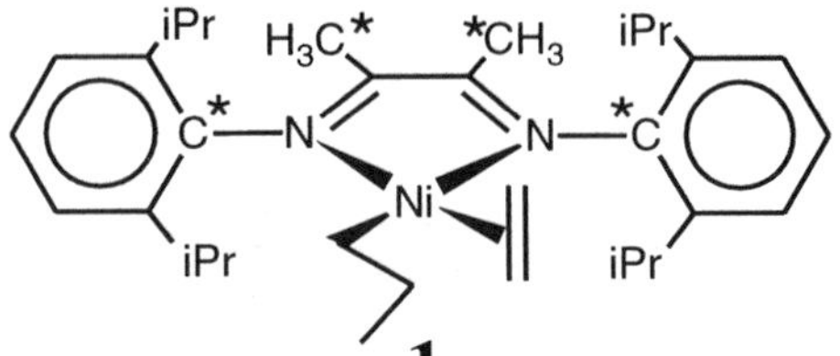

1

Figure 5. Brookhart's Ni(II) diimine olefin polymerization catalyst.

The Brookhart olefin polymerization catalyst(48,49) is a Ni(II) diimine system of the type (ArN=C(R)-C(R)=NAr)Ni(II)-R'+, **1** (Figure 5), where R=Me and Ar=2,6-C_6H_3(i-Pr)$_2$. The catalyst system is currently being developed for commercialization by DuPont(50) because it is seen as a viable alternative to traditional Ziegler-Natta catalysts and because it has the unique ability to produce branched polymer at controllable levels. In this polymerization system the bulky aryl groups play a crucial role, since without the bulky substituents the catalyst acts only as a dimerization catalyst due to the preference for the β-elimination chain termination process. In other words, without the bulky ligands the chain termination barrier is smaller than the chain propagation barrier, but with the bulky ligands this relationship is reversed. From the structure of the catalyst, it is evident that the bulky aryl substituents partially block the axial coordination sites of the Ni center and it was proposed by Johnson et al.(48) that it is likely this steric feature which impedes the termination relative to the insertion process. Our "static" QM/MM calculations(10) confirmed this notion. It was found that the termination transition state has both axial coordination sites occupied whereas the insertion transition state has only the equatorial sites occupied. Therefore, with the bulky ligands the termination is much more sterically hindered than the insertion.

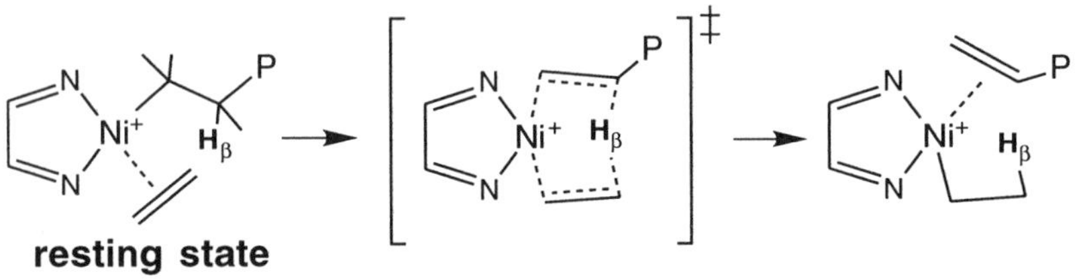

resting state

Figure 6. Proposed chain termination mechanism in the Brookhart olefin polymerization catalyst. The bulky Ar and R groups are now shown.

Chain Termination Simulation. The termination mechanism is depicted in Figure 6 whereby a β-hydrogen of the growing alkyl chain is transferred to the π-complexed monomer thereby forming a vinyl terminated polymer chain. We have performed two slow growth simulations of the process to estimate the free energy barrier of the termination - with and without the bulky ligands. The bulky R=Me and Ar=2,6-C_6H_3(i-Pr)$_2$, were treated by an augmented AMBER95(41) molecular mechanics potential whereas the Ni-diimine core including the growing chain and monomer were treated by a BP86(45,46,51) density functional potential. Full details of the simulation can be found elsewhere.(52)

For both the pure QM simulation of the system without the bulky ligands and the combined QM/MM simulation the QM calculation was performed on the 26 atom Ni diimine molecule - [(HN=C(H)-C(H)=NH)Ni-propyl + ethene]+. The link atoms which replace the dummy atoms are denoted with an asterisk in structure 1. A Nosé thermostat(35) was used to maintain an average temperature of 300 K. In the QM/MM simulation a separate thermostats were used for the QM and MM regions. Masses of the nuclei were rescaled to 50.0 amu for Ni, 2.0 amu for N and C, and 1.5 amu for H. The multiple time step procedure was not used in these simulations (it will be demonstrated in the next section). The so called "midplane" reaction coordinate(22) was utilized in both slow growth simulations. This reaction coordinate is defined as the ratio r/R where R is the length of the vector between C_2 and C_β (see Figure 6 for labeling) and r is the length of projection of the C_β-H_β bond vector onto the vector $\mathbf{R}$. This reaction coordinate constrains the transferring H atom to lie on the plane perpendicular to and passing through the endpoint of the vector $\mathbf{r}$. The midplane reaction coordinate has been previously demonstrated.(22)

Each of the two simulations presented encompassed 39000 time steps with Δt = 1.7 x 10^{-17} s. All calculations were performed on an IBM RS/6000 model 3CT workstation with 64 MB of memory. Each of the simulations required two weeks of dedicated CPU time.

Compared in Figure 7a is the free energy profile derived from the QM/MM AIMD simulation and the pure QM AIMD simulation without the bulky ligands. As suggested experimentally, there is a dramatic increase in the termination barrier when the bulky ligands are added. Moreover, the estimated free energy barrier of 14.8 kcal/mol at 300 K is in good agreement with the experimental value of ~16 kJ/mol at 272 K.(The free energy barrier of olefin insertion is experimentally estimated to be $\Delta G^\ddagger$ = 10-11 kcal/mol(53) The weight-average molecular weight, M_w, of 8.1x10^5 g/mol(48) provides an estimate for the ratio of termination events to insertion events of 1:28900. Applying Boltzmann statistics to this ratio gives a $\Delta\Delta G^\ddagger$ of 5.6 kcal/mol and an estimate for the termination barrier of 15.5-16.5 kcal/mol. In calculating $\Delta\Delta G^\ddagger$ we have assumed that every hydrogen transfer event leads to the loss of the chain and, consequently, chain termination. Plotted in Figure 7b are various structural quantities that characterize the transfer process. At the transition state, the transferred hydrogen atom is shared almost equally between the two carbon atoms. Displayed in Figure 8 is a snapshot structure of the slow growth simulation taken from the transition state region. The vertical line in Figure 7 denotes the point in the simulation where the snapshot structure in Figure 8 was extracted. Plotted in graph (c) is the relative molecular mechanics van der Waals energy for the interactions between the bulky aryl ligands and the active site propyl and ethene groups. This plot reveals that the steric interaction between the active site moieties and the bulky aryl ligands increases as the hydrogen transfer process occurs. This is a result of the expansion of the active site during the hydrogen transfer process.

The applicability of the combined QM/MM *ab initio* molecular dynamics approach to study extended systems has been demonstrated with our simulation of the chain termination process in the Brookhart polymerization catalyst. However, as is often the case, the static QM/MM calculations(10) provided similar results and insights into the chemistry at a significantly lower computational expense. The real power and applicability of the combined QM/MM AIMD method is exhibited in cases where conventional static simulations fail. One such application is to reactions which have no enthalpic barrier on the static zero K potential surface. The uptake of the monomer in the Brookhart catalyst system as shown in Figure 9 is one such case.(54)

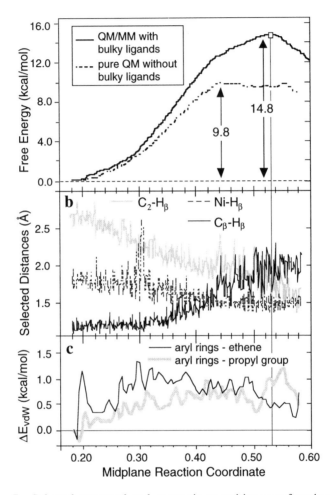

Figure 7. Selected structural and energetic quantities as a function of the reaction coordinate from the combined QM/MM simulation of the chain termination process. (a) and (b): Selected distances. (c): The molecular mechanics van der Waals interaction energies between the two bulky aryl groups and the active site ethene and propyl groups relative to the resting state value. The energies in (c) have been "smoothed" and are the running average over 500 time steps.

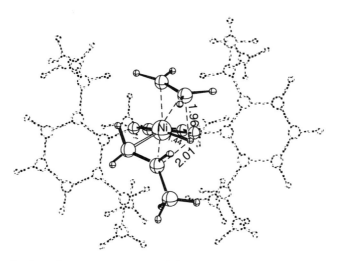

Figure 8. Snapshot structure extracted from the transition state region of the QM/MM AIMD simulation of the chain termination process in the Brookhart Ni-diimine catalyst. The snapshot structure is denoted by a vertical line in Figure 7. Bond distances are in Ångstroms.

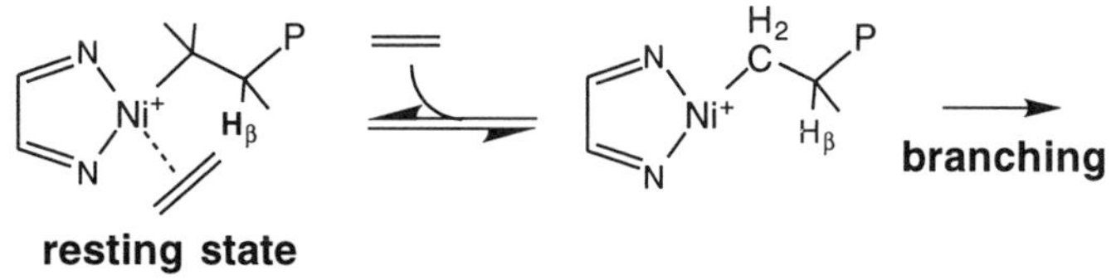

resting state   **branching**

Figure 9. Monomer uptake process with the Brookhart olefin polymerization catalyst. The bulky Ar and R groups which can be altered to tune the uptake barrier are not shown.

Monomer Uptake Simulation. Static QM/MM calculations at the zero K limit of the monomer uptake (Figure 9) show that there is no barrier. This contradicts experimental findings that suggest there is a significant barrier to the monomer capture.(48) The barrier height is in fact crucial to the polymerization chemistry because it is believed to control the level of chain branching that occurs. Furthermore, the barrier is found to depend on the ligand substitution pattern on the base Ni-diimine catalyst. For example, when the R=Me group in (ArN=C(R)-C(R)=NAr)Ni(II)-R'+, **1**, is replaced with R=H the capture free energy barrier decreases by 1.0 kcal/mol at 272 K. This is a peculiar result since the R substituent is far removed from active site and would not appear to be able to impede or promote the capture of the monomer.

In order to investigate the nature of the capture barrier at a finite temperature, we have performed a series of combined QM/MM AIMD simulations. We have calculated the free energy barrier for the uptake of ethylene into (ArN=C(R)-C(R)=NAr)Ni(II)-R'+ with (i) Ar=2,6-$C_6H_3(i$-Pr$)_2$, R=Me, (ii) Ar=2,6-$C_6H_3(i$-Pr$)_2$, R=H and (iii) Ar=H, R=H (pure QM simulation). The slow growth method was utilized where the distance between the Ni atom and the midpoint of the olefin carbons was constrained and slowly contracted during the dynamics. For each simulation the distance was varied from 4 Å to approximately 2.3 Å over 17000 time steps. The system was thermostated to maintain an average temperature 300 K. In these simulations, the multiple timestep method was used where the molecular mechanics region was over-sampled 20 times, such that the time steps in the QM and MM regions were 1.7 x 10^{-16} s and 8.5 x 10^{-18} s, respectively. Masses of the nuclei were set to 50.0 amu for Ni, 2.0 amu for N and C, and 1.5 amu for H in the QM region (including the link atom) while the masses of the C and H atoms in the MM region were scaled down 400 fold to 0.005 and 0.00375 amu, respectively.

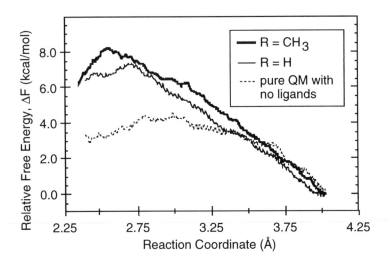

Figure 10. Free energy plots derived from a slow growth simulation of the monomer uptake with the Brookhart olefin polymerization catalyst.

The slow growth free energy profiles of the three complexes are displayed in Figure 10. The ordering of the calculated barriers follows the expected trends with the system containing no bulky ligands possessing a significantly smaller barrier than the substituted complexes. The difference in the calculated uptake free energy barrier for complex (i) with R=Me, Ar=2,6-$C_6H_3(i$-Pr$)_2$, and complex (ii) with R=H, Ar=2,6-$C_6H_3(i$-Pr$)_2$ is in remarkable agreement with the experimental value. The calculated difference is $\Delta\Delta F^{\ddagger} = 0.8$ kcal/mol at 300 K whereas the experimental difference is estimated to be $\Delta\Delta G^{\ddagger} = 1.0$ kcal/mol at 272 K. (Assuming that the uptake processes determines the amount of chain branching the relative free energy barriers of the uptake process for different complexes can be determined by comparing the level of branching observed in the polymers. For complex (i) with R=Me the level of branching has been determined to be 48 branches per 1000 carbons of the polymer at 0°C.(48) Under identical conditions, the level of branching for complex (ii) with R=H was found to be 7.0 branches per 1000 carbons. Assuming that the insertion rate for both systems is the same and applying Boltzmann statistics provides a rough estimate of $\Delta\Delta G^{\ddagger} = 1.0$ kcal/mol at 272 K.)

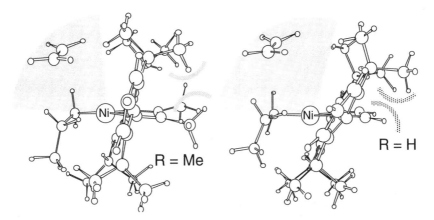

Figure 11. Representation of the active site pocket during the monomer uptake with R=Me and H. With R=Me the active site pocket is compressed compared to that with R=H because of the enhanced interaction of the R group with the i-Pr substituents on the aryl rings.

The explanation for why the substitution of R=Me with that of R=H decreases the capture barrier illustrated in Figure 11. Compared to the view in Figure 8, the view of the catalyst in Figure has been rotated 90° such that the viewer is looking down the plane of phenyl subsituents. When R=Me, the phenyl ring planes are forced to be roughly perpendicular to the Ni-diimine ring plane. This is because of the strong steric interaction between the R=Me group and the i-Pr substituents of the aryl rings. When R=Me group is replaced with R=H, the steric interaction with the i-Pr subsituents is diminished allowing the phenyl rings to relax from the perpendicular orientation relative to the Ni-diimine rings. This acts to open up the active site pocket allowing for easier uptake of the monomer. Plotted and defined in Figure 12 is the angle θ which is zero when the phenyl ring planes are perpendicular to the Ni-diimine ring plane and 90° when they are parallel to one another. The plot in Figure 12 shows that when R=Me the θ angle is confined to values between 20 and -20° whereas with R=H the variation in the θ angle is much more dramatic with values ranging from -40 to 40°. In the extreme, when all of the bulky ligands are removed the uptake of the monomer is the least impeded as evidenced by the significantly smaller uptake barrier shown in Figure 10.

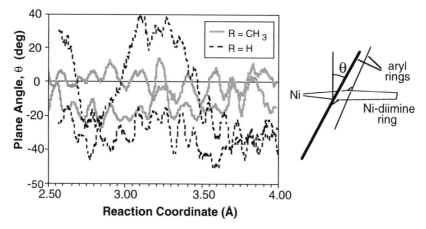

Figure 12. Plot of the θ angle during the slow growth simulation of the monomer uptake for R=Me and R=H. The definition of the angle θ is depicted to the right of the plot.

The results presented in Figure 11 are preliminary in nature and more calculations need to be performed to complete the study. It is clear from Figure 11 that at a reaction coordinate value of 4 Å the relative free energy is still dropping. Thus, in order to determine the absolute uptake barriers the simulations are currently being extended to beyond the RC value of 4 Å with a larger supercell size. Furthermore, because of the nature of the process the hysteresis may be large and therefore we are also currently performing the reverse simulations to reduce the sampling errors. Finally, the exact nature of the uptake barrier is currently being analyzed in more detail in order to provide insights into the tunability of the chain branching process, with the ultimate goal of catalyst design.

Conclusions

A new implementation to carry out Car-Parrinello *ab initio* molecular dynamics simulations of extended systems using a combined quantum mechanics and molecular mechanics potential is presented. Our implementation allows the QM/MM boundary to cross covalent bonds such that the potential surface of a single molecular system is described by a hybrid potential. Since the potential surface of the molecular mechanics region is usually much less computationally demanding to calculate than that in the QM region, we have implemented a multiple time step technique to over-sample the MM region relative to the QM region. The goal here is to provide better ensemble averaging in the MM region which is usually larger in size and therefore usually has a higher degree of configurational variability. To examine chemical reactions with the method we utilize the slow growth method to estimate reaction free energy barriers. In this method a reaction coordinate is constrained during the dynamics and slowly varied, thereby "pulling" the system over the reaction barrier from reactant to product. The "resistance" to the transformation provides an estimate of the free energy barrier via thermodynamic integration.

The applicability and efficiency of the combined QM/MM AIMD approach to study transition metal catalysis is demonstrated. In particular, we have examined the chain termination and monomer uptake processes in the Brookhart-type Ni(II) diimine olefin polymerization catalyst. In this simulation the QM/MM boundary crosses several covalent bonds, such that the Ni diimine core is treated by the DFT potential while the large bulky substituents are treated with the AMBER molecular mechanics force field. For the chain termination, the calculated free energy barrier of $\Delta F^{\ddagger} = 14.8$ kcal/mol at 25°C is in excellent agreement with the experimental result of $\Delta G^{\ddagger} \approx 16$ kcal/mol at 0°C. For the monomer uptake process, which has no enthalpic barrier on the static zero temperature potential surface, preliminary results also show good agreement with experiment. In this example, the method also provides insights into the nature of the barrier which may ultimately lead to catalyst design.

With the combined QM/MM *ab initio* molecular dynamics method, we are moving towards more realistic computational models of chemical processes. Currently finite temperature simulations of large aperiodic systems which are out of the reach of pure QM methods can be simulated with the method. In the future, with electrostatic coupling (which we are currently implementing), the combined QM/MM molecular dynamics method allows for solvent effects to be modeled. Thus, the combined QM/MM *ab initio* molecular dynamics method shows promise for building sophisticated computational models that take into account the large substituents, finite temperatures and the solvent while still remaining computationally tractable.

Acknowledgments. The authors acknowledge the invaluable contributions of Dr. Peter E. Blöchl of IBM Zürich. This work has been supported by the National Sciences and Engineering Research Council(NSERC) of Canada, as well as by the donors of the Petroleum Research Fund, administered by the American Chemical Society (ACS-PRF No. 31205-AC3) and by Novacor Research and Technology Corporation (NRTC) of Calgary. For the valuable discussions, the authors acknowledge Professor M. Brookhart. T.K.W. wishes to thank NSERC, the Alberta Heritage Scholarship Fund and the Izaak Walton Killam memorial foundation for financial support.

Literature Cited

(1) Tuckerman, M. E.; Parrinello, M. *J. Chem. Phys.* **1994**, *101*, 1316.
(2) Galli, G.; Parrinello, M. *Phys. Rev. Lett.* **1992**, *69*, 3547.
(3) Mauri, F.; Galli, G.; Car, R. *Phys. Rev. B* **1993**, *47*, 9973.
(4) Car, R.; Parrinello, M. *Phys. Rev. Lett.* **1985**, *55*, 2471.
(5) Aqvist, J.; Warshel, A. *Chem. Rev.* **1993**, *93*, 2523.
(6) Singh, U. C.; Kollman, P. A. *J. Comp. Chem.* **1986**, *7*, 718.
(7) Field, M. J.; Bash, P. A.; Karplus, M. *J. Comp. Chem.* **1990**, *11*, 700.
(8) Bash, P. A.; Field, M. J.; Davenport, R.; Ringe, D.; Petsko, G.; Karplus, M. *Biochemistry* **1991**, *30*, 5826.
(9) Tuñón, I.; Martins-Costa, M. T. C.; Millot, C.; Ruis-Lópes, M. F. *J. Chem. Phys* **1997**, *106*, 3633.
(10) Deng, L.; Woo, T. K.; Cavallo, L.; Margl, P. M.; Ziegler, T. *J. Am. Chem. Soc.* **1997**, *119*, 6177.
(11) Maseras, F.; Morokuma, K. *J. Comp. Chem.* **1995**, *16*, 1170.

(12) Gao, J. In *Reviews in Computational Chemistry*; K. B. Lipkowitz and D. B. Boyd, Ed.; VCH: New York, 1996; Vol. 7.

(13) Remler, D. K.; Madden, P. A. *Molecular Physics* **1990**, *70*, 921.

(14) Meier, R. J.; Dormaele, G. H. J. v.; Larlori, S.; Buda, F. *J. Am. Chem. Soc.* **1994**, *116*, 7274.

(15) Margl, P.; Schwarz, K.; Blöchl, P. E. *J. Chem. Phys.* **1995**, *103*, 683.

(16) Margl, P.; Schwarz, K.; Blöchl, P. E. *J. Am. Chem. Soc.* **1994**, *116*, 11177.

(17) Woo, T. K.; Margl, P. M.; Lohrenz, J. C. W.; Blöchl, P. E.; Ziegler, T. *J. Am. Chem. Soc.* **1996**, *118*, 13021.

(18) Margl, P.; Lohrenz, J. C. W.; Blöchl, P.; Ziegler, T. *J. Am. Chem. Soc.* **1996**, *118*, 4434.

(19) Margl, P.; Blöchl, P.; Ziegler, T. *J. Am. Chem. Soc.* **1995**, *117*, 12625.

(20) Curioni, A.; Sprik, M.; Andreoni, W.; Schiffer, H.; Hutter, J.; Parrinello, M. *J. Am. Chem. Soc.* **1997**, *119*, 7218.

(21) Milman, V.; Payne, M. C.; Heine, V.; Needs, R. J.; Lin, J. S.; Lee, M. H. *Phys. Rev. Lett.* **1993**, *70*, 2928.

(22) Woo, T. K.; Margl, P. M.; Blöchl, P. E.; Ziegler, T. *Organometallics* **1997**, *16*, 3454.

(23) Hartke, B.; Carter, E. A. *J. Chem. Phys.* **1992**, *97*, 6569.

(24) Hartke, B.; Carter, E. A. *Chem. Phys. Lett.* **1992**, *189*, 358.

(25) Blöchl, P. E. *Phys. Rev. B* **1994**, *50*, 17953.

(26) Blöchl, P. E.; Margl, P. M.; Schwarz, K., *ACS Symposium Series 629: Chemical Applications of Density-Functional Theory*; B. B. Laird; R. B. Ross and T. Ziegler Ed.; American Chemical Society: Washington, DC, 1996, pp 54.

(27) Blöchl, P. E. *J. Chem. Phys.* **1995**, *103*, 7422.

(28) Beveridge, D. L.; DiCapua, F. M. *Annu. Rev. Biophys. Chem.* **1989**, *18*, 431.

(29) Paci, E.; Ciccotti, G.; Ferrario, M.; Kapral, R. *Chem. Phys. Lett.* **1991**, *176*, 581.

(30) Carter, E. A.; Ciccotti, G.; Hynes, J. T.; Kapral, R. *Chem. Phys. Lett.* **1989**, *156*, 472.

(31) Fukui, K. *J. Phys. Chem.* **1970**, *74*, 4161.

(32) Fukui, K. *Acc. Chem. Res.* **1981**, *14*, 363.

(33) Straatsma, T. P.; Berendsen, H. J. C.; Postma, J. P. M. *J. Chem. Phys.* **1986**, *85*, 6720.

(34) De-Raedt, B.; Sprik, M.; Klein, M. L. *J. Chem. Phys.* **1984**, *80*, 5719.

(35) Nosé, S. *Mol. Phys.* **1984**, *52*, 255.

(36) Blöchl, P. E.; Parrinello, M. *Phys. Rev. B* **1992**, *45*, 9413.

(37) Ryckaert, J.-P.; Ciccotti, G.; Berendsen, H. J. C. *J. Comp. Phys.* **1977**, *23*, 327.

(38) Woo, T. K.; Cavallo, L.; Ziegler, T. unpublished work

(39) Baerends, E. J.; Ros, P. *Chem. Phys.* **1973**, *2*, 52.

(40) Baerends, E. J. PhD Thesis, Free University, Amsterdam, The Netherlands, 1975.

(41) Cornell, W. D.; Cieplak, P.; Bayly, C. I.; Gould, I. R.; Jr., K. M. M.; Ferguson, D. M.; Spellmeyer, D. C.; Fox, T.; Caldwell, J. W.; Kollman, P. A. *J. Am. Chem. Soc.* **1995**, *117*, 5179.

(42) Rappé, A. K.; Casewit, C. J.; Colwell, K. S.; III, W. A. G.; Skiff, W. M. *J. Am. Chem. Soc.* **1992**, *114*, 10024.

(43) Tuckerman, M. E.; Berne, B. J.; Martyna, G. J. *J. Chem. Phys.* **1992**, *97*, 1990.

(44) Gibson, D. A.; Carter, E. A. *J. Phys. Chem.* **1993**, *97*, 13429.

(45) Becke, A. *Phys. Rev. A* **1988**, *38*, 3098.

(46) Perdew, J. P. *Phys. Rev. B* **1986**, *34*, 7406.
(47) Perdew, J. P.; Zunger, A. *PRB* **1981**, *23*, 5048.
(48) Johnson, L. K.; Killian, C. M.; Brookhart, M. *J. Am. Chem. Soc.* **1995**, *117*, 6414.
(49) Johnson, L. K.; Mecking, S.; Brookhart, M. *J. Am. Chem. Soc.* **1996**, *118*, 267.
(50) Haggin, J. in *Chemical and Engineering News*, February 5 1996; pp. 6.
(51) Perdew, J. P. *Phys. Rev. B* **1986**, *33*, 8822.
(52) Woo, T. K.; Margl, P. M.; Blöchl, P. E.; Ziegler, T. *J. Phys. Chem.* **1997**, *101*, 7877.
(53) Brookhart, M., Department of Chemistry, University of North Carolina at Chapel Hill. A private communication.
(54) Woo, T. K.; Margl, P. M.; Deng, L.; Ziegler, T. in preparation.

Chapter 9

The Molecular Mechanics Valence Bond Method

Electronic Structure and Semiclassical Dynamics: Applications to Problems in Photochemistry

Michael J. Bearpark[1], Barry R. Smith[1], Fernando Bernardi[2], Massimo Olivucci[2], and Michael A. Robb[1]

[1]King's College, London WC2R 2LS, United Kingdom
[2]University of Bologna, Bologna I-40126, Italy

The Molecular Mechanics Valence Bond (MM-VB) method is a hybrid MM/QM method that uses a Heisenberg Hamiltonian (VB) for the QM part combined with force field methods. The method works with hybrid atoms that have both VB linkages to other atoms as well as classical MM interactions. The MM and QM parts are fully parametrized. The use of the VB method allows modelling of ground and excited state processes. In dynamics computations the nuclei are propagated using the gradient while the wavefunction propagation is treated via semi-classical equations which include the non-adiabatic coupling.

Applications to molecular structure and dynamics for photochemistry and photophysics will be discussed.

In recent years, there has been considerable interest in combining quantum mechanics with force field methods such as molecular mechanics (MM) in order to model large molecular systems. The purpose of this chapter is to describe our approach [1] which uses a parametrized Heisenberg Hamiltonian [2] to represent the quantum mechanical part in a valence bond (VB) space together with the MM2 [3] force field. Our objective has been to produce a modelling method, MM-VB, that reproduces the results of MC-SCF computations for ground and excited states, yet is fast enough that dynamics simulations are possible.

In a manner similar to MC-SCF methods, one identifies "active" orbitals (those involved in bond making and breaking etc.) and treats these orbitals using VB methods while the inactive atoms that form the "framework" are treated via molecular mechanics. In this way excited states can be treated on the same footing as ground states. The method is a "hybrid atom" approach. Each atom has MM links (the MM spring in Figure 1) to the surroundings as well as quantum mechanical (QM) links (the VB "orbital" in Figure 1). As we will presently show, the QM links are the exchange integrals of Heitler-London VB theory. The energy is obtained by the solution of the VB secular equation where the MM strain energy forms the "energy zero" (i.e. a constant diagonal shift). Clearly the parametrization of the VB component of the problem must change according to the "hybrization status" (sp^2-sp^3) and the MM out-of plane bending controls this.

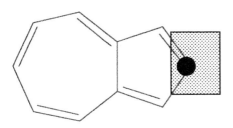 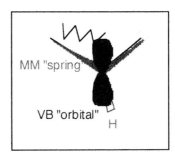

Figure 1 An MM-VB hybrid atom

For small numbers of active orbitals the VB part is very cheap. However, as the number of active orbitals becomes large, the eigenvector problem associated with the VB part of the computation dominates the computation time. The Clifford-Algebra Unitary Group Approach (CAUGA), yields an efficient implementation of MM-VB method for large scale VB problems.

THE VB PART OF MM-VB

The basis for the VB problem is the space of neutral covalent VB determinants $\{\Psi_K,...,\Psi_L\}$, in which each orbital occurs at most once with α or β spin. In a two electron problem there are only two such determinants $|1\bar{2}|$ $|\bar{1}2|$ For 4 orbitals and 4 electrons there are 6 covalent determinants.

1) $|3\ 4\bar{1}\ \bar{2}\ |$ 2) $|2\ 4\bar{1}\ \bar{3}\ |$ 3) $|2\ 3\bar{1}\ 4\ |$
4) $|1\ 4\bar{2}\ \bar{3}\ |$ 5) $|1\ 3\bar{2}\ 4\ |$ 6) $|1\ 2\bar{3}\ 4\ |$

The energy of a spin system can be determined by diagonalization of a hamiltonian [2] with matrix elements

$$H_{KL} = \delta_{KL}Q + \sum_{i,j}\Gamma_{ijji}^{KL}K_{ij} \tag{1}$$

The terms Γ_{ijji}^{KL} are the spin-free symbolic density matrix elements defined as

$$\Gamma_{ijji}^{KL} = -\langle\Psi_K|a_i^+a_j^+a_ja_i|\Psi_L\rangle \tag{2}$$

where $a_j^+a_j$ are creation annihilation operators. For example for 4 electrons the hamiltonian of equation 1 has the form.

	1	2	3	4	5	6
1	$Q - K_{12} - K_{34}$	K_{23}	$-K_{24}$	$-K_{13}$	$-K_{14}$	
2		$Q - K_{13} - K_{24}$	K_{34}	K_{12}		
3			$Q - K_{14} - K_{23}$			
4						
5						

(Matrix elements between pairs of determinants that differ by two spin interchanges i.e.. 1,6 or 2,5 or 3,4, are equal and not shown. Thus $H_{12} = H_{56}$, $H_{15} = H_{63}$ etc.)

In equation (2), the Q, K_{ij} are parameters designed to emulate the Coulomb and exchange integrals of Heitler London VB theory. These parameters are defined as

$$Q = Q_{MM} + \sum_{i,j} \left(\left[ii|jj \right] + \langle i|\hat{h}|i \rangle + \langle j|\hat{h}|j \rangle \right) \tag{3a}$$

$$K_{ij} = \left[ij|ij \right] + 2 s_{ij} \langle i|\hat{h}|j \rangle \tag{3b}$$

where $\left[ij|ij \right], \left[ii|jj \right]$ are the two-electron repulsion integrals, $\langle i|\hat{h}|i \rangle, \langle i|\hat{h}|j \rangle$ are the one-electron integrals (which will be dominated by the nuclear electron attraction term and are thus negative), Q_{MM} contains the effect of the MM frame and includes effects due to non bonded repulsions and steric effects, and s_{ij} are overlaps between the non-orthogonal atomic orbitals. Thus, from (3), it follows that the sign of K_{ij} depends mainly on the overlap between spatial orbitals: the exchange integral K_{ij} will have a positive value just if the overlap tends to zero.

This VB method is often referred to as a Heisenberg hamiltonian. The Hamiltonian given in equation 1-2 can be written as

$$H = Q - \sum_{i,j}^{N} K_{ij} \left\langle i(1)j(2) \middle| \hat{S}(1) \cdot \hat{S}(2) + \frac{1}{4}\hat{I}(1,2) \middle| i(1)j(2) \right\rangle a_i^+ a_j^+ a_j a_i \tag{4a}$$

$$= Q + \sum_{i=1}^{N} \sum_{j \neq i}^{N} K_{ij} \frac{1}{2} \left(\begin{array}{c} \hat{E}_{ij}^{\alpha\alpha} \hat{E}_{ji}^{\beta\beta} + \hat{E}_{ij}^{\beta\beta} \hat{E}_{ji}^{\alpha\alpha} \\ + \left[\hat{E}_{ij}^{\alpha\alpha} \hat{E}_{ji}^{\alpha\alpha} - \hat{E}_{ii}^{\alpha\alpha} \right] \\ + \left[\hat{E}_{ij}^{\beta\beta} \hat{E}_{ji}^{\beta\beta} - \hat{E}_{ii}^{\beta\beta} \right] \end{array} \right) \tag{4b}$$

where $\hat{E}_{ij}^{\beta\beta}$ is defined as

$$\hat{E}_{ij}^{\beta\beta} = a_{i\beta}^+ a_{j\beta}$$

and the expectation value of the term in equation 4b is just Γ_{iiji}^{KL}. Thus equation 4a is equivalent to the operator

$$\hat{H} = Q - \sum_{\mu < \eta}^{n} K_{\mu\eta}\left(2\hat{S}(\mu)\cdot\hat{S}(\eta) + \frac{1}{2}\hat{I}(\mu,\eta)\right) \tag{5}$$

acting in the spin parts of the VB determinants. Thus the VB hamiltonian is spin operator and via a comparison of equations 5 and 4a we see that we have a formal identification of electron spins $\mu\nu$ and orbitals i and j. Thus the notion of an orbital "spring" in MM-VB is associated with the coupling of electron spin (viz. equation 4a) and the strength of this coupling K_{ij}. For this reason the VB method provides an attractive formalism for the development of a hybrid approach. There are simply two types of spring on each atom: an MM one and a QM (i.e. VB) one.

The Toulouse school [2] were the first to suggest that the parameters Q, K_{ij} can be determined rigorously from effective hamiltonian theory. A detailed examination of such an approach can be found in reference 4 and there have been two successful implementations [1,5]. Since such effective hamiltonians can be constructed to reproduce the results of a CAS-SCF computation *exactly* it is clear that effects as charge-transfer or super-exchange are not neglected. Indeed modern VB methods [6] are completely equivalent to this approach if implemented without approximation. While the space of VB determinants constructed by allowing only one orbital per electron cannot represent an ionic term, this problem can be rectified (retaining the constraint of singly occupied orbitals) by the addition of additional orbitals (e.g.. 2p and 3p like orbitals for a C atom).

There are thus two problems with the implementation of such a method: parametrization of the K_{ij} from *ab initio* computation, and implentation if the VB part for large active spaces. We now give some discussion of the general strategy referring the reader to original papers for the details.

Clearly one needs a functional form for the representation of the *numerical* values of Q and K_{ij} for a given interatomic distance r_{ij}. The functional form of the representation of Q_{ij} and K_{ij} must depend on the orientation and on the hybridisation of the active orbitals (sp^n hybrids). Further the parameter Q contains the energy, Q_{MM}, of the *framework* of the molecule. Thus orientation and the hybridisation changes for an sp^3-sp^2 carbon atom must be incorporated into the functional form used to fit the data. We have used a functional of the form

$$\begin{array}{c} Q_{ij} \\ K_{ij} \end{array} = \sum_{k}^{\substack{is,ipx,ipy,ipz \\ jpx,jpy,jpz}} C_k\left\{a_k \exp\left(-b_k r_{ij}\right)\right\} \tag{7}$$

where the coefficients C_k describe orientation and the hybridisation changes. As shown in figure 2. We begin by defining a vector p_i as shown in figure 2a.. The direction of the vector p_i is taken orthogonal to the plane α containing three substituents (vectors a,b and c, assumed to be of unit length) connected to the valence centre. The coefficient C_{is} is the norm of vector p_i and is proportional to the distance q between the valence atom and the plane α. The quantities C_{ipx}, C_{py}, C_{ipz}, and C_{jpx}, C_{jpy}, C_{jpz} are the projections of unit vectors directed along p_i and p_j on two parallel local co-ordinate systems defined in figure 2b.

Since equation 6 contains two parameters (a_k and b_k), a maximum of 30 different parameters are needed for any two-center system. The coupling of the VB parameters to the MM part is thus accomplished in a simple geometric fashion.

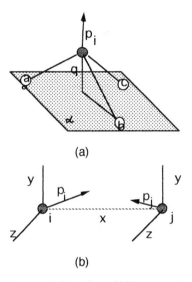

(a)

(b)

Figure 2 (adapted from reference 1) Coupling of VB parameters to the framework of the molecule.

For K_{ij} we have used 4 exponentials (ie n=4 and 8 parameters) while for Q_{ij} 11 exponentials are required (n=11 and 22 parameters). Parametrising that part of Q that contains the energy of the *framework* , Q_{MM} of the molecule would be difficult task in general. For this part of Q we shall use the standard MM potential for terms that contain the atoms that do not involve active orbitals and we shall use a modified MM potential for terms (i.e. stretching, bending and torsion) that contain atoms that have one or more active orbitals.

There are two possible strategies for the parametrization. One could parametrize the hamiltonian for the target system so that the approach becomes a sophistcated way of fitting the potential surface (i.e. one fits the hamiltonian matrix elements rather than the total energy). This approach has recently been used in a study of the dynamics of butadiene photoisomerization[7].Alternatively, one could attempt to find a universal parametrization for all sp^2-sp^2 carbon atoms. We have followed the later strategy and it has reproduced the results of CASSCF on excited state problems quite well and work is now in progress for heteroatoms and transition metals. Our parametrization was based on two electron systems and then a scaling algorithm was developed from effective hamiltonian computations on much larger systems.

In figure 3 we show the optimized geometries of two surface (S₁/S₀) crossings obtained with MM-VB for styrene together with the previous CAS-SCF results [8]. Given the strange geometries with many partly broken π bonds any pyramidalized carbon atoms the agreement is remarkable.

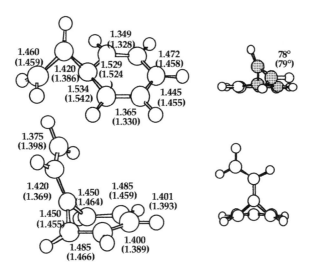

Figure 3. Two surface (S₁/S₀) crossings obtained with MM-VB for styrene together with the previous CAS-SCF results[8] in brackets.

We now focus our attention briefly on the VB problem. The Heisenberg hamiltonian is defined on a complete set of states formed by all configurations where each orbital is always singly occupied. There are several possible choices for choosing the spin coupling. The natural choice for VB wavefunctions is the Rumer functions; however, these are non-orthogonal and would destroy the simplicity of the method. The possible orthogonal basis sets are the standard Yamanochi-Kotani basis, simple determinants or the Clifford-Algebra spinors described in detail by Paldus [9] but first used in quantum chemistry by Handy and co-workers [10]. In spite of the fact that the determinantal or the Clifford-Algebra spinor basis is much larger than the standard Yamanochi-Kotani basis, the Clifford-Algebra spinor basis has two important advantages: a)the matrix elements Γ_{ijji}^{KL} (eq. 1) are 1 or -1 and thus the computation of the numerical values of the matrix elements is trivial b) the Clifford-Algebra spinor basis for states with $S_z=0$ where all the orbitals are singly occupied permits a very simple "two slope" graphical representation [1b]. We refer the reader to our original paper on this subject where full details can be found. The important point is that we can deal with up to 24 active (VB) orbitals where conventional CASSCF methods have a practical limit of about 10-12. Thus simulations can be performed on quite large systems. Shown below is the unusual optimized geometry of 18 annulene surface crossing. The CASSCF space for this problem has 449,141,836 configurations while the VB space only needs 97,240 determinants so that geometry optimizations and dynamics are feasible.

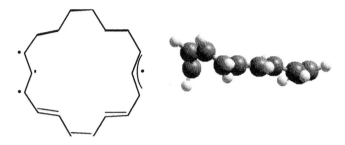

Figure 4 Geometry of [18] annulene S_0/S_1 optimized with 18 active orbitals (adapted from reference [11])

SURFACE HOPPING DYNAMICS

Our target problem for MM-VB has been surface hopping dynamics [12] (semi-classical trajectories) that traverse two or more potential energy surfaces. The theoretical background for the study of photochemistry problems has been recently reviewed [13]. A "model" with two potential energy surfaces, shown in figure 5 is typical of that encountered in many photochemical problems. The excited state surface has a minimum and transition state on the reaction path to the photochemical funnel (a conical intersection).

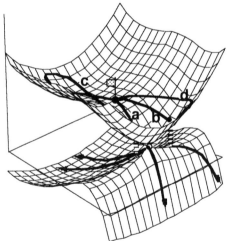

Figure 5 Examples of trajectories on a "model" potential energy surfaces for photochemistry.

Our objective has been to study the effect of various types of dynamic motion such as

a) hitting the conical intersection and emerging on the "opposite" side
b) the quantum effect of hopping in a region where the energy gap not zero
c) exploring the energy minima region and hopping on the reactant side of the surface
d) moving around the conical intersection without hitting it and then turns back

In a semi-classical trajectory one must propagate the time dependent wavefunction (equation 7)

$$\Psi_i (r,t) = \sum_\alpha a_{i\alpha} \, \phi_\alpha (r,t) \tag{7}$$

via a solution of the time dependent Schroedinger equation from equation 8

$$\sum_\alpha a_{i\alpha} \left\{ \left\langle \phi_\beta H \phi_\alpha \right\rangle - \left\langle \phi_\beta \frac{\partial}{\partial t} \phi_\alpha \right\rangle \right\} = \dot{a}_{i\beta} \tag{8}$$

where the coupling term involves the velocity $\dot{Q}$ of the classical motion of the nuclei given in equation 9

$$\left\langle \phi_\alpha \frac{\partial}{\partial t} \phi_\beta \right\rangle = -i\dot{Q} \cdot \left\langle \phi_\alpha \frac{\partial}{\partial Q} \phi_\beta \right\rangle \tag{9}$$

In contrast the nuclei are propagated using the classical equations (equation. 10) where the energy gradient is computed from the adiabatic wavefunction

$$m\ddot{Q} = -\frac{\partial E_\alpha}{\partial Q} \tag{10}$$

The surface-hopping algorithm of Tully and Preston [12] is used to allow excited state trajectories to transfer to the ground state in the conical intersection region, by monitoring the magnitude of the coefficients a in equation 8. The difference in energy between S_0 and S_1 at the hop is then redistributed along the component of the momentum parallel to the nonadiabatic coupling vector, as described by Truhlar [12c]. The MM-VB energy and gradient are used to solve the equations of motion directly [13,14] without fitting the potential energy surface and without numerical integration. Full details of our implementation have been given [15]. Thus the trajectories are propagated using a series of local quadratic approximations to the MMVB potential energy surface, as suggested by Helgaker et al [13]. The stepsize is determined by a trust radius [14]. Initial conditions (i.e. the initial velocities) are determined by random sampling of each excited state normal mode within an energy threshold , starting at the Franck-Condon geometry on S1. This generates an ensemble of trajectories - a "classical wavepacket". While one could not ordinarily contemplate semi-classical trajectories because of the necessity of evaluating the wavefunction and gradient "on the fly", this approach becomes feasible with MM-VB because the wavefunction computation is very fast indeed. Thus MM-VB yields truly hybrid dynamics with the gradient and surface hop criteria being determined quantum mechanically and the trajectory being propagated classically with the "difficult" part corresponding to the environment being relegated to the MM part.

A brief discussion of our recent results on azulene[16] gives an indication of the type of study that is possible. Femtosecond laser studies and spectroscopic linewidth measurements have now established that radiationless decay from S_1 to the ground state takes place in less than a picosecond. Accordingly, the objective of our computations was to rationalize this fact. While CASSCF computations can be used to document the critical points on the surface to provide a calibration, CASSCF dynamics with 10 active orbitals is too expensive. In reference 16 we show how such ultrafast S_1 decay can be explained by relaxation through an unavoided S_1/S_0 crossing (i.e. a conical intersection). The S_1 relaxation dynamics of azulene from its Franck-Condon structure are then modelled semiclassically with MM-VB which reproduces the structure of the ab initio CASSCF potential energy surface quite well. Two representative (types a and b in figure 5) trajectories are shown in figure 6. The relaxation from the Franck-Condon geometry involves mainly the change in the trans-annular bondlength from a single bond (ca. 1.5Å) to a double bond (ca. 1.35Å) where S_1/S_0 become degenerate. On the left hand side of figure 6 we see a trajectory that decays (almost completely diabatically with zero gap) within the first one-half oscillation from the Franck-Condon geometry. On the right-hand-side of figure 6, we see a trajectory run with ca 4 kcal mol^{-1} excess energy. In this case, the system completes an oscillation on the excited state surface before undergoing a transition to the ground state with a finite gap. Notice that in both cases, the vibration simply continues on the ground state with a weakened single bond potential. The results of the simulation suggests that the S_1->S_0 decay takes place in femtoseconds on the timescale of at most a few vibrations.

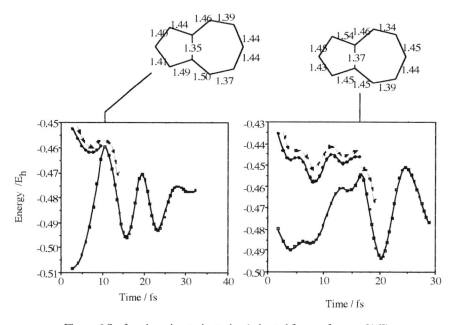

Figure 6 Surface hopping trajectories (adapted from reference [16])

CONCLUSIONS

In applications work MM-VB has been shown to yield correct surface topology for photochemistry problems [1a, 15-21]. This has tempted us to look at complicated systems such as stilbene isomerization [21] where it is impossible to explore the whole potential energy surface accurately with CASSCF.

We believe that the "hybrid atom" model combined with VB can provide the route to a useful hybrid methodology and our prototypical computations for C atom systems would appear to substantiate this claim. However, while one avoids the problem of joining MM and QM in this simple approach, one has the difficult problem of parametrization of the VB part. Dealing with ionic states (double active space), inclusion of medium effects do not seem to be insurmountable problems because these effects are easily included with the VB model and it's parameters.

REFERENCES

[1] a) Bernardi,F.; Olivucci M.; and Robb, M. A.; J. Am. Chem. Soc. **1992**, *114* 1606, b)Bearpark, M. J.; Bernardi, F.; Olivucci, M.; Robb, M. A.; Chem. Phys. Lett. 1994, 217, 513-519.

[2]Said, M.; Maynau, D.; Malrieu, J.P; Bach, MAG; *J. Amer. Chem.Soc* **1984**, *106* , 571, Maynau, D.;Durand, Ph., Duadey,J.P.; Malrieu,J.P. *Phys.Rev.* A **1983**,*28*, 3193, Durand, Ph., Malrieu,J.P. *Adv. Chem. Phys.* **1987**,*68*, 931

[3]) Boyd. D.; Lipkowitz,K *J. Chem Ed.* **1982**,*59*,269, Allinger, N.L. *Adv.Phys.Org. Chem.* **1976**, *13*,1 Allinger, N.L., Yuh, Y QCPE, **1980**, *12*, 395, Burkert U.; Allinger, N.L., *Molecular Mechanics* ACS Monograph 177; Amer. Chem.Soc. 1982, Clark,T *Handbook of Computational Chemistry;* Wiley, New York, 1985

[4] Bernardi, F. ; Olivucci, M. ; McDouall, JJW. ; Robb,M.A. *J.Chem. Phys.* **1988**, *89* ,6365

[5] Treboux G, Maynau D, Malrieu J-P J. Phys. Chem. **1995**, 99.6417

[6] Cooper, D.L.; Gerratt, J.; and Raimondi, M; *Adv. Chem. Phys. 1987*, 67, 319

[7]Ito, M and Ohmine, I, J. Chem. Phys. **1997** *106* 3159

[8]Bearpark, M. J.; Olivucci, M.; Wilsey, S.; Bernardi, F.; Robb, M. A.; *J. Am. Chem. Soc.* **1995**, 117, 6944-6953

[9] Paldus, J.; Sarma, C.R. J. Chem. Phys. **1985** *83* 5135

[10]Knowles, P. J. , Handy, N. C. Chem. Phys. Lett. **1984** *111* 315

[11] Bearpark,MJ; Bernardi, F. ; Clifford, S. ;Olivucci, M. ; Robb, M. A.; Vreven T. Mol. Phys **1996** *89* 37

[12] a) Preston, R. K.; Tully, J. C.; J. Chem. Phys. 1971, 54, 4297. b)Preston, R. K.; Tully, J. C.; J. Chem. Phys. 1971, 55, 562.c)Blais, N. C.; Truhlar, D. G.; Mead, C. A.; J. Chem. Phys. 1988, 89, 6204.

[13] a)Bernardi, F.; Olivucci, M.; Robb, M. A. *Chem. Soc. Rev.* **1996**, *25*, 321 b)Klessinger, M. *Angew. Chem. Int. Ed. Engl.* **1995**, *34*, 549-551.

[13] Helgaker, T.; Uggerud, E.; Jensen, H. J. Aa.; Chem. Phys. Lett. 1990, 173, 145-150.

[14] Chen, W.; Hase, W. L.; Schlegel, H. B.; Chem. Phys. Lett. 1994, 228, 436-442.

158

[15]Bearpark,M. J.; Robb M. A.;,Smith B.R.; Bernardi F.; Olivucci M. Chem. Phys. Lett. 1995, 242, 27-32.

[16]Bearpark,M. J.; Bernardi, F. ; Clifford, S.; Olivucci, M. ; Robb, M.A.;Smith B. R. J. Amer. Chem. Soc. **1996** *118*, 169

[17 Garavelli, M. ; Celani, P.; Fato,M.; Olivucci M.;, and Robb M.A. J.Phys. Chem. A **1997** 101, 2023

[18] Bernardi,F. Olivucci, M.; Robb,M. A.; Smith B. R.;J. Amer. Chem. Soc **1996** 118 5254

[19]Clifford, S. ; Bearpark, M. J. ; Bernardi,F.;Olivucci, M.; Robb, M. A.; Smith B. R. ;J. Amer. Chem. Soc.**1996** 118, 7353-7360

[20] Bearpark, M. J.; Bernardi, F.; Olivucci M; Robb, M A; Intern. J. Quantum Chem. **1996** *69* 505

[21]Bearpark, M. J.; Bernardi, F.; Olivucci M; Robb, M A; Clifford S.; and Vreven, T. ; J. Phys. Chem. A *101*, 3841-3847 **1997**)

Chapter 10

Density Functional Theory Ab Initio Molecular Dynamics and Combined Density Functional Theory and Molecular Dynamics Simulations

Dongqing Wei[1] and Dennis R. Salahub[1,2]

[1]Centre de Recherche en Calcul Appliqué, 5160 Boulevard Décarie, Bureau 400, Montréal, Québec, H3X 2H9, Canada
[2]Département de Chimie, Université de Montréal, C.P. 6128, Succursale Centre-ville, Montréal, Québec H3C 3J7, Canada

Theoretical studies of chemical reactions in an active site of a biological system, or other complex media such as solutions and interfaces involve two major aspects of theoretical chemistry. Firstly, we are interested in developing accurate tools to learn about the detailed dynamics of chemical reactions. One approach is the quantum dynamics calculation based on a given potential energy surface[1]. Another is *ab initio* molecular dynamics(AIMD) simulation[2, 3, 4, 5, 6] for the classical motion of nuclei, where the potential energy function in the classical dynamics is updated at each MD time step. Quantum effects for light atoms, such as, hydrogen can be included using a path integral method[7]. Secondly, proper treatment of the influence of solvent, enzyme and surface is very important because the reaction in a complex environment can be very different from that in the gas phase. It is unrealistic to treat the environment by high level quantum mechanical theory which is certainly needed for the dynamics of the reaction. Therefore various combined techniques have been developed[8, 9, 10, 11] to allow high quality calculations in the active centers and to model the environment by a fast and available computational method, such as, semiempirical quantum mechanics or empirical force fields. In its current implementation, one either studies thermodynamics and equilibrium structure of various model systems using one of the combined techniques, or studies reaction dynamics in the isolated state. Eventually, we should be able to carry out a full dynamical simulation involving both the active centers and the environment.

In this chapter, we report some of our recent results of theoretical calculations based on the Gaussian implementation of Kohn-Sham density functional theory[12,

13, 14]. Much of our efforts has been devoted to AIMD simulations[2, 3, 4, 5, 6], which were carried out by simply integrating the classical motion of the nuclei on the Born-Oppenheimer surface[3]. At each MD step, the forces on the nuclei are given by the quantum energy gradients and the molecular orbitals are updated by solving a Schrödinger type equation, the Kohn-Sham equation of DFT, in the Born-Oppenheimer approximation. This approach is often referred to as Born-Oppenheimer MD(BOMD)[4, 5]. The AIMD simulation was used to study reaction dynamics in the hypervelocity collision of NH_4^+ and NO_3^-[15], the proton transfer process in malonaldehyde[16], the dynamics of sodium clusters[17] and the conformational dynamics of alanine dipeptide analog[18].

It has become apparent now that the quantum chemical techniques based on DFT have the merit of being computationally efficient and accurate to deal with chemical reaction problems where charge and configuration re-distribution, polarization, correlation and exchange have to be treated adequately[12, 13, 14]. In parallel with our AIMD studies, we have chosen to combine quantum DFT methodologies with classical MD simulation techniques. Our first calculation studied the polarization effects of classical discrete solvents on a quantum water molecule[9]. This QM/MM technique was also applied to calculate the proton transfer barrier in aqueous solution. We also briefly mention progress on frozen density functional calculations.

AIMD simulations

In the Born-Oppenheimer AIMD simulation[3], the motion of nuclei is described by classical dynamics. In terms of Newton's equation, one has

$$m_j \ddot{r}_j = -\frac{\partial E}{\partial r_j} \tag{1}$$

where m_j are the atomic masses and E is the potential energy. The potential energy and its gradients are obtained by carrying out all-electron *ab initio* density functional calculations. In the current study, the Kohn-Sham density functional theory is implemented using the Linear Combination of Gaussian-Type Orbitals(LCGTO) approach[14]. Gradient-corrected potentials for electron correlation and exchange[19, 20] have been used in our calculations. The Verlet integrator[21] was used to propagate the equation of motion(Eq. 1). The time step varies depending on the system studied, which is chosen considering factors, such as, atomic mass, initial velocity and the properties of interest. Constant energy simulations are performed unless otherwise stated[3].

Solid state explosions

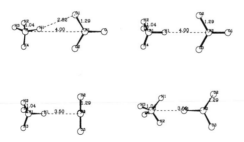

Fig. 1. Schematic diagram of initial geometries for the NH_4^+ - NO_3^- collision, where the smaller ball represents hydrogen, nitrogen is located in the center, oxygen is represented by a larger ball bonded to the central nitrogen; the distances are in $\overset{\circ}{A}$. Geometries (a), (b), (c) and (d) are placed from left to right and followed by second row.

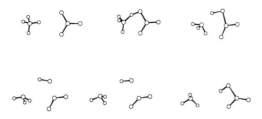

Fig. 2. Snapshot of the collision with the initial velocity 6.37 ks/s and the initial geometry is shown in 1(a), where the time sequence is: 0.0 fs, 24 fs, 39 fs, 51 fs, 63 fs, 75 fs

As an example, our AIMD simulation was applied to study the hypervelocity collision dynamics of NH_4^+ and NO_3^-. Ultimately, such a simulation will help to understand the complex chemical reactions behind the shock wave front in a condensed phase explosion. The shock front separates the unreacted material by only a few lattice spacings from the shocked material which is flowing at several kilometers per second[22, 23, 24]. Indeed exactly what sequence of states the material experiences as it transforms from the initial state in front of the shock to the final high pressure, high temperature state well behind the shock has been termed one of the outstanding questions in shock physics[25]. Our studies were carried out including initial conditions chosen to mimic conditions expected to be present at a detonation front. The time step is taken to be 10 a.u, i.e, 0.242 fs which is chosen to accommodate the hypervelocity and light particles involved. A typical simulation is started with a nitrogen in NH_4^+ $3.5 - 4.0 A^0$ ($N - N$ distance) apart from the nitrogen in NO_3^-, the former with an initial velocity, say, 6.37 km/s, the latter with a zero initial velocity. A collision process takes 100-300 time steps to complete, i.e., 24-72 fs depending on initial velocity and starting geometry. The product and intermediate also depend on the initial geometry and velocity.

The starting geometries are displayed in Fig. 1. Results were obtained for a system with starting geometry (a) and initial velocity 6.37 km/s on the direction defined by the $N - N$ bond. The energy profiles of collision processes are shown in Fig. 3. Clearly the total energy is well conserved, which indicates that our choice of time step is adequate. As the two particles approach each other the potential energy becomes more negative such that the attractive interaction between these two particles increases. The collision starts at $\sim$ 18 fs, reaches maximum penetration at t $\sim$ 27 fs. The height of the peak is closely related to the initial velocity, i.e., the magnitude of the initial kinetic energy. Then new species are generated and the system starts to release the kinetic energy it absorbed during the collision process.

A snap shot of this collision is used to illustrate the details of the chemical reaction. As NH_4^+ moves towards NO_3^- it picks up some velocity towards the oxygen close to the hydrogen in the front, driven by the attraction force. As a result, the hydrogen first collides with the oxygen which is followed by a proton transfer from NH_4^+ to NO_3^- generating NH_3 and NHO_3 fragments. This is expected knowing that in the optimized geometry at zero temperature the proton is quite close to NO_3^-[26]. The proton carries sufficient kinetic energy to break the $N - O$ bond so that an OH radical is formed and tries to leave NO_2. At t $\sim$ 60 fs, three species are observed, NH_3, OH and NO_2. However the kinetic energy of OH is not large enough to escape from the attraction of NO_2. Eventually OH re-joins NO_2 to form HNO_3.

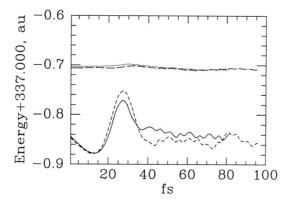

Fig. 3. The solid and dotted curves are the potential and total energy, respectively, for the initial geometry defined by Fig 1(a); The short and long dashed curves are the corresponding plots for the initial geometry defined by Fig 1(b).

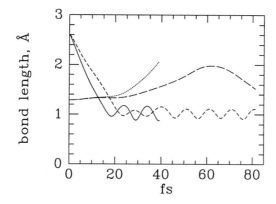

Fig. 4. Bond profile of $O - H$ and $N - O$; The solid and dotted curves are for the initial geometry defined by Fig. 1(a) and an initial velocity of 10 km/s, short and long dashed curves are for the same initial geometry but with an initial velocity of 6.37 km/s.

The reaction dynamics can also be viewed on a plot of the bond profile(Fig. 4). The short dashed line shows the change of the bond length between the reactive hydrogen and oxygen, and the long dashed line shows the change of the $N - O$ bond. The proton transfer occurs at t $\sim$ 22 fs, $\sim$ 5 fs after the collision starts. The $O - H$ bond undergoes vibration due to the kinetic energy carried by the proton. The $N - O$ profile clearly illustrates the break and re-formation of the $N - O$ bond. We would expect that a larger initial velocity could produce a free OH radical. Results for an initial velocity of 10 km/s are represented by the solid line and the dotted lines in Fig. 4. The proton transfer occurs at a much earlier time, $\sim$ 17 fs. OH is clearly breaking away from NO_2.

The initial geometry should have a significant effect on the final product and the collision process. If the simulation starts from the initial geometry (b) in Fig. 1 since the front hydrogen is pointing to the nitrogen of NO_3 along the direction of initial velocity both oxygens in the front have an equal probability to react. The hydrogen reaches the nitrogen first, it delays the proton transfer about 20 fs. Even starting from a geometry shown in (c) the proton transfer also occurs, but again is delayed due to the indirect collision between hydrogen and oxygen. If we start from geometry (d) no new species are generated, the collision is a non-reactive one.

Our result is consistent with the ideas of Dremin and Klimenko[23, 27], whose studies suggest that hypervelocity collisions between molecules in a condensed phase shock front in the undisturbed material can be sufficient to cause the formation of radicals and other active species immediately behind the shock front. Although our study is clearly of limited scope, only a few trajectories have been examined, it does demonstrate the possibility of using our AIMD techniques to obtain detailed mechanistic insight for this type of reaction.

Metal clusters

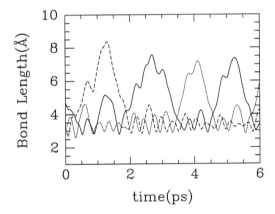

Fig. 5. Bond profile of Na_3 cluster, the solid, dotted, and dashed curves represent the three bonds of the cluster

The structure and dynamics of metal clusters have received much recent attention. We applied our DFT AIMD simulation to study the polarization at finite temperature, the dynamics and the vibrational spectrum of the sodium cluster Na_3, which has been studied by other groups using HF based methods[6]. The quantum calculations were done using a model core potential[28] along with a gradient-corrected functional due to Becke and Perdew[19, 20]. The time step is 200 a.u., i.e., 4.84 fs and the simulation is a constant temperature one at 300K. Some preliminary results have been obtained. We plot the bond profile in Fig. 5. The sodium undergoes a structural change from a triangular state to linear and then to another triangular one. At T=0, the ground state[28] is a triangle(planar) with different bond distances ranging from 6.0 to 8.5 bohr[28]. Our simulation shows that at 300K, the various conformations can be easily explored by the AIDM simulation. The linear configuration is reached at each maximum of the bond distance. Over about 4 picosecond(ps), we observed three maxima, and also every bond reached the maximum once, i.e., achieved a linear configuration. Our future goal is to calculate the polarizability, and optical properties at finite temperature and compare with experimental data. Currently more simulation data is accumulating to obtain better statistics.

164

Malonaldehyde

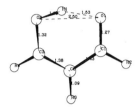

Fig. 6. Schematic diagram of malonaldehyde

Malonaldehyde has become a prototype system for studying intramolecular proton transfer[29, 30]. The most stable configuration at room temperature is found to be an intramolecularly hydrogen-bonded cis-enolform. Theoretical studies adopting two- and multi-dimensional approaches[1] have made much progress, especially, works on proton tunneling. AIMD simulation has the advantage of being able to sample phase space efficiently, and dynamical properties can be calculated without pre-constructed energy surfaces based on either theoretical calculations or experimental data[31]. A good quality AIMD simulation could give the exact proton transfer rate constant in the limit of the BO approximation(adiabatic) and the classical treatment of the proton, i.e., neglecting proton tunneling effect, which is a task nearly impossible to handle using the potential energy surface approach because of the number of degrees of freedom involved.

Calculations were carried out using the DZVP basis set[32] and the Becke-Perdew gradient-corrected functional[19, 20]. The time step is 20 a.u., about 0.484 fs. The ground state structure is shown in Fig. 6. Compared with experimental and other theoretical results, the $O_2 - H_1$ "co-valent" bond is too long(1.05 Å vs 0.97 Å(exp)), and the $O_1 - H_1$ hydrogen bond is too short(1.53 Å vs 1.68 Å(exp)). As a result, the $O_2 - H_1 - O_1$ angle is too large and the $O_1 - O_2$ distance is too small(2.5 Å vs 2.55 - 2.58 Å (exp)). This is consistent with our previous works on other hydrogen bonded systems, such as hydrated proton clusters[9]. This leads to a small proton transfer energy barrier which is estimated to be about 2.0 kcal/mol at 0 K. It is certainly too low compared with the $G2$[29] and Lap[33] calculations(about 4.0 Kcal). It would be ideal to use the Lap functional[34]. However, we have difficulty to maintain energy conservation during the simulation due to unsolved problems in obtaining energy gradients with sufficient accuracy for MD(In the current LAP implementation, a gradientless approximation is used for the correlation potential, $v_c = \partial E_c/\partial \rho$. While this yields sufficient accuracy for geometry optimizations, in MD, the small errors accumulate, leading to the energy fluactuations). Nevertheless, as a first trial of all-electron simulation using Gaussian-based DFT, the AIMD simulation is capable of demonstrating the usefulness of first principle's simulation.

Fig.7- 8 shows the bond profile at an average temperature of 290 K and of 637 K as a function of time. Clearly we can distinguish two regions, one with the proton undergoing consecutive transfer which is referred as proton "shuttling"[31] and each transfer takes about 0.1-0.2 ps to complete, another with the proton staying on one side for quite a long period without transferring. The period varies significantly with temperature. Let us call this period the "waiting period". At 290 K, the waiting period can be as long as 1 ps. At 637 K, it is usually about 0.5 ps. Another obvious feature is that there is a clear correlation between a large O-O oscillation and the occurrence of the proton

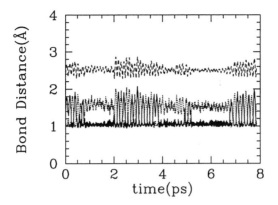

Fig. 7. Bond profile of malonaldehyde at 290 K, the solid, dotted, and dashed curves are for $O_2 - H_1$, $O_1 - H_1$ and $O_1 - O_2$, respectively

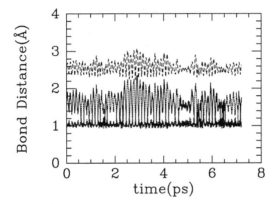

Fig. 8. Bond profile of malonaldehyde at 637 K, the solid, dotted, and dashed curves are for $O_2 - H_1$, $O_1 - H_1$ and $O_1 - O_2$, respectively

shuttling. We have plotted a few other profiles of relevant quantities, such as, angles, dihedral angles and other bond lengths. No other properties correlate as well as the O-O distance. This phenomena was not reported in previously published work[31]. It is apparent from Fig. 8 that the O-H hydrogen bond shows a much larger vibration amplitude than the O-H covalent bond as expected. One can calculate the proton transfer rate constant by taking a proper average. But it is apparent that a 7 ps trajectory is not sufficient; there are only up to three shuttling and 4 waiting periods. We are currently still accumulating data to be able to obtain reliable statistics.

Alanine dipeptide

The largest system we have so far treated with the AIMD is an alanine dipeptide analogue[35]. The calculation is a all-electron implementation with the DZVP basis set and Becke-Perdew gradient-corrected functional[19, 20]. The time step is 50 a.u., i.e., 1.12 fs, and the average temperature is about 300 K[18]. Smaller time steps(10 and 20 a.u.) were also tested but no significant difference is observed.

As a model system for simulations of peptides, alanine dipeptide has been studied extensively[35, 36, 37, 38, 39, 40, 41, 42]. High-level *ab initio* studies [35, 43] have shown that the internally hydrogen-bonded conformations, the cyclic hydrogen-bonded $C7_{eq}$ structure and the extended C5 structure, are of lowest energy. Those structures analogous to α and β helical conformation are of higher energy. One important question that needs to be answered concerns the time scales and forces activating and stabilizing secondary structure formation (i.e., formation of α- and β-sheet). AIMD simulation of this model system could give valuable insight into the conformational dynamics for peptides. But, the main dynamical transformation should occur between $C7_{eq}$ and C5, which are the most populated in phase space.

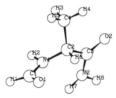

Fig. 9. Schematic diagram of alanine dipeptide analogue.

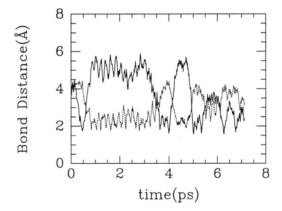

Fig. 10. Time evolution of the H-bond lengths. The solid and dashed lines are for the O_1-H_7 and O_2-H_2, respectively.

In Fig. 10, we plot the bond lengths as a function of time. The dotted and dashed lines correspond to the O-H bond lengths for the two internally hydrogen bonded conformers $C7_{eq}$ and C5, respectively. If the O-H bond length is about 2.0Å or less we define a hydrogen bond to be formed. By monitoring the forming and breaking of the two types of hydrogen bonds we can observe the conformational dynamics in the simulation. We clearly see the conversion from the $C7_{eq}$ to the C5 conformation. The frequency of transition is on the order of a few picoseconds(ps). The O-H bond length for the $C7_{eq}$ conformation reaches a lower value than the C5 conformation. The C5 hydrogen bond tends to have a longer duration. Other conformations(α-like and β-like) are also visited in this short simulation.

This time scale of a few ps is an interesting discovery. We have also carried out classical simulations using the CHARMM, MM3, AMBER and ESFF force fields[36, 44] We did not see any transition between $C5$ and $C7_{eq}$ conformation at 300K. There are two factors to consider, 1, the energy barrier in these force fields is higher than that of deMon, 2, there are parameters in the force fields that are not properly defined, for example, the dihedral angle rotational barrier may be too high, or kept constant. If we increase the temperature to, say, 375 K, a force field, such as, CHARMM, shows conformational transformation between $C5$ and $C7_{eq}$(with dominant population), and the time scale is about 20 ps. Such a conformational transformation is also found in a simulation using the GROMOS force field[45] at 300K.

Combined DFT and MD Simulation

Our main focus is to estimate how a classical solvent influences a quantum center. Solvation is chosen because it does not involve cutting a chemical bond in dividing quantum and classical regions[9]. The solvent, such as, water is modeled as sphere particles, with point charges and effective diameters. The quantum classical interaction is described by point charge-electron density and also Lennard-Jones terms. The complete solvent dynamics would require an instantaneous energy gradient(force) of the potential energy function containing quantum-classical solvent(involves one-electron integrals), solvent-solvent interaction. However, this "complete" dynamics does not lead to good liquid structure(i.e., pair correlation function) because some important factors of the quantum-classical interaction(D. Berard, D. Wei and D.R. Salahub unpublished, also see[46]) are missing. The main source of error may be the exclusion effect due to Pauli's principle which accounts for the orthogonality constraint of the wave function of the quantum water with that of the solvent and the exchange effect which is a consequence of the antisymmetrization of the wave function[9]. However, classical water models give very good liquid structure. Therefore, in our early study, we used classical solvent dynamics to achieve the ensemble average.

We first calculated the dipole moment of water molecules in liquid water. Both the Local Density Approximation(LDA) and Perdew[19] nonlocal density functional calculations were carried out. Two basis sets have been used in our quantum calculation, DZVP[32] which has the double-zeta quality with polarization functions on heavy atoms, and $TZVP+$[47]. The geometry of the quantum water is that of experiment and is fixed in all our calculations. We studied two water models, SPC[48] and TIP3P[49]. Table I summarizes the results for SPC water. The experimental dipole moment of water in the liquid phase is estimated to be 2.6 Debye(D)[50]. Our result is certainly in this range.

Table 1. Energy(a.u.) and Dipole Moment(Debye)

	Gas				Liquid			
	LDA		Perdew		LDA		Perdew	
	DZVP	$TZVP+$	DZVP	$TZVP+$	DZVP	$TZVP+$	DZVP	$TZVP+$
-E	75.8795	75.9018	76.5104	76.5344	75.9134	75.9328	76.5430	76.5644
D	2.28	1.88	2.17	1.81	2.71	2.54	2.62	2.47

We also calculated the proton transfer energy and free energy barrier in aqueous solution[9]. The barrier is found to be 3 kcal/mol higher than in the gas phase. Very large solvent fluctuations are observed which may have a significant influence on the reaction rate. A frozen density functional theory[10, 18] has been developed to include the "quantum" nature of solvent. We are currently trying a fluctuating central charge model to see if solvent dynamics can be better described.

Conclusions

We report the results of two techniques involving density functional theory that show promise for applications aimed at understanding reactions in complex media. Both the AIMD method and the CDFMD (and their eventual combination) are undergoing extensive testing and development that will be reported in the near future.

Acknowledgments

The financial support of the Natural Sciences and Engineering Research Council of Canada is gratefully acknowledged. We are grateful to Brett Dunlap, Carter White, Daniel Berard, Hong Guo, Mark Casida, Jingang Guan and Martin Leboeuf for their participation and contributions.

References

1. Makri N.; Miller W.H. J. Chem. Phys. **1989**, 91, 4026; Swell T.D.; Guo Y.; Thompson D.L. J. Chem. Phys. **1995**, 103, 8557, and reference therein.
2. Car R.; Parrinello M. Phys. Rev. Lett. **1985**, 50, 55.
3. Wei D.; Salahub D. R J. Chem. Phys. **1997**, 106, 6086.
4. Barnett R.; Landman U. Phys. Rev. **1993**, B48, 2081.
5. Jing X.; Troullier N.; Dean D.; Binggeli N.; Chelikowsky J. R.; Wu K.; Saad Y. Phys. Rev. **1994**, B50, 122234.
6. Gibson D.; Ionova I.; Carter E. Chem. Phys. Lett. **1995**, 240, 261.
7. Tuckerman M. E.; Ungar P. J.; von Rosenvinge T.; Klein M. L. J. Phys. Chem. **1996**, 100, 12878 and references therein.
8. Field M. J.; Bash P. A.; Karplus M. J. Comput. Chem. **1990**, 11, 700.
9. Wei D.; Salahub D. R. Chem. Phys. Lett. **1994**, 224, 291; J. Chem. Phys., **1994**, 101, 7633.
10. Wesolowski T. A.; Warshel A. J. Phys. Chem. **1993**, 97, 8050; Wesolowski T. A.; Weber J. Chem. Phys. Lett. **1996**, 248, 71.
11. Gao J.; Xia X. Science **1992**, 258, 631.
12. Kohn W.; Sham L. J. Phys. Rev. **1965**, 140A, 1133; Hohenberg P.; Kohn W. Phys. Rev. **1964**, 136B, 864.
13. Parr R. G.; Yang W. *Density functional theory of atoms and molecules*; Oxford University: Oxford, 1989.
14. (a) St-Amant A; Salahub D. R. Chem. Phys. Lett. **1990**, 169, 387; (b) St-Amant A. Ph.D. Thesis, Université de Montréal (1992); (c) **deMon-KS** version 1.2, Andzelm J.W.; Casida M.E.; Koester A.; Proynov E.; St-Amant A.; Salahub D.R.; Duarte H.; Godbout N.; Guan J.; Jamorski C.; M. Leboeuf, V. Malkin, O. Malkina, F. Sim, and A. Vela, deMon Software, University of Montreal, 1995.
15. Wei D.; Salahub D.R.; Dunlap B.; White C. to be submitted.
16. Wei D.; Salahub D.R. to be submitted.
17. Wei D.; Guan J.; Casida M.; Salahub D.R. to be submitted.
18. Berard D.; Wei D.; Salahub D.R. *Proceeding of Pacific Symposium on Biocomputing*, Jan 6-9, 1997.
19. Perdew J.P. Phys. Rev. **1986**, B33, 8800; **1986**, B34, 7406.
20. Becke A. D. Phys. Rev. **1988**, A38, 3098.
21. Verlet L. Phys. Rev. **1967**, 159, 98.
22. Holian B. L.; Hoover W. G.; Moran W.; Straub G. K. Phys. Rev., **1980**, A22, 2798.

23. Dremin A. N.; Klimenko V. Yu. Prog. Astronaut. Aeronaut. **1981**, 75, 253.
24. Robertson D. H.; Brenner D. W.; White C. T. Phys. Rev. Lett. **1991**, 67, 3132; White C. T.; Robertson D. H.; Brenner D.W. Physica A **1992**, 188, 357.
25. Belak J. *High–Pressure Science and Technology*, Schmidt S.C.; Shaner J.W.; Samara G.A.; Ross M. Eds; AIP Press: New York, 1994, P.1063.
26. Doyle R.J. Jr; Dunlap B.I. J. Phys. Chem. **1994**, 98, 8261.
27. Brenner D. W.; Robertson D. H.; Elert M. L.; White C. T. Phys. Rev. Lett. **1993**, 70, 2174.
28. Guan J.; Casida M. E.; Köster A.M.; Salahub D.R.; Phys. Rev. B **1995**, 52, 2184.
29. Barone V.; Adamo C. J. Chem. Phys. **1996**, 105, 11007.
30. Luth K.; Scheiner S. J. Phys. Chem. **1994**, 98, 3582; Wiberg K.B.; Ochterski J. Streitwieser A. J. Am. Chem. Soc. **1996**, 118, 8291; Sim S.; St-Amant A.; Salahub D.R. J. Am. Chem. Soc. **1992**, 114, 4391, reference therein.
31. Wolf K.; Mikenda W.; Nusterer E.; Schwartz K.; Ulbricht C. preprint.
32. Godbout N.; Salahub D.R.; Andzelm J.; Wimmer E. Can. J. Chem., **1992**, 70, 560.
33. Sirois S.; Proynov E.I.; Nguyen D.T.; Salahub D.R. J. Chem. Phys., submitted.
34. Proynov E.I; Vela A.; Ruiz E.; Salahub D.R. *New Methods in Quantum Theory*, NATO ASI series, Avery J.; Herschbach D. Tsipis C. Eds, Kluwer: Dordrecht, 1995.
35. Head-Gordon T.; Head-Gordon M.; Frisch M. J.; Brooks C. L. III; Pople J. A. J. Am. Chem. Soc. **1991**, 113, 5989.
36. Brooks C. L. III; Pettit M.; Karplus M. Proteins, A Theoretical Perspective of Dynamics, Structure, and Thermodynamics, in *Advances in Chemical Physics*; John Wiley and Sond: New York, 1988; Vol. 71.
37. Tobias D. J.; Brooks C. L. III, J. Phys. Chem. **1992**, 96, 3864.
38. Smith P.E.; Pettit M.; Karplus M. J. Phys. Chem. **1993**, 97, 6907.
39. Fraternali F.; Van Gunsteren W.F.; Biopolymers, **1994**, 34, 347.
40. Gould I. R.; Cornell W. D.; Hillier I. H. J. Am. Chem. Soc. **1994**, 116, 9250.
41. Loncharich R. J.; Brooks B. R.; Pastor R. W. Biopolymers, **1992**, 32, 523,
42. MacArthur M.W.; Thornton J. W. J. Mol. Biol. **1996**, 264, 1180.
43. Wei D.; Guo H.; Salahub D. R., to be submitted.
44. *User Guide*, Biosym/MSI, 9685 Scranton Road, San Diego, CA 92121-3752.
45. Rommel-Möhle K.; Hofmann H. J. J. Mol. Struct. **1993**, 285,211.
46. Wang J.; Boyd R.J. J. Chem. Phys., **1996**, 104, 7261.
47. Guan J.; Duffy P.; Carter J.T.; Chong D.P; Casida K.; Casida M.E.; Wrinn M. J. Chem. Phys. **1993**, 98, 4753.
48. Berendsen H.J.C; Postma J.P.M; von Gunsteren W.F.; Hermans J. in Intermolecular Forces, Pullman B. Eds; Reidel: Dordrecht, Holland, 1981, P. 331.
49. Jorgensen W.L. J. Am. Chem. Soc. **1981**, 103, 335.
50. Coulson C.A.; Eisenberg D. Proc. R. Soc. London, Ser A **1966**, 291, 445; Whalley W. Chem. Phys. Lett. **1978**, 53, 449.

SOLVATION

Chapter 11

Generalized Molecular Mechanics Including Quantum Electronic Structure Variation of Polar Solvents: An Overview

Hyung J. Kim[1], Badry D. Bursulaya[1], Jonggu Jeon[1], and Dominic A. Zichi[2]

[1]Department of Chemistry, Carnegie Mellon University, 4400 Fifth Avenue, Pittsburgh, PA 15213–2683
[2]NeXstar Pharmaceuticals, Inc., 2860 Wilderness Place, Boulder, CO 80301

A brief account is given of recent theoretical work on the computationally-efficient quantum mechanical description of the solvent electronic structure variation via a truncated adiabatic basis-set representation. By the inclusion of both linear and non-linear solvent electronic response, this goes beyond many existing classically polarizable solvent descriptions, widely used in simulation studies. Its implementation with Molecular Dynamics (MD) simulation techniques and application to liquid water are described.

Due to its essential role in condensed-phase processes, solvation has received extensive theoretical and experimental attention for quite some time. Among others, the computer simulation methods have been very instrumental in revealing molecular-level details of solvation (*1 –3*). In recent years, considerable theoretical efforts have been focused on the construction of electronically polarizable solvent descriptions (*4 –23*). With a few exceptions (*19 ,22 ,23*), classical electrostatics are employed for polarizability, so that the induced dipole moment for each molecule is proportional to the local electric field arising from its surroundings. While this accounts for the induction effects to the leading order in the electric field, the higher-order effects are not included. Another difficulty is the electronic transitions and accompanying relaxation. Even though ab initio MD couched in density functional theory provides significant improvement in the solvent electronic description (*24*), intensive CPU demand restricts its application to the simulations of small systems for a short period of time (*25 –30*). Also, its generic difficulty with the excited states does not allow for electronic transitions, relevant for various electronic spectroscopies. Therefore it seems worthwhile to develop a theoretical description that captures important electronic features in a quantum mechanical way and yet is computationally efficient to allow for extended simulation studies of macroscopic systems.

In this contribution, we give a brief account of our recent efforts to address the above-mentioned solvent electronic aspects in the context of MD simulations. By employing a truncated adiabatic basis-set (TAB) description, we incorporate both the linear and nonlinear solvent electronic response into our theory. This is applied to liquid water to study its solvation properties via MD. Here we give only enough description to make our accounts reasonably self-contained; for further details, the reader is referred to the original papers (*31* –*33*).

Theoretical Formulation

In the TAB formulation, the electronic wave function of each solvent molecule is represented as a linear combination of a few vacuum energy eigenstates

$$\psi^i_\alpha = \sum_\mu c^i_{\alpha\mu}\,\phi^i_\mu \; ; \qquad \hat{h}^i\,\phi^i_\mu = \epsilon^{(0)}_\mu\,\phi^i_\mu \; , \qquad (1)$$

where i labels the molecules, $\hat{h}^i$ is the vacuum electronic Hamiltonian for i, ϕ^i_μ's and $\epsilon^{(0)}_\mu$'s are its energy eigenfunctions and eigenvalues and α and μ denote the single-molecule (1M) electronic states in solution and in vacuum, respectively. Due to Coulombic interactions, $c^i_{\alpha\mu}$ vary with the solvent configurations. Thus the evolving electronic character of each constituent molecule is effected via a differing mixing of a few relevant electronic configurations of TAB. In this sense, the theory presented here is somewhat similar to the valence-bond description (*34* –*40*), widely used for various reaction processes in solution. A major difference is that ϕ^i_μ's are real electronic states in our theory. Thus equation 1 reasonably describes not only the ground but also low-lying excited states in solution. This allows for electronic transitions and thus spectroscopy in solution. By contrast, the valence-bond functions are designed to describe the lowest energy state, so that they often do not yield reliable information on the excited states. Compared with classically polarizable solvent descriptions, both the linear and nonlinear electronic polarizabilities are included in equation 1 through the solvation-dependent mixing of the basis functions.

Since the TAB representation is couched in the electrically-neutral 1M basis functions, the resulting excited states are of excitonic character. While this does not allow for charge transfer and electrical conduction, it will be of minor importance for bulk properties (*4*). More serious is the absence of electronic exchange that is responsible for the short-range repulsive interactions. This is due to the implicit treatment of the electrons through ϕ^i_μ's in the TAB. To remedy this, we incorporate the short-range repulsive interactions into our description via the Lennard-Jones (LJ) potentials (see equation 2 below).

In a point-dipole approximation, the effective electronic Hamiltonian $\hat{H}$ for the solvent system consisting of N molecules is given by

$$\hat{H} = \hat{H}_{\text{el}} + \sum_j\sum_{i(>j)} v^{ij}_{\text{LJ}} \; ; \quad \hat{H}_{\text{el}} = \sum_i \hat{h}^i + \sum_j\sum_{i(>j)} \hat{p}^i\cdot\boldsymbol{T}_{ij}\cdot\hat{p}^j \; ; \quad \boldsymbol{T}_{ij} = -\frac{3\hat{n}\hat{n} - \boldsymbol{I}}{|\vec{r}_i - \vec{r}_j|^3} \; , \quad (2)$$

where $\hat{p}^i$ is the electric dipole operator for i, $\vec{r}_i$ is its position vector, v^{ij}_{LJ} and $\boldsymbol{T}_{ij}$ are, respectively, the LJ interaction and dipole tensor between i and j, $\hat{n}$

is a unit vector in the direction of $\vec{r}_i - \vec{r}_j$ and $\boldsymbol{I}$ is a unit matrix. In the TAB representation consisting of M basis functions, $\hat{h}^i$ is a diagonal $M \times M$ matrix. By contrast, $\hat{p}^i$ is, in general, not diagonal; its off-diagonal elements correspond to the transition dipole moments between different ϕ_μ^i's. For clarity, we add a remark here: in contrast to repulsion arising from electron exchange, the dispersion can be, to a large degree, accounted for in the TAB description by including the intermolecular electronic correlation effects via, e.g., the MP2 method (31). However, in this initial study with low-level electronic structure calculations, we take into account the dispersion through the LJ potentials instead. For simplicity, we further assume that v_{LJ}^{ij} do not depend on the electronic states.

We determine the system electronic wave function and energy eigenvalue in the self-consistent field (SCF) approximation

$$|\Psi_{SC}> = \prod_{i=1}^{N} |\psi^i> ; \quad \hat{H}_{el} |\Psi_{SC}> = E_{SC} |\Psi_{SC}> ; \quad U = E_{SC} + \sum_{j}\sum_{i(>j)} v_{LJ}^{ij} , \tag{3}$$

where E_{SC}^G and U are, respectively, the ground-state SCF and total potential energies of the system. This involves solving a set of "Fock" equations

$$\hat{f}^i |\psi^i> = \left[\hat{h}^i + \hat{v}_{1M}^i\right] |\psi^i> = \epsilon^i |\psi^i> ; \quad \hat{v}_{1M}^i \equiv \sum_{j(\neq i)} <\psi^j |\hat{p}^j| \psi^j> \cdot \boldsymbol{T}_{ji} \cdot \hat{p}^i , \tag{4}$$

where $\hat{v}_{1M}^i$ is an effective 1M operator representing the electrostatic interaction between molecule i and the self-consistent field arising from all other molecules. The structure of equation 4 is very similar to that of the Hartree-Fock quantum chemistry method (41). Except for the absence of exchange terms in equation 4, these two theories are nearly isomorphic to each other (molecule $\leftrightarrow$ electron and TAB $\leftrightarrow$ electronic basis set). We arrange the 1M SCF solutions for each molecule in the order of increasing energy and denote their states and energies as $|\alpha^i>$ and ϵ_α^i ($\alpha = 0, 1, \ldots, M-1$). Then the ground electronic state $|\Psi_{SC}>$ for the entire solvent system in the SCF approximation results when every molecule is in its ground state $|0^i>$

$$|\Psi_{SC}> = \prod_{i=1}^{N} |0^i> ; \quad \hat{f}^i |\Psi_{SC}> = \epsilon_0^i |\Psi_{SC}> ; \tag{5}$$

$$\mathcal{E}^G = \sum_i \epsilon_0^i ; \quad E_{SC}^G = <\Psi_{SC}|\hat{H}_{el}|\Psi_{SC}> ,$$

where $\mathcal{E}^G$ is the sum of the 1M ground state energies ϵ_0^i's.

The ground-state dipole moment $\vec{\mu}^i$ and electronic polarizability tensor $\boldsymbol{\alpha}^i$ of molecule i in solution are given by (31)

$$\vec{\mu}^i = <0^i |\hat{p}^i| 0^i> ; \quad \boldsymbol{\pi}^i \equiv 2 \sum_{a(\neq 0)} \frac{\vec{p}_{a0}^i \vec{p}_{a0}^i}{\epsilon_a^i - \epsilon_0^i} ; \quad \vec{p}_{a0}^i = <0^i |\hat{p}^i| a^i> ;$$

$$\boldsymbol{\alpha}^i = \boldsymbol{\pi}^i + \sum_j \boldsymbol{\pi}^i \cdot \boldsymbol{T}_{ij} \cdot \boldsymbol{\pi}^j \cdot \boldsymbol{T}_{ji} \cdot \boldsymbol{\pi}^i - \sum_{j,k} \boldsymbol{\pi}^i \cdot \boldsymbol{T}_{ij} \cdot \boldsymbol{\pi}^j \cdot \boldsymbol{T}_{jk} \cdot \boldsymbol{\pi}^k \cdot \boldsymbol{T}_{ki} \cdot \boldsymbol{\pi}^i + \ldots , \tag{6}$$

where the terms involving T_{ij} arise from the interaction-induced effects; i.e., a change in $\vec{\mu}^i$ induces a dipole change for j through electrostatic interactions. Thus π^i is the polarizability of i that would result if the interaction-induced dependence of the electronic structure of the surrounding molecules is neglected. Even though it is a 1M quantity, π^i depends on the intermolecular interactions through ϵ_α^i and $\vec{p}_{a0}^i$. Thus, π^i generally differs from the gas-phase polarizability and so does α^i. Also, both π^i and α^i vary with i because different molecules are subject to differing local solvation environments. This is closely related to nonlinear hyperpolarizability, absent in many classically polarizable potential models.

The total electronic polarizability tensor Π of the solvent is given by

$$\Pi = \sum_i \pi^i - \sum_{i,j} \pi^i \cdot T_{ij} \cdot \pi^j + \sum_{i,j,k} \pi^i \cdot T_{ij} \cdot \pi^j \cdot T_{jk} \cdot \pi^k + \dots . \tag{7}$$

We note that an expression formally identical to equation 7 is widely used to describe the interaction-induced polarizability changes in liquid and related light scattering spectroscopy (42, 43). One major difference is that 1M polarizability is a fixed quantity in many existing approaches, whereas it fluctuates dynamically with the solvent configuration in our theory. As illustrated in Figure 4 below, this has an important consequence for nonlinear electronic spectroscopy.

We can also study the excited-state electronic properties by modifying the configuration interaction (CI) method of quantum chemistry. Analogous to gas-phase electronic structure theory (41), 1M excited states are not directly coupled to the SCF ground state [Brillouin's theorem] (31). Thus we can perform the CI-type calculations among the former (1MCI) independent of the latter. This will provide a size-consistent method for the excited states, which takes into account to a certain degree electronic relaxation associated with excitations. This involves a diagonalization of a CI matrix

$$< a^i |\hat{H}_{el}| b^j > = (\mathcal{E}^G + \epsilon_a^i - \epsilon_0^i)\,\delta_{ab}\,\delta_{ij} + \vec{p}_{a0}^i \cdot T_{ij} \cdot \vec{p}_{0b}^j\,(1-\delta_{ij}) \quad (a,b \neq 0) . \tag{8}$$

Water Electronic Structure

Ab Initio Calculations in Vacuum. We now consider the application of the theoretical formulation above to liquid water. In order to determine a suitable TAB representation for water, we first computed the ground and first five dipole-allowed singlet excited states of an isolated water molecule with the CASSCF method. We used the augmented cc-pVDZ basis of Dunning et al.; this is of double zeta quality, extended by the addition of the polarization and diffuse functions for each atom including hydrogen (44, 45). The effects of the Rydberg states were included via the Rydberg basis of Dunning and Hay for oxygen (46). The resulting basis (aug-cc-pVDZ + DH-Rydberg) yields 53 AO's for water. The CASSCF calculations were carried out with the GAMESS program (47) with 1 core and 12 active MO's (i.e., 5 a_1, 3 b_1, 3 b_2, and 1 a_2 orbitals) for each state. The SPC geometry of water (48) was employed [Figure 1]. The CASSCF results compiled in Table I are used as a reference for parametrizing the TAB functions for liquid water (see below).

Table I. CASSCF Results for an Isolated Water Molecule

A. Vacuum Energy Gap[a]

Water Electronic State	CASSCF Results	Experiments[b]
$\tilde{A}\,^1B_1$	7.58	7.3, 7.4, 7.49
$\tilde{B}\,^1A_1$	9.80	9.69, 9.70, 9.73
$\tilde{C}\,^1B_1$	10.45	9.99, 10.01, 10.10
$\tilde{D}\,^1A_1$	10.52	10.14, 10.15, 10.172
$^1B_2$	11.33	11.4

[a] In units of eV.
[b] For references on experimental studies, see, e.g., ref. *49* .

B. Electric Dipole Matrix Elements[a],[b]

Electronic State	$\tilde{X}\,^1A_1$	$\tilde{A}\,^1B_1$	$\tilde{B}\,^1A_1$	$\tilde{C}\,^1B_1$	$\tilde{D}\,^1A_1$	$^1B_2$
$\tilde{X}\,^1A_1$	$1.837\hat{z}$					
$\tilde{A}\,^1B_1$	$-1.329\hat{x}$	$-1.659\hat{z}$				
$\tilde{B}\,^1A_1$	$1.686\hat{z}$	$-0.220\hat{x}$	$-0.982\hat{z}$			
$\tilde{C}\,^1B_1$	$-0.322\hat{x}$	$5.355\hat{z}$	$0.011\hat{x}$	$5.778\hat{z}$		
$\tilde{D}\,^1A_1$	$0.988\hat{z}$	$4.874\hat{x}$	$1.876\,\hat{z}$	$0.848\hat{x}$	$-0.380\hat{z}$	
$^1B_2$	$-0.851\hat{y}$	0	$-7.482\hat{y}$	0	$4.134\hat{y}$	$-1.624\hat{z}$

[a] In units of D.
[b] All unit vectors refer to the water molecular frame [Figure 1].

TAB Description for Simulation. A small basis set consisting of a few vacuum states is usually not sufficient to describe diversified electronic structural aspects of water due to the neglect of a large number of excited states. To improve this, we expand the TAB by introducing four auxiliary states—referred to as surrogate states, which effectively represent the truncated states. The parameters of the surrogate states as well as those of the six real states are adjusted, so that they reasonably reproduce the average internal energy in liquid and vacuum polarizability. In ref. *32* , we studied two different models, TAB/10 and TAB/10D. Here we consider only the former whose vacuum Hamiltonian and dipole operators are given in Table II. Due to the adjustment, the parameters of the six real states are somewhat different from the ab initio results in Table I.

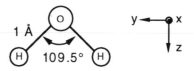

Figure 1: The geometry of SPC water (*48*).

In the simulations, we employed the interaction site model description for the solvent charge distribution, rather than the point-dipole approximation. The charge distributions of the TAB functions and their transition dipole moments were represented by partial point charges centered on a pair of "fictitious" sites, as well as the three atomic sites corresponding to O and two H's. The relative displacement of the former two sites [Table III] is perpendicular to the molecular plane, so that they can describe the out-of-plane polarizability. As for the real atomic sites, we used the SPC geometry [Figure 1] (48·).

Table II. Parameters for TAB/10 potential

A. Vacuum Hamiltonian

Diagonal elements (eV) $= [0, 7.6, 9.6, 10.5, 11.0, 11.5, 14.0, 16.8, 18.0, 25.0]$

B. Electric Dipole Operator[a]

$$\hat{p} = \begin{pmatrix} 1.85\hat{z} & -2.15\hat{x} & 1.60\hat{z} & -0.50\hat{x} & 1.20\hat{z} & -2.58\hat{y} & -3.00\hat{y} & 2.26\hat{z} & -3.00\hat{x} & 2.65\hat{z} \\ -2.15\hat{x} & -1.60\hat{z} & -0.42\hat{x} & 2.70\hat{z} & 4.20\hat{x} & 0 & 0 & 0 & 0 & 0 \\ 1.60\hat{z} & -0.42\hat{x} & -0.70\hat{z} & -2.52\hat{x} & 1.80\hat{z} & -5.00\hat{y} & 3.92\hat{y} & 2.66\hat{z} & -2.40\hat{x} & 0 \\ -0.50\hat{x} & 2.70\hat{z} & -2.52\hat{x} & 5.00\hat{z} & 0.84\hat{x} & 0 & 0 & 0 & 0 & 0 \\ 1.20\hat{z} & 4.20\hat{x} & 1.80\hat{z} & 0.84\hat{x} & -0.80\hat{z} & 4.00\hat{y} & 3.92\hat{y} & 2.66\hat{z} & -2.40\hat{x} & 0 \\ -2.58\hat{y} & 0 & -5.00\hat{y} & 0 & 4.00\hat{y} & -1.00\hat{z} & 0 & 0 & 0 & 0 \\ -3.00\hat{y} & 0 & 3.92\hat{y} & 0 & 3.92\hat{y} & 0 & 4.00\hat{z} & 2.35\hat{y} & 0 & 0 \\ 2.26\hat{z} & 0 & 2.66\hat{z} & 0 & 2.66\hat{z} & 0 & 2.35\hat{y} & 3.00\hat{z} & 0 & 0 \\ -3.00\hat{x} & 0 & -2.40\hat{x} & 0 & -2.40\hat{x} & 0 & 0 & 0 & -1.60\hat{z} & 0 \\ 2.65\hat{z} & 0 & 0 & 0 & 0 & 0 & 0 & 0 & 0 & -1.60\hat{z} \end{pmatrix}$$

[a] In units of D.

The vacuum electronic properties of TAB/10 water are summarized in Table III. It is characterized by the correct gas-phase dipole moment, $\mu^0 = 1.85$ D. Also, due to the out-of-plane charge distributions, its quadrupole moments Q^0_{ii} compare well with the experimental values. Since the majority of highly excited states are not included in the TAB description, the TAB/10 model in vacuum is less polarizable and more anisotropic than an isolated real water molecule.

Simulation Methods

The simulations were performed in the canonical ensemble of 128 water molecules at the temperature of 298 K and density of 0.997 g cm^{-3}, using the extended system method of Nosé (50). The trajectories were integrated with a time step of 2 fs with the Verlet algorithm (51). At each time step the 1M wave functions were calculated by solving equation 4 iteratively; the SCF solutions were converged with a relative tolerance of 10^{-8} in total electrostatic energy (32, 33). The Coulombic interactions were computed with the Ewald method (52) with account of the self-consistency condition between the central and image molecule charges. The intermolecular forces were evaluated using the Hellmann-Feynman theorem that is exact in the SCF regime. An excellent energy conservation without drift was obtained. The simulations were carried out with 10 ps of equilibration, followed by a 400 ps trajectory from which equilibrium averages were computed.

Table III. Properties of TAB/10 Water Potential in Vacuum[a]

	TAB/10	SPC/E	Experiments
$\vec{r}_{fict}$[b]	$(\pm 0.50, 0, 0.25)$	-	-
μ^0	1.85	2.35	1.85[c]
Q^0_{xx}	-2.32	-1.88	-2.50[d]
Q^0_{yy}	2.45	2.19	2.63[d]
Q^0_{zz}	-0.13	-0.31	-0.13[d]
α^0_{xx}	1.413	0	1.415[e]
α^0_{yy}	1.528	0	1.528[e]
α^0_{zz}	1.226	0	1.468[e]
$\bar{\alpha}^0$	1.39	0	1.470[e], 1.429[f]
σ_{OO}	3.240	3.160	-
ϵ_{OO}	0.152	0.1554	-
σ_{OH}	1.960	0	-
ϵ_{OH}	0.070	0	-
σ_{HH}	0.680	0	-
ϵ_{HH}	0.032	0	-

[a] Units: μ^0 (D), Q^0_{ii} (10^{-26} esu cm^2), polarizability α^0_{ii} (Å^3) and LJ parameters, $\sigma_{\alpha\beta}$ (Å) and $\epsilon_{\alpha\beta}$ (kcal mol^{-1}). Q^0_{ii} here refer to the center of mass of water.
[b] The coordinates of the two out-of-plane fictitious sites in the molecular frame [Figure 1] (units: Å). The origin is at the center of the oxygen atom.
[c] Ref. 53 .
[d] Ref. 54 .
[e] Ref. 55 .
[f] Ref. 56 .

Results

The MD results for the TAB/10 water model are summarized in Table IV and displayed in Figures 2–4. We begin with the solvent structure.

Equilibrium Structure. The results for the radial distribution functions are shown in Figure 2. The SPC/E results and the deconvolution of the elastic neutron scattering experiments (57) are also displayed. For the oxygen-oxygen distribution g_{OO}, the TAB/10 model yields a better first peak position than SPC/E, while the second peak is less distinctive for the former than for the latter. We parenthetically remark here that different water geometry, for example, TIP4P (58), could enhance the tetrahedral structure of g_{OO}, compared to the current TAB/10 result with SPC geometry. The coordination number n determined by integrating $g_{OO}(r)$ up to $r = 3.5$ Å is essentially the same for both SPC/E and TAB/10: $n = 4.9$ for the former and 5.0 for the latter. This is in good accord with the experimental result 5.2 ± 0.4 (*59*). As for the O–H distribution g_{OH}, the TAB/10 prediction for the first peak is in better accord with experiments than that of the SPC/E. For the hydrogen-hydrogen distribution g_{HH}, TAB/10 water yields excellent agreement with the measurements, whereas SPC/E predicts an overstructured first peak.

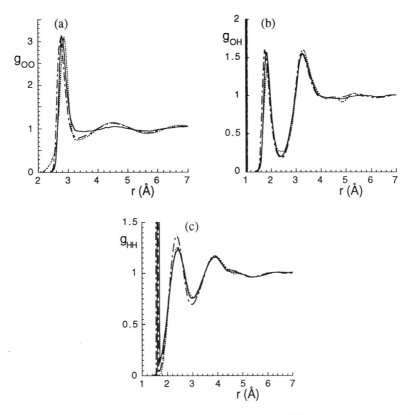

Figure 2: The radial distribution functions (a) $g_{OO}(r)$, (b) $g_{OH}(r)$ and (c) $g_{HH}(r)$: TAB/10 (——); SPC/E (– · –); experimental results (*57*) (· · ·).

Dynamics. The translational diffusion coefficient D_{tr} is given by

$$D_{tr} = \lim_{t \to \infty} \frac{1}{6t} < | \vec{r}_{CM}^{\,i}(t) - \vec{r}_{CM}^{\,i}(0) |^2 >_{eq} \qquad (9)$$

where $\vec{r}_{CM}^{\,i}(t)$ is the position of the center of mass for i at time t and $< \ldots >_{eq}$ denotes an equilibrium ensemble average. The MD yields $D_{tr} = 2.4 \pm 0.1$ Fick. This is in excellent agreement with the experimental value $D_{tr} = 2.3$ Fick (*60*).

Dielectric Properties. The electronic polarizability tensor in solution can be determined via equation 6. With the neglect of the interaction-induced coupling effects, viz., $\boldsymbol{\alpha^i} \approx \boldsymbol{\pi^i}$, its equilibrium average and trace simplify to

$$\bar{\alpha}^s \approx \bar{\pi}^s = N^{-1} \sum_i < \boldsymbol{\pi^i} >_{eq} ; \qquad \bar{\alpha}^s = \frac{1}{3}\mathrm{Tr}\ \bar{\boldsymbol{\alpha}}^s \approx \frac{1}{3}\mathrm{Tr}\ \bar{\boldsymbol{\pi}}^s . \qquad (10)$$

The MD results with equation 10 are given in Table IV. [To test the validity of this approximation, we have compared exact numerical results for $\bar{\alpha}^s$ and those of equation 10 for several different MD configurations. It was found that their relative difference is only a few %.] Comparison with Table III shows that the solution-phase polarizability is different from that in vacuum; the TAB/10 model is more polarizable in vacuum than in solution. Also, its polarizability anisotropy becomes reduced in solution. This arises from water electronic structure change induced by solvation. Also, average polarizability anisotropy of TAB/10 in solution is very close to the gas-phase measurements (55).

Table IV. Simulation Results for Liquid Water[a]

	TAB/10	SPC/E	Experiments
U/N	-10.07	-9.9	-9.92[b]
$\bar{\mu}^s$	2.58 ± 0.30	2.35	-
$\bar{\alpha}^s_{xx}$	1.28	-	-
$\bar{\alpha}^s_{yy}$	1.40	-	-
$\bar{\alpha}^s_{zz}$	1.34	-	-
$\bar{\alpha}^s$	1.34	-	-
ε_∞[c]	$1.56\ (1.69)$	1	1.79[d]
ε_0	79 ± 10	67 ± 10	78[d]
D_{tr}	2.4 ± 0.1	2.4 ± 0.4	2.3[e]

[a] Units: average potential energy per molecule U/N (kcal mol^{-1}), dipole moment $\bar{\mu}^s$ (D), polarizability (Å^3) and D_{tr} (Fick) [1 Fick $= 0.1$ Å^2 ps^{-1}].
[b] Ref. 58.
[c] The value in parentheses is determined with the Clausius-Mossotti equation 12.
[d] Ref. 61.
[e] Ref. 60.

We can evaluate the optical dielectric constant via (62)

$$\frac{\varepsilon_\infty - 1}{3} = \frac{1}{3}\frac{4\pi}{3V} \mathrm{Tr} < \boldsymbol{\Pi} >_{eq} \approx \frac{4\pi N}{3V} \bar{\pi}^s , \qquad (11)$$

where we have again neglected the interaction-induced effects in passage to the last expression. Here V is the volume of the simulation cell. With equation 11, TAB/10 yields $\varepsilon_\infty = 1.56$. If we instead employ the Clausius-Mossotti equation

$$\frac{\epsilon_\infty - 1}{\epsilon_\infty + 2} = \frac{4\pi N}{3V} \bar{\alpha}^s , \qquad (12)$$

we obtain $\varepsilon_\infty = 1.69$. While this is in fair agreement with the experimental value $\varepsilon_\infty = 1.79$ (61), the TAB description underestimates ε_∞ due to the truncation of the excited states. This state of affairs could be, in principle, improved by increasing the number of TAB functions. The static dielectric constant ε_0, on the other hand, can be determined in the linear response regime via (63 –65)

$$\varepsilon_0 = \varepsilon_\infty + \frac{4\pi}{3k_B TV}\left(<\vec{M}^2>_{eq} - <\vec{M}>_{eq}^2\right) \; ; \qquad \vec{M} = \sum_i \vec{\mu}^i \; , \qquad (13)$$

where k_B is Boltzmann's constant, T is the temperature and $\vec{M}$ is the total dipole moment of water. We found $\varepsilon_0 = 79 \pm 10$ for TAB/10, which is in excellent accord with the experimental result.

Absorption Spectrum. We now proceed to the distribution of excitation energies to gain insight into absorption spectroscopy. In this initial attempt, we neglect the electronic relaxation effects and consider only the 1M SCF excited states. [The electronic relaxation could be partially accounted for via the 1MCI method in equation 8.] The results are displayed in Figure 3. The first peak near 8.2 eV shows reasonable agreement with the absorption band of liquid water around 8.2–8.5 eV (66 –70). Also the broad character of the band, arising from the inhomogeneous distribution of differing local solvent environments, is well described by the TAB/10 model. Compared with the lowest electronic transition $\tilde{X}\,^1A_1 \to \tilde{A}\,^1B_1$ with energy 7.6 eV in vacuum [cf. Table IIA], the corresponding transition in solution is blue-shifted by ~ 0.6 eV, consonant with experiments. While the inclusion of electronic relaxation may shift the absorption band to a lower energy compared to that in Figure 3, the qualitative and even semiquantitative features observed here will remain essentially unchanged because of the nonequilibrium solvation effects arising from the solvent permanent dipoles (71). To the best of our knowledge, this is the first simulation study that yields semiquantitative agreement with photoabsorption measurements [cf. refs. 19 and 26].

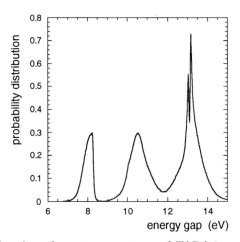

Figure 3: The photoabsorption spectrum of TAB/10 water with the neglect of the electronic relaxation effects.

Optical Kerr Effect Spectroscopy. Finally, we consider optical Kerr effect (OKE) spectroscopy, characterized by third-order nuclear response (72)

$$R^{(3)}(t) = -\frac{1}{k_B T}\frac{\partial}{\partial t} < \Pi_{ab}(t)\,\Pi_{ab}(0) > \qquad (t \geq 0)\,, \qquad (14)$$

where Π_{ab} $(ab = xy, xz, yz)$ is an off-diagonal component of $\boldsymbol{\Pi}$ in the lab frame. The MD result for the TAB/10 Kerr response function, Also presented there is a model response function, $\bar{R}^{(3)}$, obtained with *fixed* nonfluctuating electronic polarizability. $\bar{R}^{(3)}$ was evaluated by replacing fluctuating $\boldsymbol{\pi}^i$ in equation 7 with its solution-phase ensemble average, $\bar{\boldsymbol{\pi}}^s$ [equation 10]; this will be referred to as the nonfluctuating (NF) model. To separate out the polarizability effects on dynamics, we used the TAB/10 trajectory to compute $\bar{R}^{(3)}$; thus the underlying nuclear dynamics for both the TAB/10 and NF models are exactly the same. Therefore, the only difference between the two is that the polarizability of TAB/10 fluctuates dynamically with the solvent configuration, whereas that of NF remains unchanged. As noted above, their anisotropy is very close to Murphy's measured value (55). Thus TAB/10 and NF are nearly isotropically polarizable in solution.

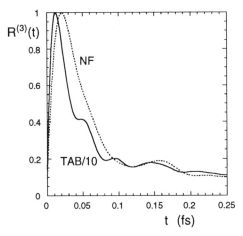

Figure 4: The third-order nuclear Kerr response function $R^{(3)}(t)$ [equation 14] for TAB/10 water (—). For comparison, the Kerr response $\bar{R}^{(3)}(t)$ of the NF model with fixed electronic polarizability $\bar{\boldsymbol{\pi}}^s$ is also shown ($\cdots$).

One of the most pronounced features in Figure 4 is that Kerr response of the TAB/10 and NF models differs markedly. After the initial peak around 15 fs, the latter yields no distinctive structures in $\bar{R}^{(3)}$ up to ~ 150 fs; this is in striking contrast with the $R^{(3)}$ peaks around 50 and 95 fs. Also the short-time behavior of NF is considerably slower than that of TAB/10. The reader should note that this dramatic difference in Kerr response arises solely from the polarizability *fluctuations* (or lack thereof). As shown in ref. *33*, this is due to the instantaneous adjustment of the TAB/10 molecular polarizability to the fluctuating local electric field. This nonlinear solvent electronic response enhances the librational character of, and introduces the short-time oscillatory structures into, Kerr response of liquid water (*33*). We also point out that the TAB/10 response function $R^{(3)}$ compares well with recent OKE measurements (*73 -76*). The structures around 20 and 200 fs determined from experiments are in reasonable accord with those of $R^{(3)}$ at $\sim$ 12 and $\sim$ 150 fs. Also, the existence of a peak near 50 fs in the latter is in good agreement with the experimental findings (*74 -76*). Therefore, despite its nearly isotropic polarizability in solution, the TAB/10 potential reproduces the experimental results on nuclear Kerr response fairly well due to the inclusion of polarizability fluctuations. For a more detailed description, the reader is referred to ref. *33* . [For previous MD studies on nonlinear electronic spectroscopy on water, see, for example, refs. *77 -80.*]

Concluding Remarks

In this article, we outlined our recent efforts to include the solvent electronic polarizability in MD simulations. By employing an efficient quantum mechanical description in a truncated adiabatic basis-set representation, both linear and nonlinear aspects of the solvent electronic structure variation were accounted for. Its application to liquid water was effected via the TAB/10 potential model with 10 basis functions. The parametrization of the basis functions was couched in terms of the CASSCF ab initio results for an isolated water molecule with the aid of experimental information. In the simulations, the diagonal and overlap charge distributions of the TAB functions were represented by partial charges on the five interaction sites of water; two of them are fictitious sites located off the molecular plane to describe out-of-plane polarizability. The ground electronic wave function and energy for the entire solvent system were determined at the SCF level and the intermolecular forces were evaluated using the Hellmann-Feynman theorem.

It was found that the TAB/10 potential provides a good overall description for bulk water. Its predictions for structural, dynamic, spectroscopic and dielectric properties of water are in good agreement with experiments. Its translational diffusion coefficient and static dielectric constant were found to be, respectively, $D_{tr} \approx 2.4$ Fick and $\varepsilon_0 \approx 79$; the corresponding experimental results are $D_{tr} = 2.3$ Fick and $\varepsilon_0 = 78$. The first absorption band centered around 8.2 eV and associated blue-shift compare well with the experimental findings. Also, despite its small polarizability anisotropy, TAB/10 water correctly captures experimentally-observed short-time behavior of nuclear Kerr response—in particular, the structure around 50–60 fs; this is usually not reproduced with classically polarizable solvent descriptions.

Finally, compared with a nonpolarizable model with 5 interaction sites, the performance of TAB/10 at the SCF level was found to be slow by a factor of $\sim$ 8. This is mainly due to the repeated diagonalizations of the Fock matrices [equation 4] to determine the solvent electronic structure. One promising aspect, though, is that our generalized molecular mechanics algorithm would allow a time step larger than 2 fs since the electronic degrees of freedom are optimized accurately by solving the Fock equations. Also in view of its various advantages, e.g., nonlinear electronic response and spectroscopy, we believe that the TAB formulation provides a promising and viable alternative to classical polarizable solvent descriptions and density functional MD.

Acknowledgments

This work was supported in part by NSF Grant No. CHE-9412035.

Literature Cited

[1] Allen, M. P.; Tildesley, D. J. *Computer Simulation of Liquids*; Clarendon: Oxford, 1987.

[2] For a collection of papers on various aspects of computer simulation studies, see *Simulation of Liquids and Solids*, Ciccotti, G.; Frenkel, D.; McDonald, I. R., Eds.; North-Holland: Amsterdam, 1987.

[3] For a recent review on MD studies of assorted chemical reactions in solution, see Whitnell, R. M.; Wilson, K. R. *Rev. Comp. Chem.* **1993**, *4*, 67.

[4] Barnes, P.; Finney, J. L.; Nicholas, J. D.; Quinn, J. E. *Nature* **1979**, *282*, 459.

[5] Rullman, D.; van Duijnen, P. T. *Mol. Phys.* **1988**, *63*, 451.

[6] Sprik, M.; Klein, M. L. *J. Chem. Phys.* **1988**, *89*, 7556.

[7] Ahlström, P.; Wallqvist, A.; Engström, S.; Jönsson, B. *Mol. Phys.* **1989**, *68*, 563.

[8] Kuwajima, S.; Warshel, A. *J. Phys. Chem.* **1990**, *94*, 460.

[9] Niesar, U.; Corongiu, G.; Clementi, E.; Kneller, G. R.; Bhattacharya, D. K. *J. Phys. Chem.* **1990**, *94*, 7949.

[10] Cieplak, P.; Kollman, P.; Lybrand, T. *J. Chem. Phys.* **1990**, *91*, 6755.

[11] Zhu, S.-B.; Yao, S.; Zhu, J.-B.; Singh, S.; Robinson, G. W. *J. Phys. Chem.* **1991**, *95*, 6211.

[12] Kozack, R. E.; Jordan, P. C. *J. Chem. Phys.* **1992**, *96*, 3120.

[13] Dang, L. X. *J. Chem. Phys.* **1992**, *97*, 2659.

[14] Smith, D. E.; Dang, L. X. *J. Chem. Phys.* **1994**, *100*, 3757.

[15] Halley, J. W.; Rustad, J. R.; Rahman, A. *J. Chem. Phys.* **1993**, *98*, 4110.

[16] Bernardo, D. N.; Ding, Y.; Krogh-Jespersen, K; Levy, R. M. *J. Phys. Chem.* **1994**, *98*, 4180.

[17] Rick, S. W.; Stuart, S. J.; Berne, B. J. *J. Chem. Phys.* **1994**, *101*, 6141.

[18] Åstrand, P.-O.; Linse, P.; Karlström, G. *Chem. Phys.* **1995**, *191*, 195.

[19] Borgis, D.; Staib, A. *Chem. Phys. Lett.* **1995**, *238*, 187.

[20] Svishchev, I. M.; Kusalik, P. G.; Wang, J.; Boyd, R. J. *J. Chem. Phys.* **1996**, *105*, 4742.

[21] Chialvo, A. A.; Cummings, P. T. *J. Chem. Phys.* **1996**, *105*, 8247.

[22] Gao, J. *J. Phys. Chem.* **1997**, *B101*, 657.

[23] Ortega, J.; Lewis, J. P.; Sankey, O. F. *J. Chem. Phys.* **1997**, *106*, 3696.

[24] Car, R.; Parrinello, M. *Phys. Rev. Lett.* **1985**, *55*, 2471.

[25] Hartke, B.; Carter, E. A. *J. Chem. Phys.* **1992**, *97*, 6569.

[26] Laasonen, K.; Sprik, M.; Parrinello, M.; Car, R. *J. Chem. Phys.* **1993**, *99*, 9080.

[27] Tuckerman, M.; Laasonen, K.; Sprik, M.; Parrinello, M. *J. Phys. Chem.* **1995**, *99*, 5749.

[28] Martyna, G. J.; Deng, Z.; Klein, M. L. *J. Chem. Phys.* **1993**, *98*, 555.

[29] Hammes-Schiffer, S.; Andersen, H. C. *J. Chem. Phys.* **1993**, *99*, 523.

[30] Margl, P.; Schwarz, K.; Blöchl, P. E. *J. Chem. Phys.* **1994**, *100*, 8194.

[31] Bursulaya, B. D.; Kim, H. J., *Generalized Molecular Mechanics Including Quantum Electronic Structure Variation of Polar Solvents. I. Theoretical Formulation via a Truncated Adiabatic Basis Set Description, J. Chem. Phys.* In press.

[32] Bursulaya, B. D.; Jeon, J.; Zichi, D. A.; Kim, H. J. *Generalized Molecular Mechanics Including Quantum Electronic Structure Variation of Polar Solvents. II. A Molecular Dynamics Simulation Study of Water, J. Chem. Phys.* In press.

[33] Bursulaya, B. D.; Kim, H. J., *J. Phys. Chem.* **1997**, *B101*, 10994.

[34] Warshel, A.; Weiss, R. M. *J. Am. Chem. Soc.* **1980**, *102*, 6218.

[35] Warshel, A. *Computer Modeling of Chemical Reactions in Enzymes and Solutions*; Wiley: New York, 1991.

[36] Pross, A.; Shaik, S. S. *Acc. Chem. Res.* **1983**, *16*, 363.

[37] Kochi, J. K. *Angew. Chem., Int. Ed. Engl.* **1988**, *27*, 1227.

186

[38] Kim, H. J.; Hynes, J. T. *J. Am. Chem. Soc.* **1992**, *114*, 10508, 10528.

[39] Bursulaya, B. D.; Zichi, D. A.; Kim, H. J. *J. Phys. Chem.* **1995**, *99*, 10069.

[40] Bursulaya, B. D.; Kim, H. J. *J. Phys. Chem.* **1996**, *100*, 16451.

[41] See, e.g., A. Szabo and N. S. Ostlund, *Modern Quantum Chemistry* (McGraw-Hill, New York, 1982).

[42] Madden, P. A. In *Spectroscopy and Relaxation of Molecular Liquids*, Steele, D., Yarwood, J., Eds.; Elsevier: Amsterdam, 1991.

[43] Ladanyi, B. M. In *Spectroscopy and Relaxation of Molecular Liquids*, Steele, D., Yarwood, J., Eds.; Elsevier: Amsterdam, 1991.

[44] Dunning, Jr., T. H. *J. Chem. Phys.* **1989**, *90*, 1007.

[45] Kendall, R. A.; Dunning, Jr., T. H.; Harrison, R. J. *J. Chem. Phys.* **1992**, *96*, 6769.

[46] Dunning, Jr., T. H.; Hay, P. J. In *Methods of Electronic Structure Theory*, Schaefer III, H. F., Ed.; Plenum: 1977, Vol. 3.

[47] Schmidt, M. W.; Baldridge, K. K.; Boatz, J. A.; Elbert, S. T.; Gordon, M. S.; Jensen, J. H.; Koseki,S.; Matsunaga, N.; Nguyen, K. A.; Su, S. J.; Windus, T. L., Dupuis, M.; Montgomery, J. A. *J. Comp. Chem.* **1993**, *14*, 1347.

[48] Berendsen, H. J. C.; Gigera, J. R.; Straatsma, T. P. J. *J. Phys. Chem.* **1987**, *91*, 6269.

[49] Durand, G.; Chapuisat, X. *Chem. Phys.* **1985**, *96*, 381.

[50] Nosé, S. *J. Chem. Phys.* **1984**, *81*, 511.

[51] Verlet, L. *Phys. Rev.* **1967**, *159*, 98.

[52] Heyes, D. M. *J. Chem. Phys.* **1982**, *74*, 1924.

[53] Clough, S. A.; Beers, A.; Klein, G. P.; Rothman, L. S. *J. Chem. Phys.* **1973**, *59*, 2254.

[54] Verhoeven, J.; Dynamus, A. *J. Chem. Phys.* **1970**, *52*, 3222.

[55] Murphy, W. F. *J. Chem. Phys.* **1977**, *67*, 5877.

[56] Zeiss, G. D.; Meath, W. J. *Mol. Phys.* **1977**, *33*, 1155.

[57] Soper, A. K.; Phillips, M. G. *Chem. Phys.* **1986**, *107*, 47.

[58] Jorgensen, W. L.; Chandrasekhar, J.; Madura, J. D.; Impey, R. W.; Klein, M. L. *J. Chem. Phys.* **1983**, *79*, 926.

[59] Soper, A. K. *J. Chem. Phys.* **1994**, *101*, 6888.

[60] Mills, R. *J. Phys. Chem.* **1973**, *77*, 685.

[61] Buckingham, A. D. *Proc. Royal Soc. London* **1956**, *A238*, 235.

[62] Neumann, M.; Steinhauser, O. *Chem. Phys. Lett.* **1984**, *106*, 563.

[63] Kirkwood, J. G. *J. Chem. Phys.* **1939**, *7*, 911.

[64] De Leeuw, S. W.; Perram, J. W.; Smith, E. R. *Proc. Royal Soc. London* **1983**, *A388*, 177.

[65] Neumann, M. *J. Chem. Phys.* **1986**, *85*, 1567.

[66] Onaka, R.; Takahashi, T. *J. Phys. Soc. Jpn.* **1968**, *24*, 548.

[67] Shibaguchi, T.; Onuki, H.; Onaka, R. *J. Phys. Soc. Jpn.* **1997**, *42*, 152.

[68] Verrall, R. E.; Senior, W. A. *J. Chem. Phys.* **1969**, *50*, 2746.

[69] Painter, L. R.; Birkhoff, R. D.; Arakawa, E. T. *J. Chem. Phys.* **1969**, *51*, 243.

[70] Popova, S. I.; Alperovich, L. I.; Zolotarev, V. M. *Opt. Spect.* **1972**, *32*, 288.

[71] Kim, H. J.; Bursulaya, B. D.; Jeon, J.; Zichi, D. A. *Computer Simulation Study of Electronic Spectroscopy in Water*, submitted to *Proceedings of SPIE*.

[72] See, e.g., Mukamel, S. *Principles of Nonlinear Optical Spectroscopy*; Oxford: New York, 1995.

[73] Chang, Y. J.; Castner, Jr., E. W. *J. Chem. Phys.* **1993**, *99*, 7289.

[74] Castner, Jr., E. W.; Chang, Y. J.; Chu, Y. C.; Walrafen, G. E. *J. Chem. Phys.* **1995**, *102*, 653.

[75] Palese, S.; Schilling, L.; Miller, R. J. D.; Staver, P. R.; Lotshaw, W. T. *J. Phys. Chem.* **1994**, *98*, 6308.

[76] Palese, S.; Buontempo, J. T.; Schilling, L.; Lotshaw, W. T.; Tanimura, Y.; Mukamel, S.; Miller, R. J. D. *J. Phys. Chem.* **1994**, *98*, 12466.

[77] Impey, R. W.; Madden, P. A.; McDonald, I. R. *Mol. Phys.* **1982**, *46*, 513.

[78] Bosma, W. B.; Fried, L. E.; Mukamel, S. *J. Chem. Phys.* **1993**, *98*, 4413.

[79] Saito, S.; Ohmine, I. *J. Chem. Phys.* **1995**, *102*, 3566.

[80] Saito, S.; Ohmine, I. *J. Chem. Phys.* **1997**, *106*, 4889.

Chapter 12

RISM-SCF Study of Solvent Effect on Electronic Structure and Chemical Reaction in Solution: Temperature Dependence of pKw

Fumio Hirata[1], Hirofumi Sato[1], Seiichiro Ten-no[1], and Shigeki Kato[2]

[1]Division of Theoretical Study, Institute for Molecular Science, Okazaki, 444, Japan
[2]Department of Chemistry, Graduate School of Science, Kyoto University, Kyoto 606, Japan

The RISM-SCF method, or a hybridized method of the electronic structure theory and statistical mechanics of liquids, is applied to the auto-ionization process in water putting special stress on its temperature dependence. It is found that changes in the solvation free energies with increasing temperature drive the chemical reaction toward neutral species while energies associated with solvent-induced reorganization of electronic structure make opposite contribution. The actual temperature dependence is dominated by the latter contribution, which gives rise to good agreement with experimental results. It is suggested that the observed temperature dependence of pKw is related to the great sensitivity of the electronic structure of OH$^-$ on solvent effect.

A chemical reaction is undoubtedly the most important issue in the theoretical chemistry, and the electronic structure is a key to solve the problem. As long as molecules in the gas phase are concerned, theories for the electronic structure have been enjoying its great success. However, when it comes to molecules in solution, the stage of theories is still an infant. Since it is anyway impossible to solve the Schrödinger equation for entire system including about 10^{23} solvent molecules, a most promising approach so far is any type of hybrid between classical solvent and quantum solute. Roughly three types of hybrid approaches have been appeared depending on how to treat the solvent degrees of freedom: molecular simulations, continuum models, and the integral equations. The continuum models employ the electrostatic theory in order to evaluate the reaction field exerted on the solute from the continuum dielectric regarded as a solvent. A variety of different methodologies including Onsager's reaction field, Born's model and the image charge models have been implemented for the reaction field. An obvious advantage of the method is its handiness, while well-documented disadvantages are the lack of molecular picture of solvent and an artifact introduced at the boundary between solute and solvent. The integral equation method is free from such disadvantages, and gives a microscopic picture for the solvent effect on the electronic structure of molecule with a reasonable amount of computation time. We have recently proposed a new method refereed to as RISM-SCF based on the integral equation theory of molecular liquids (RISM) (*1*),(*2*)and the ab initio electronic structure theory(*3*). The integral equation approach replaces the reaction field in the continuum models by a microscopic expression in terms of the site-site radial distribution functions between solute and solvent. The statistical solvent distribution around solute is determined by the electronic structure or the partial charges of solute, while the electronic structure of solute is influenced by the solvent distribution. Therefore, the Hartree-Fock equation and the RISM equation should be solved in a self-consistent manner. It is the self-consistent determination of the solute electronic structure and the solvent distribution around the solute that features the RISM-SCF procedure. It has been further extended to the electronic structure with multiconfigurations, which also enables us calculation of optimized geometry in solutions(*4*). The method has been successfully applied to a variety of chemical processes in solution including the coupled solvent and substitution effects on proton affinity and tautomerization reaction in organic solvents. (*3-6*) As an example of the applications, here, we report a study for autoionization of liquid water putting special stress on its temperature dependence. (Sato, H. ; Hirata, F., *J. Phys. Chem.* in press)

A water molecule has *amphoteric* character, which means it can act as both an acid and a base. The autoionization equilibrium process in water,

$$H_2O+H_2O \rightleftarrows H_3O^++OH^-, \tag{1}$$

is one of the most important and fundamental reactions in a variety of fields in chemistry. The ionic product defined by

$$K_w=[H_3O^+][OH^-], \tag{2}$$

and its logarithm

$$pK_w=-\log K_w, \tag{3}$$

are measures of the autoionization. The quantity can be related to the free energy change (ΔG) associated with the reaction of Eq. (1) by the standard thermodynamic relations(*7*).

$$\Delta G=2.303RT\,pK_w. \tag{4}$$

It is experimentally known that the pK_w value shows significant temperature dependence; i.e., it decreases with increasing temperature (*8*). The observation apparently contradicts an intuitive consideration with respect to the effect of hydrogen-bonding on intramolecular O-H bond and its temperature dependence. Rising temperature must intensify the rotational and translational motion of each water molecule and the hydrogen bondings are weakened. If the above autoionization process is only governed by the hydrogen bond strength alone, it is feasible that the equilibrium leans toward the left hand side of the equation.

The free energy change consists of various contributions including changes in the electronic energy and "solvation" free energy of the molecular species concerning the reaction, which are related to each other. A microscopic description of the reaction, therefore, requires a theory that accounts for both the electronic and liquid structures of water. Many theoretical attempts have been made to evaluate pKw and pKa values in aqueous solutions, which involve determination of molecular parameters for H_3O^+ and OH^- ions in the liquid(9-14). Most of them have employed the molecular simulation techniques combined with the standard *ab initio* molecular orbital method. However, the prediction of pKw is usually very difficult since it requires high accuracy in the description of the electronic and liquid structures as well as their coupling. In the present report, we apply the RISM-SCF / MCSCF method to investigate the temperature dependence of pKw.

RISM-SCF/MCSCF Theory

Since details of the RISM-SCF / MCSCF method are reported elsewhere (3,4), here we make a brief sketch of the theory. We define the total free energy $\mathcal{A}$ as the sum of the electronic energy of the solute molecule E_{solute} and the excess chemical potential $\delta\mu$:

$$\mathcal{A} = E_{solute} + \Delta\mu, \tag{5}$$

E_{solute} is defined in usual *ab initio* electronic structure method

$$E_{solute} = \langle \Psi | \mathcal{H} | \Psi \rangle + E_{nuc}, \tag{6}$$

where $\mathcal{H}$ is electronic hamiltonian of solute, $|\Psi\rangle$ is wave function, and E_{nuc} is nuclear repulsion energy. The Fock operator F of the isolated molecule is derived by imposing the constrains to the orthonormality of configuration state functions and one particle orbital.

$$\delta\left(E_{solute} - [constraint\ for\ orthonormality]\right) = 0 \rightarrow F. \tag{7}$$

For the chemical potential term we adopted the free energy derived by Singer and Chandler (15).

$$\delta\mu = -\frac{\rho}{\beta}\sum_{\alpha,s}\int dr\left(c_{\alpha,s} - \frac{1}{2}h_{\alpha,s}^2 + \frac{1}{2}c_{\alpha,s}h_{\alpha,s}\right), \tag{8}$$

where $\beta = 1/k_BT$, k_B is Boltzmann's constant and ρ is number of density. The Greek subscripts refer to the interaction sites of solute molecules, and the Roman to the sites on the solvent. $c_{\alpha s}$ and $h_{\alpha s}$ are the direct and total correlation functions, respectively.

The quantity $\mathcal{A}$ can be regarded as a functional of the correlation functions in molecular liquid theory as well as solute molecular wave functions described by molecular orbital and CI coefficients in molecular orbital theory. Imposing the constrains, new Fock operator F^{new} can be derived by variations with respect to these functions.

$$\delta\left(E_{solute} + \delta\mu - [constraint\ for\ orthonormality]\right) = 0. \tag{9}$$

Thus the Fock operator in the case of Hartree Fock approximation is

$$F^{new} = F - f\sum_{\lambda} V_{\lambda} b_{\lambda}, \tag{10}$$

where f is the occupation number, and V_{λ} is the electrostatic potential on λ - site in solute molecule and expressed as a mean field,

$$V_{\lambda} = \rho \sum_{\alpha} q_{\alpha} \int 4\pi r^2 \frac{g_{\lambda j}}{r} dr, \tag{11}$$

where q_{α} is a partial charge on the solvent site α. It is noted that electronic structure of a solute molecule and statistical solvent structure can be calculated in a self - consistent manner. Since we used self - consistent scheme to derive the Fock operator, energy gradient with respect to the nuclear coordinates are readily obtained.

Computational Detail

In order to facilitate the calculation of the free energy change ΔG^{aq} associated with the reaction in Eq. (1),we use a thermodynamic cycle illustrated in Scheme, regarding the species concerning the reaction as "solute" in infinitely dilute aqueous solution. (We use "Δ" for changes of quantities associated with the chemical reaction and "δ" for changes due to solvation. Superscripts "aq" and "vac" are used to distinguish aqueous and vacuum environments.) Then, ΔG^{aq} can be written in terms of the energy change

Scheme: Thermodynamic Cycle of Autoionization of Water

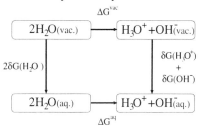

associated with the reaction in vacuo ΔG^{vac} and the free energy change of the reacting species due to solvation as

$$\Delta G^{aq} = \Delta G^{vac} + \delta G(H_3O^+) + \delta G(OH^-) - 2\delta G(H_2O), \tag{12}$$

where $\delta G(H_3O^+)$, $\delta G(OH^-)$ and $\delta G(H_2O)$ are respectively the free energy changes of H_3O^+, OH^- and H_2O upon solvation. The free energy change δG for each species can be decoupled into intra and inter-molecular contributions as

$$\delta G = \delta G_{intra} + \delta\mu, \tag{13}$$

where δG_{intra} is the intra-molecular contribution consisting of the changes in the kinetic (translational, rotational and vibrational) terms and the electronic reorganization energy (16),

$$\delta G_{intra} = \delta G_{kin} + \delta E_{elec}. \tag{14}$$

The electronic reorganization energy is defined as difference of the electronic energies between

$$\delta E_{elec} = E_{elec}^{aq} - E_{elec}^{vac}, \tag{15}$$

where E_{elec}^{aq} and E_{elec}^{vac} are the electronic energy in solution and in vacuo, respectively (6). The inter-molecular contribution, $\delta\mu$, is the excess chemical potential, or the solvation free energy. From Eqs. (12) - (15), we have

$$\Delta G^{aq} = \Delta E_{elec}^{vac} + \Delta\delta G_{kin} + \Delta\delta E_{elec} + \Delta\delta\mu, \tag{16}$$

where

$$\Delta E_{elec}^{vac} = E_{elec}^{vac}(H_3O^+) + E_{elec}^{vac}(OH^-) - 2E_{elec}^{vac}(H_2O),$$
$$\Delta\delta G_{kin} = \delta G_{kin}(H_3O^+) + \delta G_{kin}(OH^-) - 2\delta G_{kin}(H_2O),$$
$$\Delta\delta E_{elec} = \delta E_{elec}(H_3O^+) + \delta E_{elec}(OH^-) - 2\delta E_{elec}(H_2O)$$
$$\Delta\delta\mu = \delta\mu(H_3O^+) + \delta\mu(OH^-) - 2\delta\mu(H_2O).$$

In the above equations, ΔE_{elec}^{vac}, which concerns the electronic structures of the species only in the gas phase, can be calculated from the standard ab initio MO method. $\Delta\delta E_{elec}$ and $\Delta\delta\mu$ are our major concerns in the present study, which requires a self-consistent determination of the electronic structure and the statistical solvent distribution. We employ the RISM-SCF procedure to calculate those quantities. $\Delta\delta G_{kin}$ can be obtained from the elementary statistical mechanics given the optimized molecular geometry for the reacting species in gas and solution phases. Optimizing molecular geometry in the solution phase, however, is not a standard problem, which requires calculation for energy gradients of a solvated molecule. The RISM-MCSCF method, which has been proposed recently, enables us the geometry optimization of a molecule in solution phases.

Here we make brief comments concerning approximations and models used in the present study. For the electronic structure calculations, the standard triple-ζ basis set augmented by p polarization functions on hydrogen atoms, d polarization function and diffusive p function (α_p=0.059 (17)) on oxygen atom is employed in the Hartree - Fock (HF) method . Molecular geometries of all the species (H_2O, H_3O^+, and OH^-) are optimized in both gas and solution phases under a constraint that H_2O and H_3O^+ are assumed to possess C_{2v} and C_{3v} symmetry, respectively.

In solving the RISM equation, we employ a SPC - like water model (18) which has been successful in liquid phase calculations. The same Lennard - Jones parameters are used for H_3O^+ and OH^-. The experimental density of water and its temperature dependence, which are taken from the literature (19), are used in the calculation.

Our primary concern in the present study is the temperature dependence of pK_w in the aqueous phase. So, we concentrate our attention on the relative value of pK_w at temperature T to that at $T = 273.15$ K chosen as a standard, that is,

$$\Delta_T pK_w(T) = pK_w(T) - pK_w(273.15)$$
$$= \frac{1}{2.303R} \left\{ \frac{\Delta G^{aq}(T)}{T} - \frac{\Delta G^{aq}(273.15)}{273.15} \right\} \cdot \qquad (17)$$

The difference of the electronic energies in vacuo, ΔE_{elec}^{vac}, makes a dominant contribution to determine the "absolute" value of pK_w. However, ΔE_{elec}^{vac} is essentially independent of T, and it contributes to temperature dependence of pK_w in aqueous phase only through $1/T$ in Eq. (17).

Results and discussion

Solvent Induced Molecular Polarization and Solvation Structure. Temperature dependence of effective charges on oxygen atoms q_O, which represents a change in the electronic structure induced by solvent effect, is shown in Figure 1. The effective charges were determined in such a way that electrostatic potential produced by the charges fits best with those arising from the electronic cloud of the solute (3). Compared with the gas phase values, -0.766 (water) , -0.633 (hydronium ion), and -1.128 (hydroxide ion), charges in all the species are largely enhanced by the interaction with surrounding solvent molecules. The change in the electronic structure or the polarization is appreciably reduced as temperature increases. The behavior can be explained in terms of the increased molecular motion such as rotation. As the motions increase, the hydrogen-bond between the solute species and solvent molecules is weakened, and the polarizing effects becomes less and less. Consequently, the electronic structure is relaxed toward that in the isolated molecule as temperature increases. Among three species, the greatest relaxation of polarization is observed in the hydroxide ion, which can be understood in terms of the temperature dependence of the hydration structure of "solute" molecules.

In Figure 2, we show three sets of pair correlation functions (PCF) between (a) solute oxygen and solvent hydrogen, and (b) solute hydrogen and solvent oxygen, at 293.15 K. It is readily seen that sharp peaks are found around 2.0 Å in a pair of the ionic species and atoms with opposite charges in water molecules. An oxygen atom in the hydroxide ion has a large negative charge and attracts hydrogen sites of solvent water acutely. In contrast, a hydrogen atom in the hydronium ion attracts oxygen sites of solvent, but in a moderate manner. Water has an intermediate character of these two ions. The highest peak in PCF between hydroxide O and water H concerns the greatest effective charge of this ion shown in Figure 1. The results indicate that the electronic structure of the hydroxide ion is most sensitive to the electrostatic field of solvent and that it is most easily to be polarized. Shown in Figure 3 is alteration of PCF due to temperature change, $\Delta_T g(T) = g(T) - g(273.15K)$. The oxygen site of hydronium ion has the least negative charge among the three species, and $\Delta_T g(T)$ is rather loose and broad. But, changes in the others are extremely sharp and large. Especially, a negative deviation around 2.0 Å corresponds to change in height of the first peak in PCF, being indicative of a drastic change in hydrogen bonds as temperature increases. It should be noted that $\Delta_T g$ in hydroxide ion is about ten times greater than that in water. The positive deviation around 1.5 Å is a consequence of the increased probability of

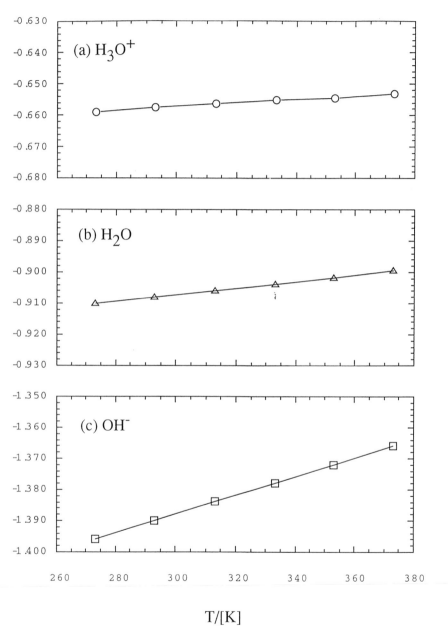

Figure 1. Temperature dependence of electronic structure represented by effective charge on oxygen site: (a) hydronium ion; (b) water; (c) hydroxide ion.

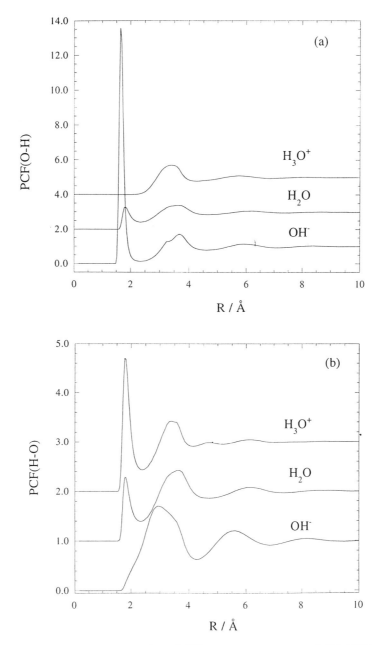

Figure 2. Pair correlation function (PCF) at room temperature (293.15K): (a) solute oxygen and solvent hydrogen; (b) solute hydrogen and solvent oxygen.

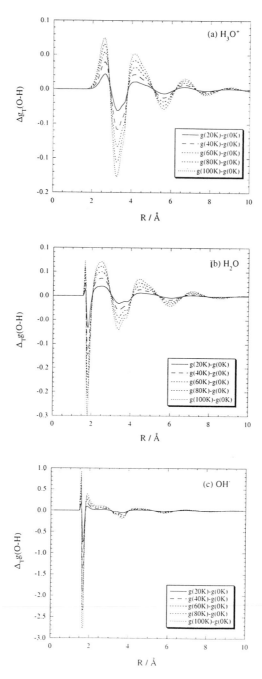

Figure 3. Temperature dependence of PCF: (a) hydronium ion; (b) water; (c) hydroxide ion.

repulsive configurations with increasing temperature and can be compared with the results reported for the temperature derivatives of the ion-water PCF(20).

Temperature Dependence of Free Energy and pK_W. Free energy components in aqueous solution at selected temperatures are listed in Table I. Three upper groups in the table include energy components of each species. The kinetic term G_{kin} contributed from the kinetic energies decreases with increasing temperature, which can be understood by the elementary statistical mechanics. The temperature dependence of δE_{elec} can be explained in terms of the polarization of solute (*i.e.*; effective charge) and solvation structure discussed in the previous section. As has been discussed there, the electronic structure on each species is relaxed toward that in the gas phase upon rising temperature. Thus, δE_{elec} decreases as temperature increases. It is noteworthy that δE_{elec} of hydroxide ion is estimated 2 or 3 times larger than those of any other species, being consistent with the finding in the previous section, that the anion is more polarizable than the other species. Our results for the excess chemical potential, $\delta\mu$, or the hydration free energy are in good agreement with those reported by other authors based on the dielectric continuum models.

Table I. Free Energy Component at Selected Temperature

	273.15[K]	293.15[K]	373.15[K]
H_2O			
δE_{elec}	2.38	2.27	1.87
$\delta\mu$	-5.17	-4.58	-2.94
δG_{kin}	7.45	6.89	4.52
ΔG	4.67	4.58	3.45
H_3O^+			
δE_{elec}	3.74	3.59	3.11
$\delta\mu$	-73.49	-72.33	-68.38
δG_{kin}	15.38	14.77	12.23
ΔG	-54.37	-53.97	-53.04
OH^-			
δE_{elec}	9.43	8.93	7.15
$\delta\mu$	-140.02	-137.95	-130.17
δG_{kin}	0.20	-0.29	-2.33
δG	-130.39	-129.31	-125.35
Energy Difference			
ΔE^{vac}_{elec}	229.54	229.54	229.54
$\Delta\delta E_{elec}$	8.41	7.98	6.53
$\Delta\delta\mu$	-203.17	-201.13	-192.67
$\Delta\delta G_{kin}$	0.67	0.71	0.87
δG^{aq}	35.45	37.10	44.27

All the energies are given in kcal/mol.

For the hydration energy of hydronium ion, for instance, Emsley et al. (*12*) and Chipot et al.(*10*) reported -74.6 to -89.2 kcal/mol and -78.8 kcal / mol, respectively, which are very close to our estimation at the room temperature (293.15K), -72.33 kcal / mol. The hydration free energy of each species increases, or changes toward more positive side, with increasing temperature, which reflects reduced electrostatic interactions including the hydrogen-bonding due to increased thermal motions. The change is much greater for OH⁻ (~ 9.9 kcal/mol) than for H_2O (~ 2.2 kcal/mol) and H_3O^+ (~ 5.1 kcal/mol) in this range of temperatures. The behavior can be again ascribed to the characteristics of the anion which is more polarizable than the other species. The total free energy of each species is determined by a subtle balance among these components, and only H_2O shows a decrease with rising temperature.

Changes in the free energy components, associated with the autoionization reaction (Eq. (1)), are listed in the bottom group of Table 1. Since $\Delta\delta\mu$'s and ΔE_{elec}^{vac} are greater in magnitudes by a few order compared to the other components, and since they largely compensate each others, we plot the sum of those contributions. It is important to note that the sum ($\Delta E_{elec}^{vac} + \Delta\delta\mu$) shows rather large positive temperature dependence. Compared to these contributions, $\Delta\delta G_{kin}$ is almost negligible in the present level of estimations both in magnitudes and in temperature dependence. $\Delta\delta E_{elec}$ shows a small and negative dependence on temperature, to which the main contribution comes from the hydroxide ion.

Finally, we show a resultant $\Delta_T pK_w$ and their components in Figure 4. $\Delta_T pK_w$ is decomposed into the following four terms:

$$\Delta_T pK_w(T) = \Delta_T pK^{vac}_{w, elec} + \Delta_T pK_{w, reorg} (T)$$
$$+ \Delta_T pK_{w, \delta\mu}(T) + \Delta_T pK_{w, kin}(T)$$

where $\Delta_T pK^{vac}_{w, elec}$, $\Delta_T pK_{w, reorg}$, $\Delta_T pK_{w, \delta\mu}$, and $\Delta_T pK_{w, kin}$ are related, respectively, to the corresponding free energy changes, ΔE^{vac}_{elec}, $\Delta\delta E_{elec}$, $\Delta\delta\mu$, and $\Delta\delta E_{kin}$ by Eq. (16). In order to facilitate the inspection, a sum of $\Delta_T pK_{w, \delta\mu}$ and $\Delta_T pK^{vac}_{w, elec}$ is also plotted in the figure. Those contributions are very large but they compensate each other, and the sum shows slight and positive dependence on temperature. It is important to note that the components in $\Delta_T pK_w$ show different temperature dependence from corresponding terms in $\Delta G^{aq}(T)$ due to "T" which appears in the denominator in Eq. (17). It is readily seen from the figure that the temperature dependence of pK_w is determined by an interplay of several contributions with different physical origins. It is also clear that the temperature dependence is dominated by $\Delta_T pK_{w, reorg}$ after the largest contributions compensated each other. As a consequence, the theoretical results for the temperature dependence of the ionic product show good agreement with experiments.

To summarize, so-called solvation free energies, or $\delta\mu$, contribute to a positive temperature dependence on $\Delta_T pK_w$, while the energies associated with reorganization of electronic structure give rise to the opposite temperature dependence. In other words, the solvation free energies contribute to make a water molecule less dissociative with increasing temperature, which is in accord with the intuitive consideration in terms of the reduced hydrogen-bonding. On the other hand, the electronic reorganization energy contributes to facilitate the dissociation as temperature increases. The latter dominates the temperature dependence of the ionic product. The temperature dependence is in turn largely determined by characteristics of the electronic structure of OH⁻, because the temperature dependence of the electronic reorganization energy is dominated by that species.

Improvements of the theoretical calculations, especially in descriptions of electronic structures, can be expected for ΔpK_w. However, as long as the temperature dependence is concerned, the qualitative feature of our results seems to be insensitive

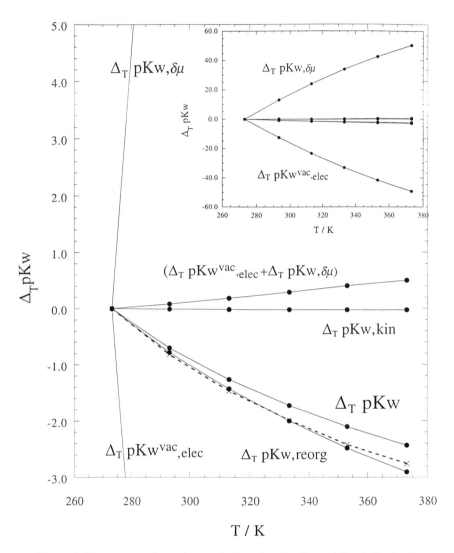

Figure 4. Temperature dependence of pK$_w$ of water. Dashed line indicates the experimentally observed value, while all other solid lines are computed by RISM - SCF / MCSCF procedure.

to such elaboration in the quantum description according to our preliminary calculations based on the second - order Møller - Plesset perturbation.

Acknowledgments

The present study is supported by Grants-in-Aid for Scientific Research form the Japanese Ministry of Education, Science, Sports, and Culture.

References

(1) Chandler, D.; Andersen, H. C. *J. Chem. Phys.* **1972**, *57*, 1930 .
(2) Hirata, F.; Rossky, P. J. *Chem. Phys. Lett.* **1981**,*84*, 329 .
(3) Ten-no, S.; Hirata, F.; Kato, S. *Chem. Phys. Lett.* **1993**, *214*, 391:
Ten-no, S.; Hirata F.; Kato, S. *J. Chem. Phys.* **1994**, *100*, 7443.
(4) Sato, H.; Hirata, F.; Kato, S. *J. Chem. Phys.* **1996**, *105*, 1546.
(5) Kawata, M.; Ten-no, S.; Kato, S.; Hirata, F. *Chem. Phys.* **1996**, *203*, 53.
(6) Kawata, M.; Ten-no, S.; Kato, S.; Hirata, F. *J. Phys. Chem.* **1996**, *100*, 1111.
(7) Pearson, R. G. *J. Am. Chem. Soc.* **1986**, *108*, 6109.
(8) Dobos, D. *Electrochemical data* ; a handbook for electrochemists in industry and universities, elsevier scientific publishing Inc., Amsterdam, 1975.
(9) Tuckerman, M.; Laasonen, K.; Sprik, M.; Parrinello, M. *J. Phys. Chem.* **1995**, *99*, 5749.
(10) Chipot, C.; Gorb, L. G.; Rivail, J.-L. *J. Phys. Chem.* **1994**, *98*, 1601.
(11) Tortonada, F. R.; Pascual - Ahuir, J.-L.; Silla, E.; Tuñón, I. *J. Phys. Chem..* **1993**, *97*, 11087.
(12) Emsley, J.; Lucas, J.; Overill, R. E. *Chem. Phys. Lett.* **1981**, *84*, 593.
(13) Gao, J. *Acc. Chem. Res.* **1996**, *29*, 298.
(14) Tanaka, Y.; Shiratori, Y.; Nakagawa, S. *Chem. Phys. Lett.* **1990**, *169*, 513.
(15) Singer, S. J.; Chandler, D. *Mol. Phys.* **1985**, *55*, 621.
(16) Maw, S.; Sato, H.; Ten-no, S.; Hirata, F. *Chem. Phys. Lett.* **1997**, *276*, 20.
(17) Poirier, R.; Kari, R.; Csizmadia, I. G. *Physical science data 24, Handbook of gaussian basis set*, Elsevier: New York, 1985.
(18) Berendsen, H. J. C.; Postma, J. P. M.; van Gunstern, E. F. J. Hermans, in *Intermolecular Forces*, ed. Pullman, B., Ed; Reidel, Dordrecht, 1981.
(19) *CRC Handbook of chemistry and physics,* 68th ed.; CRC Press: Boca Raton, FL, 1987.
(20) Chong, S.-H.; Hirata, F. *J. Phys. Chem. B* **1997**, *101*, 3209.

Chapter 13

Universal Solvation Models

Gregory D. Hawkins[1], Tianhai Zhu[1], Jiabo Li[1], Candee C. Chambers[1,3],
David J. Giesen[1,4], Daniel A. Liotard[2], Christopher J. Cramer[1],
and Donald G. Truhlar[1]

[1]Department of Chemistry and Supercomputer Institute, University of Minnesota,
Minneapolis, MN 55455
[2]Laboratoire de Physico-Chimie Theorique, Universite de Bordeaux 1, 351 Cours
de la Liberation, 33405 Talence Cedex, France

This chapter presents an overview of the SM5 suite of universal solvation models for computing free energies of solvation in water and nonaqueous solvents. After a general review of the theoretical components of all the SM5 solvation models, we specifically compare the performance of those that have been parameterized for both aqueous and organic solvents. These are called the universal solvation models, and they include models based on semiempirical neglect of diatomic differential overlap molecular orbital theory, density functional theory, and *ab initio* Hartree-Fock theory, and also a model with implicit electrostatics.

The combination of quantum mechanics (QM) for a solute and classical mechanics (CM) for a solvent can be achieved in two main ways. In the first (*1,2*), the solvent is represented atomistically, solvent-solvent interactions are represented by an analytical force field, and solvent-solute interactions are treated by some combination of analytical force field terms (e.g., Lennard-Jones potentials) and addition of a classical potential perturbing the solute Hamiltonian and thereby the electronic structure. Analytical force fields are usually called a molecular mechanics treatment, and this kind of QM/CM treatment is called QM/MM. An alternative classical treatment of the solvent is to treat it as a continuum (*3–5*). Most continuum solvent calculations include only electrostatic interactions, in which case the solvent is treated as a dielectric medium, and the relevant classical equation is the Poisson equation or, for nonzero ionic strength, the Poisson-Boltzmann equation. However, we have proposed a series of QM/CM solvation models (*6–22*) in which, in addition to electrostatics, we include microscopic analogs of classical thermodynamic surface tension terms. These are called atomic surface tensions, and they account for short-range effects of a continuum solvent, especially those due to the first solvation shell.

Just as QM/MM calculations require a protocol for joining the QM solute and the MM solvent, the appropriate formalism for joining a solute to a dielectric continuum is the

[3]Current address: Departments of Chemistry and Physics, Mercyhurst College, 501 East 38th Street, Erie, PA 16546
[4]Eastman Kodak Company, Rochester, NY 14650

much older reaction field concept. For a polarizable solute this requires self-consistent reaction field (SCRF) theory (*3–5*). The formalism for a continuum treatment of first-solvation-shell effects is the concept of solvent-accessible surface area (SASA) (*23,24*). When a realistic solvent radius is employed to calculate the SASA, it equals the area of a surface through the center of the first solvation shell, and thus—in a continuum sense—it is proportional to the "average" number of solvent molecules in this shell.

The explicit treatment of solvent molecules in QM/MM methods leads to three significant difficulties. First is the fact that conventional MM methods (*25–29*) do not account for solvent electronic polarizability. Second is the slow convergence of solute-solvent interactions with respect to distance (*30–32*); this requires careful convergence of sums of long-range Coulomb forces. Third is the necessity for Monte Carlo or molecular dynamics averaging over the myriad of configurations that the explicit solvent molecules may assume with respect to the solute and each other. None of these issues is a difficulty for continuum models. First of all solvent bulk polarizability is included through the solvent's dielectric constant, and short-range non-bulk polarizability effects are included in the atomic surface tensions. Second, long-range electrostatic effects are not truncated because they are much more easily included (*10*) in a continuum theory than in a many-body simulation. Third, conformational averaging is implicit in the parameters that characterize the continuum, and explicit averaging over solvent positions is not required. Of course there are also disadvantages. For example, one cannot examine details of the solvent structure, and it is less obvious how to include nonequilibrium solvation effects. However, we shall concentrate here on the calculation of free energies of solvation under equilibrium conditions, so those disadvantages will not concern us.

A critical element in modern continuum models is that the boundary between the atomic solute and the continuum solvent is not required to have a simple shape, e.g., spherical, spheroidal, or ellipsoidal, as in the early days of Onsager, Kirkwood, and the 1970s and early 1980s (*33–36*). Rather the shape of the solute is determined by a superposition of partially overlapping atomic spheres, as in space filling models. A critical element in all our solvation models to date is that we use the generalized Born approximation (GBA) to the Poisson equation (*6,37-42*) for the electrostatic SCRF term. The solute shape enters this term through a dielectric descreening algorithm based on the work of Still *et al.* (*4*). We have improved this algorithm numerically (*10*), but the physical idea is unchanged. In particular, electric polarization of the solvent screens the solute's intramolecular interactions because those field lines pass through the dielectric medium representing the volume occupied by solvent (*43,44*). However, the interactions in one part of the solute may be descreened from the solvent by another part of the solute; thus the solvent screening and solute descreening depend on solute shape. We take account of this in one of two ways. The original approach (*6,10,42*) is to integrate over the electric polarization free energy density in each volume element of space occupied by solvent. This is called volume descreening. The second approach (*13,15,44*) is to reduce the descreening effect of solute atom A on solute atom B to an effective pair potential; this is called pairwise descreening (PD).

We originally developed a series of solvation models labeled variously SM1 through SM4 (*6–13*). These models combined the GBA with volume descreening for the SCRF term with the AM1 (*45–48*) and PM3 (*49,50*) general parameterizations of neglect-of-diatomic-differential-overlap (NDDO) (*51*) semiempirical molecular orbital theory and used an evolving series of functional forms for atomic surface tensions. The atomic surface tensions were written in terms of surface tension coefficients whose values were fit to training sets of experimental free energies of solvation. In some cases, the atomic surface tensions were functions of bond orders computed from the SCRF electronic density. Models SM1–3.1 (*6–10,13*) were parameterized for aqueous solution, and model SM4 (*11,12*) was parameterized for alkane solvents. In order to make the electrostatic portions of the models more realistic, we also developed a class IV charge model called Charge Model 1 (CM1) (*52*), and this charge model was used to create the reaction field in the SM4 models. Solvation models SM1–SM4 were successful in their own right, especially SM2–SM4. In addition they provided insight into the theoretical underpinnings required for a successful solvation model.

With this experience as background, we embarked on a project to create universal solvation models, by which we mean solvation models that may be applied consistently in water and also in any organic solvent. Furthermore, motivated by the continuing usefulness of a variety of "levels" in gas-phase electronic structure theory, e.g., the NDDO level, the *ab initio* Hartree-Fock (HF) level (*53*), and density functional theory (DFT) (*54*), and motivated by the continuing usefulness of a variety of basis sets (*55*) and charge models (*52,56–60*), we have created several levels of universal solvation models based on different charge models and descreening algorithms and parameterized them for a variety of solute-electronic-structure levels and basis sets. In particular we have created universal solvation models with class IV charges, class II charges, and implicit charges and with volume and pairwise descreening, and we parameterized them for the MNDO (*61*) and MNDO/d (*62–64*) general parameterizations of NDDO theory as well as AM1 and PM3 and also for *ab initio* HF and DFT. For HF and DFT, the surface tension coefficients depend on basis set, and we have begun to find the parameters for several useful basis sets. (The universal models, i.e., those parameterized for both water and organic solvents, all use volume descreening; models based on pairwise descreening (*15*) have been parameterized only for aqueous solution.) Finally we note we have distinguished two modes of treating the solute geometry: the "flexible" mode in which the solute geometry is re-optimized in solution and the "rigid" mode in which it is not; in the latter (*19–22*) the effect of solute geometry relaxation in the vicinity of equilibrium structures is included implicitly in the parameters. The purpose of this chapter is to summarize and compare these universal solvation models (*14–22*), which are called as a group the SM5 suite of solvation models.

Overview of SM5 Solvation Models

The distinguishing characteristic of SM5 solvation models is the set of functional forms used for the atomic surface tensions. They depend on solute bond distances, in contrast to the earlier functional forms that depended on solute bond orders. They also depend on a small set of solvent descriptors. All SM5 solvation models use the same basic approach to atomic surface tensions, although there has been some evolution from the earlier SM5 models (*14–19*) to the later ones (*20–22*) in which these forms have become finalized.

A great deal of the effort involved in creating the SM5 suite of solvation models has been the development of a broad training set of experimental data. The training set includes a wide representation of solute functional groups and solvents, and the solutes selected for the training set are essentially free of complicating conformational issues. Furthermore we have made an attempt to eliminate incorrect and unreliable experimental data. The training set has evolved (*6–22*); in the present article all statistical characterization of SM5 models will be carried out using the final SM5 training set, and the reader should realize that this data set is larger than the training set actually used to parameterize the earlier (*14–19*) models. Although the sources of experimental data for the final SM5 training set are diverse (full references are given in our original papers; see especially Ref. (*20*)), we note that five sources of data proved especially useful (*65–69*).

A critical component in our ability to parameterize the models for general organic solvents is the use of a minimal number of descriptors to specify a solvent. In the final version of the SM5 models (*20–22*), the atomic surface tension coefficients depend on six solvent descriptors for organic solvents:

n the index of refraction for visible light

α Abraham's hydrogen bond acidity parameter $\Sigma\alpha_2^H$ (*71*) β   Abraham's hydrogen bond basicity parameter $\Sigma\beta_2^H$ (71)

γ   the macroscopic surface tension at a solvent-air interface

φ   the number of aromatic carbon atoms in the solvent divided by the total number of non-hydrogenic atoms in the solvent

ψ   the number of F, Cl, and Br atoms in the solvent divided by the total number of nonhydrogenic atoms in the solvent.

In addition the SCRF term involves:

ε   the dielectric constant

For aqueous solution, we again use ε in the SCRF term, but we do not require the functional dependence of the atomic surface tensions on n, α, β, γ, φ, and ψ to hold for water. Instead we parameterize water separately because of its unique properties. Furthermore, because the water training set is more sensitive to electrostatic effects than is the organic one, we parameterize water first in order to determine acceptable values for the intrinsic Coulomb radii (which is what we call the atomic radii used in the SCRF calculation; we also call them electrostatic radii); these intrinsic Coulomb radii are then fixed at the same values for calculations in organic solvents.

The calculations also require atomic radii for the calculation of first-solvation-shell effects. For these radii, all SM5 models use Bondi's van der Waals radii (70) without any adjustment.

Beginning with Ref. (20), all functional forms, radii, and training sets are frozen, and only the solute wave function and charge model and the surface tension coefficients change in Refs. (21) and (22).

We checked for two different SM5 models whether including functions of ε in the solvent descriptors for surface tensions would reduce the mean errors. Interestingly, it does not, at least to any meaningful extent This may indicate that our electrostatic radii are reasonable. If they were not reasonable, then effects associated with the "first solvation shell" would be expected to have a significant electrostatic component.

The SM5 family

The SM5 family now consists of seven solvation models, some of which have been parameterized for more than one method for obtaining the solute wave function. The models and their distinguishing features are

SM5.42R	rigid model based on class IV charges obtained by Charge Model 2 (CM2) (60) and on volume descreening
SM5.4	flexible model based on class IV charges obtained by CM1 (52) and on volume descreening
SM5.4PD	flexible model based on class IV charges obtained by CM1 (52) and on pairwise descreening
SM5.2R	rigid model based on class II charges (56,57) and on volume descreening
SM5.2PD	flexible model based on class II charges (56) and on pairwise descreening
SM5.0R	rigid model employing only atomic surface tensions (i.e., all the charges are set equal to zero, and electrostatic effects are included only implicitly though the atomic surface tensions)
SM5.05R	like SM5.0R except that class I charges are added to some of the atoms in ions and zwitterions, and electrostatic interactions are calculated using volume descreening.

Table 1. Parameterizations of universal solvation models

model	solute wave function	Ref.
SM5.42R	BPW91/MIDI!(6D)	(21)
	BPW91/DZVP	(21)
	HF/MIDI!(6D)	(22a)
	AM1	(22b)
	PM3	(22b)
SM5.4	AM1	(14,16–18)
	PM3	(14,16–18)
SM5.2R	AM1	(20)
	PM3	(20)
	MNDO	(20)
	MNDO/d	(20)
SM5.0R	–	(19)

Note that .4, .2, or .0 denotes the choice of charge model (in .0 models, the charges are zero, or, more precisely, they are implicit in the surface tensions), R or its absence denotes rigid or flexible, and PD or its absence denotes the choice between pairwise or volume descreening. The highest possible level would be an SM5.42 flexible model with volume descreening, but we have not created such a model yet.

For levels employing class II or higher charges, the surface tension parameters are reoptimized for each level of solute wave function (because these empirical parameters can make up for any systematic deficiencies in a given choice of solute level). Table 1 gives a complete list of the solute wave functions for which each of the five levels that involve either class II and class IV charges has been optimized to date. The notation indicates the wave function after a slash (solidus). Thus SM5.42/HF/MIDI!(6D) indicates that the wave function is calculated at the HF/MIDI!(6D) level. The basis sets examined so far are MIDI!(6D) (60,72) and DZVP (73).

The rigid models are designed to be used with any reasonable gas-phase geometry, and some attempt was made to parameterize them in such a way as to not be overly sensitive to small changes in geometries (although some residual sensitivity remains, especially in SM5.42R models due to the dependence of CM2 charges on geometry). As a matter of completeness, we note that *all* rigid models are parameterized using HF/MIDI! geometries. (Note: The MIDI! basis set (72) uses five spherical harmonic d functions in each d shell, and the MIDI!(6D) basis (60, 72) differs only in that it uses six Cartesian d functions in each d shell. The results obtained with these two basis are very similar. The MIDI! basis is therefore preferred because it is smaller, but some program packages do not support spherical harmonic d functions, so the MIDI!(6D) basis set can be used with a larger number of electronic structure programs.)

Comparison of Universal SM5 Solvation Models

In this section we compare the universal SM5 solvation models. There are twelve universal parameterizations, and they are listed in Table 1.

In the rest of this paper, all results for flexible models are based on separate geometry optimizations in the gas phase and in solution, and all results for rigid models are based on HF/MIDI! geometries, with one exception. The exception is the SM5.0R model, where results are shown for HF/MIDI! gas-phase geometries, AM1 gas-phase geometries, and MM3* gas-phase geometries, where MM3* is the MACROMODEL (74) implementation of the MM3 (29,75–77) force field. Ordinarily these would be denoted SM5.0R//HF/MIDI!, SM5.0R//AM1, and SM5.0R//MM3*. In the present paper, to simplify the discussion and tables, we will use the convention that //HF/MIDI! is implied for rigid models when no geometry is indicated.

All SM5 models are parameterized for solutes containing H, C, N, O, F, S, Cl, Br, and I, and—in addition—some of them are parameterized for solutes containing P. The final SM5 training set has 2135 solvation free energies for 275 neutral solutes in 91 solvents and 43 free energies of solvation for ions in water. In this section we will not discuss phosphorus-containing compounds or ions. For neutral solutes without phosphorus, the final SM5 training set has 2084 solvation free energies for 260 solutes in 91 solvents. The non-phosphorus solutes are classified into 31 classes and the solvents into 19.

Tables 2 and 3 show mean unsigned errors for each of the solvent classes. The bottom row gives the overall mean unsigned error, which ranges from 0.38 to 0.48 kcal/mol. The models with class II charges and those lacking charges altogether perform better than the models with class IV charges. Nevertheless, based on our extensive validation for gas-phase dipole moments (60), we believe that the models with class IV charges are more physical and will be more reliable for solute functionalities that are not represented in the training set and transition states. The robust nature of the SM5 suite is indicated by the remarkably even spread of errors for individual solvent classes. Tables 4 and 5 show the breakdown by solvent class of the mean *signed* errors. Here there is more variation in the errors, with free energies of solvation systematically overestimated or underestimated for certain solvent classes. The most poorly treated solvent classes are nitriles, nitro compounds, and haloaromatics. Nevertheless, the results are quite encouraging; the median absolute value of the solvent-class mean signed errors in Tables 4 and 5 is only 0.12 kcal/mol, and the mean absolute value of these 247 mean signed errors is only 0.17 kcal/mol. The SM5.2R models, especially those based on MNDO and MNDO/d, do the best at minimizing these systematic errors, and the SM5.42R/AM1 and SM5.42R/PM3 models do the least well.

Tables 6 and 7 show the breakdown of mean unsigned errors by solute class, and Tables 8 and 9 show the breakdown of mean signed errors by solute class. The single most significant conclusion that one can draw from these tables is that the various models and parameterizations do about equally well. This is extremely encouraging; apparently the empirical surface tensions do an excellent job of making up for any systematic deficiencies in the wave functions and partial atomic charges.

The magnitudes of the class II charges appear to be somewhat smaller, on average, than the magnitudes of class IV charges. In general, we see that there is a tendency for models that underestimate the accurate partial atomic charges or treat them implicitly to be slightly easier to parameterize for free energies of solvation. This may be due to reducing instability in the electrostatic portion of the calculation.

Table 10 is of special interest for this symposium because it demonstrates the use of a rigid SM5 model with gas-phase geometries that were calculated using the molecular mechanical method MM3*. Since the MM3* model does not have sufficient types to determine geometries for H_2 or solutes containing nitro groups the 7 such molecules in our testing set were omitted. In addition, we limited our examination to solvation free energies in organic solvents (not water), for a total of 220 neutral solutes and 1802 data points. For this set of solutes, Table 10 shows that the SM5.0R model applied with MM3* gas-phase geometries produces extremely similar results to the SM5.0R model applied to HF/MIDI! geometries. The overall mean unsigned error is 0.38 kcal/mol using either set of geometries. The mean signed error over the whole set of molecules increases only from 0.00 to 0.02 when using the MM3* geometries versus the HF/MIDI! geometries with which the SM5.0R model was parameterized. For comparison, results for the SM5.4/AM1 model, which is based on self-consistently relaxed solute geometries, are also shown.

A question often asked by referees is: how well would the model do if the test set included molecules not in the training set? To address this, we performed a series of systematic tests for the SM5.2R/MNDO/d model in which we omitted one quarter of the

Table 2. Mean Unsigned Error (kcal/mol) of Selected SMx Models by Solvent Functional Group Class

| Solvent Class | Number of | | | SM5.42R/BPW91/... | | SM5.42R/HF/ MIDI!(6D) | SM5.4/... | |
	Solvents[a]	Solute classes[b]	Data[c]	DZVP	MIDI!(6D)		AM1	PM3
Aqueous	1	31	248	0.41	0.43	0.46	0.55	0.49
Alkanes	11	30	475	0.32	0.33	0.32	0.33	0.31
Cycloalkanes	2	24	106	0.38	0.38	0.40	0.41	0.39
Arenes	12	16	256	0.38	0.38	0.39	0.42	0.38
Aliphatic alcohols	12	31	299	0.57	0.55	0.52	0.57	0.53
Aromatic alcohols	2	7	12	0.59	0.55	0.53	0.71	0.68
Ketones	4	10	35	0.43	0.49	0.45	0.48	0.40
Esters	2	8	36	0.53	0.60	0.54	0.44	0.42
Aliphatic ethers	4	19	99	0.49	0.52	0.52	0.57	0.50
Aromatic ethers	3	5	15	0.53	0.59	0.53	0.32	0.30
Amines	2	6	12	0.32	0.36	0.37	0.66	0.56
Pyridines	3	5	15	0.46	0.49	0.40	0.33	0.34
Nitriles	2	5	10	0.69	0.76	0.59	0.54	0.46
Nitro compounds	4	8	27	0.56	0.63	0.58	0.68	0.60
Tertiary amides	2	5	10	0.37	0.38	0.26	0.33	0.31
Haloaliphatics	12	27	269	0.46	0.47	0.47	0.58	0.48
Haloaromatics	6	11	106	0.44	0.40	0.38	0.50	0.45
Miscellaneous acidic solvents	3	5	15	0.49	0.48	0.40	0.41	0.37
Miscellaneous non-acidic solvents	4	12	39	0.41	0.42	0.43	0.37	0.35
Total:	91	31	2084	0.43	0.43	0.43	0.48	0.43

[a]Number of solvents in this solvent class

[b]Number of solutes classes for which data exists in this solvent class. The solute classes are listed in Tables 6–9.

[a]Total number of solute/solvent data involving this solvent class

Table 3. Mean Unsigned Error (kcal/mol) of Selected SMx Models by Solvent Functional Group Class

Solvent Class	SM5.42R/...		SM5.2R/...				SM5.0R//...	
	AM1	PM3	MNDO/d	MNDO	AM1	PM3	HF/MIDI!	AM1
Aqueous	0.42	0.43	0.47	0.47	0.44	0.51	0.53	0.57
Alkanes	0.31	0.32	0.29	0.29	0.32	0.30	0.29	0.31
Cycloalkanes	0.39	0.39	0.36	0.36	0.37	0.35	0.35	0.36
Arenes	0.40	0.39	0.30	0.30	0.34	0.33	0.28	0.29
Aliphatic alcohols	0.53	0.54	0.46	0.47	0.53	0.49	0.44	0.45
Aromatic alcohols	0.54	0.54	0.49	0.49	0.52	0.51	0.48	0.48
Ketones	0.53	0.53	0.35	0.35	0.42	0.40	0.42	0.41
Esters	0.52	0.54	0.50	0.50	0.59	0.56	0.51	0.53
Aliphatic ethers	0.52	0.53	0.46	0.46	0.50	0.47	0.44	0.43
Aromatic ethers	0.52	0.54	0.43	0.43	0.49	0.46	0.37	0.34
Amines	0.39	0.41	0.45	0.45	0.43	0.40	0.55	0.55
Pyridines	0.42	0.42	0.30	0.30	0.43	0.38	0.32	0.35
Nitriles	0.76	0.73	0.45	0.45	0.60	0.56	0.50	0.47
Nitro compounds	0.58	0.64	0.55	0.55	0.47	0.55	0.47	0.49
Tertiary amides	0.34	0.32	0.23	0.23	0.36	0.34	0.43	0.47
Haloaliphatics	0.48	0.48	0.45	0.45	0.45	0.45	0.53	0.51
Haloaromatics	0.54	0.51	0.27	0.27	0.34	0.29	0.29	0.28
Miscellaneous acidic solvents	0.48	0.47	0.38	0.38	0.52	0.48	0.31	0.36
Miscellaneous non–acidic solvents	0.39	0.40	0.42	0.42	0.46	0.44	0.58	0.59
Total:	0.43	0.44	0.38	0.39	0.41	0.40	0.40	0.41

Table 4. Mean Signed Error (kcal/mol) of Selected SMx Models by Solvent Functional Group Class

Solvent Class	Number of			SM5.42R/BPW91/...		SM5.42R/HF/MIDI!(6D)	SM5.4/...	
	Solvents[a]	Solute classes[b]	Data[c]	DZVP	MIDI!(6D)		AM1	PM3
Aqueous	1	31	248	-0.01	-0.03	-0.03	0.04	0.04
Alkanes	11	30	475	-0.03	-0.06	-0.04	-0.10	-0.11
Cycloalkanes	2	24	106	0.14	0.07	0.01	-0.02	-0.06
Arenes	12	16	256	0.25	0.19	0.24	0.20	0.19
Aliphatic alcohols	12	31	299	0.00	-0.05	-0.08	-0.11	-0.12
Aromatic alcohols	2	7	12	0.25	0.07	0.07	0.51	0.44
Ketones	4	10	35	-0.14	-0.20	-0.15	-0.35	-0.17
Esters	2	8	36	0.41	0.37	0.36	0.27	0.29
Aliphatic ethers	4	19	99	0.07	-0.02	-0.05	0.02	0.07
Aromatic ethers	3	5	15	-0.44	-0.58	-0.51	-0.02	-0.01
Amines	2	6	12	0.19	0.03	0.01	0.56	0.46
Pyridines	3	5	15	-0.17	-0.31	-0.24	0.10	0.13
Nitriles	2	5	10	-0.66	-0.76	-0.59	-0.54	-0.41
Nitro compounds	4	8	27	-0.36	-0.30	-0.22	-0.65	-0.52
Tertiary amides	2	5	10	-0.03	-0.14	-0.04	-0.12	-0.08
Haloaliphatics	12	27	269	-0.10	-0.12	-0.11	0.01	-0.01
Haloaromatics	6	11	106	-0.34	-0.29	-0.24	-0.30	-0.23
Miscellaneous acidic solvents	3	5	15	0.16	0.02	-0.05	0.02	-0.04
Miscellaneous non-acidic solvents	4	12	39	-0.08	-0.10	-0.08	-0.07	-0.13
Total:	91	31	2084	0.00	-0.04	-0.03	-0.03	-0.03

[a] Number of solvents in this solvent class
[b] Number of solutes classes for which data exists in this solvent class. The solute classes are listed in Tables 6–9.
[c] Total number of solute/solvent data involving this solvent class

Table 5. Mean Signed Error (kcal/mol) of Selected SMx Models by Solvent Functional Group Class

Solvent Class	SM5.42R/...		SM5.2R/...				SM5.0R//...	
	AM1	PM3	MNDO/d	MNDO	AM1	PM3	HF/MIDI!	AM1
Aqueous	-0.04	-0.04	-0.05	-0.07	-0.03	-0.01	0.06	0.11
Alkanes	-0.03	-0.04	-0.05	-0.05	-0.06	-0.05	-0.05	-0.06
Cycloalkanes	0.13	0.12	0.00	0.00	0.07	0.04	-0.07	-0.08
Arenes	0.27	0.27	0.08	0.08	0.16	0.13	-0.05	-0.07
Aliphatic alcohols	-0.06	-0.06	-0.07	-0.09	-0.03	-0.04	-0.03	0.00
Aromatic alcohols	0.17	0.15	0.16	0.16	0.16	0.16	0.14	0.16
Ketones	-0.27	-0.24	0.02	0.02	-0.07	-0.02	0.25	0.27
Esters	0.36	0.35	0.42	0.42	0.50	0.46	0.43	0.46
Aliphatic ethers	-0.02	-0.01	0.02	0.02	0.08	0.07	0.09	0.11
Aromatic ethers	-0.48	-0.49	-0.40	-0.40	-0.43	-0.43	-0.33	-0.29
Amines	0.22	0.19	-0.04	-0.04	0.10	0.03	-0.21	-0.18
Pyridines	-0.24	-0.23	-0.09	-0.09	-0.13	-0.11	0.03	0.10
Nitriles	-0.76	-0.73	-0.27	-0.27	-0.53	-0.40	0.06	0.11
Nitro compounds	-0.41	-0.37	-0.02	-0.01	-0.23	-0.14	0.25	0.24
Tertiary amides	-0.10	-0.07	0.13	0.13	0.03	0.09	0.28	0.37
Haloaliphatics	-0.16	-0.15	-0.05	-0.05	-0.10	-0.07	0.04	0.02
Haloaromatics	-0.43	-0.41	-0.06	-0.06	-0.22	-0.15	0.14	0.15
Miscellaneous acidic solvents	0.19	0.19	0.16	0.16	0.17	0.18	0.10	0.15
Miscellaneous non-acidic solvents	-0.05	-0.05	-0.13	-0.13	-0.21	-0.18	-0.21	-0.15
Total:	-0.03	-0.03	-0.02	-0.03	-0.02	-0.01	0.01	0.02

Table 6. Mean Unsigned Error of Selected SMx Models by Solute Functional Group Class

Solute Class	Number of			SM5.42R/BPW91/...		SM5.42R/HF/ MIDI!(6D)	SM5.4/...	
	Solutes[a]	Solvent classes[b]	Data[c]	DZVP	MIDI!(6D)		AM1	PM3
Unbranched alkanes	9	19	84	0.45	0.40	0.38	0.42	0.44
Branched alkanes	5	3	12	0.40	0.42	0.45	0.50	0.51
Cycloalkanes	5	6	18	0.35	0.52	0.43	0.81	0.83
Alkenes	9	4	27	0.33	0.28	0.25	0.48	0.35
Alkynes	5	3	14	0.26	0.21	0.16	0.26	0.21
Arenes	9	19	134	0.50	0.57	0.51	0.31	0.29
Alcohols	17	19	385	0.37	0.37	0.36	0.38	0.36
Ethers	12	19	93	0.42	0.48	0.44	0.47	0.43
Aldehydes	7	8	38	0.55	0.41	0.41	0.44	0.45
Ketones	12	18	203	0.44	0.44	0.40	0.42	0.42
Carboxylic acids	5	14	124	0.41	0.45	0.43	0.64	0.62
Esters	14	8	249	0.33	0.33	0.34	0.49	0.45
Bifunctional compounds containing H, C, O	5	8	28	1.16	1.07	1.11	0.87	0.86
Inorganic compounds containing H and O	2	9	22	0.60	0.57	0.63	1.28	1.06
Aliphatic Amines	15	10	168	0.36	0.34	0.31	0.40	0.34
Aromatic Amines	11	12	81	0.30	0.33	0.35	0.46	0.44
Nitriles	4	6	22	0.48	0.44	0.90	0.42	0.38
Nitrohydrocarbons	6	8	38	0.55	0.43	0.40	0.53	0.23
Amides & ureas	4	6	11	1.08	1.24	1.29	1.93	0.93
Bifunctional compounds containing N	6	3	11	0.61	0.69	0.70	1.76	1.03
Inorganic compounds containing N	2	8	15	0.64	0.70	0.69	1.92	0.58
Thiols	4	5	14	0.31	0.33	0.42	0.34	0.29
Sulfides	6	6	23	0.61	0.59	0.68	0.72	0.68
Disulfides	2	3	5	0.23	0.25	0.19	0.10	0.10
Fluorinated hydrocarbons	9	5	19	0.35	0.49	0.54	0.49	0.33
Chloroalkanes	13	5	35	0.49	0.34	0.32	0.29	0.27
Chloroalkenes	5	4	16	0.65	0.65	0.61	0.46	0.38
Chloroarenes	8	6	37	0.33	0.43	0.61	0.32	0.28
Brominated hydrocarbons	14	6	50	0.34	0.36	0.35	0.25	0.28
Iodinated hydrocarbons	9	6	28	0.18	0.34	0.31	0.18	0.20
Multifunctional halogenated solutes	26	9	80	0.75	0.70	0.67	0.62	0.67
Total:	260	19	2084	0.43	0.43	0.43	0.48	0.43

[a] Number of solutes in this solute class [b] Number of solvent classes for which there are data for this solute class. The solvent classes are listed in Tables 2–5. [c] Total number of solute/solvent data involving solutes in this solute class

Table 7. Mean Unsigned Error (kcal/mol) of Selected SMx Models by Solute Functional Group Class

Solute Class	SM5.42R/...		SM5.2R/...				SM5.0R/...	
	AM1	PM3	MNDO/d	MNDO	AM1	PM3	HF/MIDI!	AM1
Unbranched alkanes	0.40	0.41	0.38	0.38	0.45	0.42	0.41	0.43
Branched alkanes	0.41	0.41	0.45	0.45	0.39	0.42	0.41	0.44
Cycloalkanes	0.33	0.31	0.34	0.34	0.31	0.30	0.38	0.37
Alkenes	0.31	0.28	0.21	0.21	0.31	0.25	0.20	0.19
Alkynes	0.15	0.14	0.13	0.13	0.18	0.17	0.11	0.15
Arenes	0.50	0.53	0.39	0.39	0.53	0.47	0.39	0.38
Alcohols	0.37	0.38	0.33	0.33	0.35	0.34	0.35	0.36
Ethers	0.51	0.49	0.39	0.39	0.40	0.39	0.43	0.45
Aldehydes	0.47	0.54	0.46	0.46	0.41	0.43	0.51	0.54
Ketones	0.42	0.44	0.33	0.33	0.34	0.35	0.36	0.37
Carboxylic acids	0.46	0.47	0.35	0.35	0.39	0.38	0.40	0.40
Esters	0.40	0.39	0.29	0.29	0.33	0.31	0.28	0.31
H, C, O Bifunctional compounds	1.00	1.00	0.86	0.86	1.07	0.99	0.85	0.83
H and O Inorganic compounds	0.68	0.66	0.58	0.58	0.57	0.58	0.65	0.65
Aliphatic Amines	0.34	0.31	0.27	0.27	0.32	0.30	0.26	0.32
Aromatic Amines	0.35	0.33	0.43	0.43	0.35	0.39	0.50	0.52
Nitriles	0.54	0.54	0.35	0.35	0.39	0.26	0.51	0.49
Nitrohydrocarbons	0.24	0.29	0.23	0.23	0.57	0.40	0.24	0.25
Amides & ureas	1.08	1.01	1.49	1.49	1.48	1.32	1.86	1.89
N Bifunctional compounds	0.66	0.64	0.96	0.96	0.88	0.91	1.02	1.02
N Inorganic compounds	0.69	0.57	0.57	0.57	0.62	0.53	0.71	0.73
Thiols	0.32	0.34	0.48	0.45	0.43	0.40	0.38	0.43
Sulfides	0.53	0.67	0.98	0.84	0.90	0.83	0.66	0.67
Disulfides	0.26	0.22	0.16	0.17	0.16	0.17	0.19	0.27
Fluorinated hydrocarbons	0.46	0.50	0.67	0.68	0.42	0.46	0.70	0.69
Chloroalkanes	0.40	0.48	0.27	0.36	0.37	0.51	0.38	0.41
Chloroalkenes	0.64	0.58	0.65	0.54	0.73	0.88	0.78	0.82
Chloroarenes	0.45	0.39	0.56	0.64	0.33	0.37	0.29	0.29
Brominated hydrocarbons	0.37	0.44	0.33	0.35	0.42	0.33	0.37	0.37
Iodinated hydrocarbons	0.26	0.37	0.41	0.48	0.49	0.37	0.43	0.41
Multifunctional halogenated solutes	0.71	0.72	0.64	0.69	0.58	0.79	0.68	0.68
Total:	0.43	0.44	0.38	0.39	0.41	0.40	0.40	0.41

Table 8. Mean Signed Error of Selected SMx Models by Solute Functional Group Class

Solute Class	Number of			SM5.42R/BPW91/...		SM5.42R/HF/ MIDI!(6D)	SM5.4/...	
	Solutes[a]	Solvent classes[b]	Data[c]	DZVP	MIDI!(6D)	MIDI!(6D)	AM1	PM3
Unbranched alkanes	9	19	84	0.21	0.09	0.10	0.13	0.17
Branched alkanes	5	3	12	0.08	0.13	0.18	-0.08	-0.10
Cycloalkanes	5	6	18	-0.31	-0.47	-0.32	0.71	0.77
Alkenes	9	4	27	0.19	-0.02	-0.04	0.39	0.27
Alkynes	5	3	14	-0.03	-0.01	0.02	-0.05	-0.04
Arenes	9	19	134	-0.41	-0.53	-0.47	-0.10	-0.12
Alcohols	17	19	385	-0.01	0.01	0.02	0.04	0.03
Ethers	12	19	93	0.11	0.04	0.10	0.02	0.11
Aldehydes	7	8	38	0.38	-0.08	-0.21	0.08	-0.03
Ketones	12	18	203	-0.25	-0.28	-0.19	-0.07	-0.07
Carboxylic acids	5	14	124	-0.05	-0.20	-0.11	0.30	0.26
Esters	14	8	249	0.08	0.13	0.10	-0.30	-0.27
Bifunctional compounds containing H, C, O	5	8	28	0.94	0.83	0.75	0.38	0.33
Inorganic compounds containing H and O	2	9	22	0.00	-0.01	0.03	-1.27	-1.05
Aliphatic amines	15	10	168	0.10	0.08	0.07	-0.10	0.00
Aromatic amines	11	12	81	0.00	-0.01	0.07	-0.33	-0.18
Nitriles	4	6	22	0.01	0.00	-0.90	-0.03	0.02
Nitrohydrocarbons	6	8	38	0.03	0.01	0.01	-0.05	-0.01
Amides & ureas	4	6	11	0.73	0.86	0.83	1.01	0.29
Bifunctional compounds containing N	6	3	11	-0.43	-0.43	-0.23	0.62	-0.69
Inorganic compounds containing N	2	8	15	-0.35	-0.42	-0.41	0.14	0.46
Thiols	4	5	14	0.24	0.31	0.39	0.05	0.08
Sulfides	6	6	23	-0.09	-0.16	-0.14	-0.43	-0.43
Disulfides	2	3	5	0.00	-0.01	0.00	-0.02	0.10
Fluorinated hydrocarbons	9	5	19	-0.22	-0.36	-0.42	0.45	0.26
Chloroalkanes	13	5	35	-0.45	-0.09	0.11	-0.19	-0.18
Chloroalkenes	5	4	16	0.65	0.65	0.62	0.45	0.37
Chloroarenes	8	6	37	0.11	-0.41	-0.59	-0.10	-0.13
Brominated hydrocarbons	14	6	50	-0.20	-0.27	-0.23	-0.10	-0.20
Iodinated hydrocarbons	9	6	28	0.01	-0.05	-0.02	-0.03	-0.04
Multifunctional halogenated solutes	26	9	80	0.30	0.29	0.21	0.03	-0.03
Total:	260	19	2084	0.00	-0.04	-0.03	-0.03	-0.03

[a] Number of solutes in this solute class [b] Number of solvent classes for which there are data for this solute class. The solvent classes are listed in Tables 2–5. [c] Total number of solute/solvent data involving solutes in this solute class

Table 9. Mean Signed Error (kcal/mol) of Selected SMx Models by Solute Functional Group Class

Solute Class	SM5.42R/...		SM5.2R/...				SM5.0R/...	
	AM1	PM3	MNDO/d	MNDO	AM1	PM3	HF/MIDI!	AM1
Unbranched alkanes	0.10	0.14	0.05	0.05	0.16	0.16	0.06	0.18
Branched alkanes	0.08	0.13	0.07	0.07	0.01	0.08	0.21	0.30
Cycloalkanes	-0.27	-0.19	-0.14	-0.14	-0.25	-0.12	-0.16	-0.07
Alkenes	0.20	0.17	-0.05	-0.05	0.21	0.11	-0.05	0.03
Alkynes	-0.01	-0.01	0.04	0.04	-0.01	0.00	0.04	0.09
Arenes	-0.44	-0.49	-0.27	-0.27	-0.42	-0.39	-0.13	-0.07
Alcohols	-0.04	-0.03	-0.01	-0.01	-0.02	-0.01	-0.01	0.02
Ethers	0.16	0.18	0.09	0.09	0.09	0.11	0.12	0.16
Aldehydes	-0.39	-0.49	-0.27	-0.27	-0.15	-0.23	-0.14	-0.21
Ketones	-0.13	-0.14	-0.09	-0.09	-0.18	-0.15	-0.07	-0.14
Carboxylic acids	-0.03	-0.04	-0.04	-0.04	-0.06	-0.06	0.05	-0.06
Esters	0.03	0.04	0.02	0.02	0.04	0.04	-0.04	-0.16
H, C, O Bifunctional compounds	0.83	0.81	0.41	0.41	0.76	0.62	0.24	0.19
H and O Inorganic compounds	0.00	0.00	0.00	0.00	-0.02	0.00	-0.08	-0.09
Aliphatic amines	0.06	0.05	0.01	0.01	0.07	0.02	0.00	0.14
Aromatic amines	0.04	0.00	0.14	0.14	-0.02	0.14	0.18	0.25
Nitriles	0.00	0.00	0.09	0.09	0.32	-0.09	0.11	0.12
Nitrohydrocarbons	0.00	0.00	0.00	0.00	0.02	-0.06	-0.02	-0.05
Amides & ureas	0.64	0.63	1.38	1.38	1.38	1.26	1.51	1.49
N Bifunctional compounds	-0.36	-0.27	-0.89	-0.89	-0.83	-0.84	-0.96	-0.82
N Inorganic compounds	-0.40	-0.24	-0.18	-0.18	-0.28	-0.08	-0.15	-0.18
Thiols	0.28	0.33	0.34	0.30	0.31	0.29	0.28	0.37
Sulfides	-0.08	-0.01	-0.13	-0.10	-0.12	-0.06	-0.24	-0.10
Disulfides	0.00	0.00	0.00	0.00	0.00	0.00	0.00	0.18
Fluorinated hydrocarbons	-0.26	-0.29	-0.41	-0.44	-0.11	-0.10	0.14	0.23
Chloroalkanes	-0.34	-0.39	-0.01	-0.16	-0.27	0.43	0.09	0.03
Chloroalkenes	0.64	0.58	0.65	0.52	0.73	0.88	0.78	0.82
Chloroarenes	-0.40	-0.33	-0.54	-0.62	-0.18	-0.32	-0.08	-0.02
Brominated hydrocarbons	-0.31	-0.39	-0.19	-0.26	-0.10	-0.21	0.05	0.05
Iodinated hydrocarbons	-0.05	-0.09	-0.06	-0.04	-0.03	-0.09	-0.02	0.06
Multifunctional halogenated solutes	0.28	0.31	0.08	0.15	0.10	0.03	0.07	0.09
Total:	-0.03	-0.03	-0.02	-0.03	-0.02	-0.01	0.01	0.02

Table 10. Application of the SM5.0R model to molecular mechanics (MM3*) geometries in organic solvents. (Note nitrohydrocarbons and hydrogen were left out because not enough types in MM3* to run these molecules.)

Solute Class	Number of			SM5.4/AM1		SM5.0R		SM5.0R /MM3*	
	Solutes[a]	Classes[b]	Data[c]	MSE[d]	MUE[e]	MSE[d]	MUE[e]	MSE[d]	MUE[e]
Unbranched alkanes	9	18	76	0.21	0.40	0.12	0.40	0.22	0.43
Branched alkanes	5	2	7	0.35	0.37	0.49	0.52	0.56	0.57
Cycloalkanes	4	5	13	1.04	1.04	-0.06	0.35	0.01	0.35
Alkenes	8	3	18	0.42	0.49	-0.04	0.19	0.02	0.21
Alkynes	5	2	9	-0.07	0.27	0.02	0.12	0.08	0.14
Arenes	9	18	126	-0.11	0.32	-0.17	0.38	-0.11	0.36
Alcohols	17	18	369	0.03	0.38	-0.02	0.35	-0.02	0.35
Ethers	12	18	81	-0.04	0.43	0.12	0.41	0.13	0.41
Aldehydes	7	7	32	0.15	0.46	-0.12	0.54	-0.13	0.54
Ketones	12	17	191	-0.09	0.42	-0.10	0.36	-0.09	0.35
Carboxylic acids	5	13	119	0.28	0.64	0.03	0.39	0.03	0.39
Esters	13	7	236	-0.30	0.49	-0.02	0.28	-0.03	0.29
Bifunctional compounds containing H, C, O	4	7	23	0.44	0.98	0.41	0.92	0.41	0.92
Inorganic compounds containing H and O	1	8	18	-1.38	1.38	0.01	0.68	0.01	0.68
Aliphatic amines	11	9	153	-0.12	0.36	0.00	0.24	0.09	0.26
Aromatic amines	11	11	71	-0.29	0.42	0.16	0.43	0.23	0.44
Nitriles	4	5	18	-0.11	0.40	0.14	0.47	0.17	0.47
Amides & Ureas	2	5	7	1.01	1.93	1.32	1.66	1.38	1.73
Bifunctional compounds containing N	4	2	6	1.46	2.49	-0.88	0.97	-0.88	0.96
Inorganic compounds containing N	2	7	13	0.10	1.79	-0.21	0.66	-0.24	0.66
Thiols	3	4	10	0.07	0.35	0.37	0.40	0.40	0.42
Sulfides	6	5	17	-0.51	0.76	-0.21	0.71	-0.18	0.70
Disulfides	2	2	3	-0.04	0.04	0.00	0.27	0.10	0.28
Fluorinated hydrocarbons	5	4	13	0.39	0.45	0.02	0.35	0.07	0.51
Chloroalkanes	7	4	22	-0.21	0.30	0.05	0.35	0.06	0.37
Chloroalkenes	4	3	11	0.36	0.37	0.70	0.70	0.74	0.74
Chloroarenes	6	5	29	-0.11	0.35	-0.15	0.30	-0.10	0.29
Brominated hydrocarbons	14	5	36	-0.11	0.25	-0.02	0.36	0.01	0.36
Iodinated hydrocarbons	9	5	20	-0.01	0.15	-0.02	0.50	0.03	0.49
Multifunctional halogenated solutes	19	8	55	-0.01	0.61	0.13	0.64	0.13	0.63
Total:	220		1802	-0.03	0.48	0.00	0.39	0.05	0.39

[a]Number of solutes in this solute class [b]Number of solvent classes for which there are data for this solute class [c]Total number of solute/solvent data involving solutes in this solute class [d]Mean signed error over data in solute class [e]Mean unsigned error over solute class

data (over 500 data) from the training set which has 2082 data. Thus the model was trained on 1561–1562 data, but tested on the full 2082 data. The mean unsigned error increased by less than 0.01 kcal/mol, which is less than 3% of the mean unsigned error of 0.38 kcal/mol in Table 3. This is consistent with our general experience. We believe that the most robust model is obtained by parameterizing with a large, broad training set.

Although the present manuscript is primarily concerned with neutral solutes, we also tested the models for ions in aqueous solution. The SM5.2R/MNDO/d model yields a mean unsigned error of 3.8 kcal/mol for 43 ions without phosphorus, and the results for other models are similar. Considering that the experimental data are uncertain by about ±5 kcal/mol for ions (as compared to about ±0.2 kcal/mol for neutrals), this appears to be quite acceptable.

Platforms

The SM5.4/AM1 and /PM3 parameterizations, all the parameterizations of PD models, the SM5.0R model, and the SM5.05R model are available in the latest released version of AMSOL (78) and SM5.4R/AM1 and /PM3 and SM5.2R/AM1, /PM3, and /MNDO will be in the next version of AMSOL (79).

SM5.0R and SM5.05R are also available in OMNISOL (80).

SM5.2R/AM1, /PM3, /MNDO, and /MNDO/d have been implemented in AMPAC (81).

SM5.42R/BPW91/MIDI!(6D) and /BPW91/DZVP have been implemented in DGAUSS (82) and GAUSSIAN94 (83).

SM5.4R/HF/MIDI!(6D) has been implemented in GAUSSIAN94 (83) and GAMESS (84,85).

Additional implementations involving other density functionals and other basis sets are in progress.

Concluding remarks

The SM5 suite of solvation models provides reasonable accuracy for free energies of solvation over a broad range of solute functionality and solvent type. The results are just as accurate on the average (sometimes even more accurate!) when affordable low levels of electronic structure theory are used. This approach therefore provides a convenient way to mix a quantum mechanical treatment of the solute with a classical treatment of the solvent. Careful attention to the first solvation shell, which "links" the quantal and classical regions, allows one to obtain quantitative accuracy for both ions and neutrals of arbitrary shape.

Acknowledgments

This work was supported in part by the National Science Foundation and under the general computational methodology provisions of an Advanced Technology Project supported by the National Institutes of Science and Technology.

References

(1)   Bash, P. A.; Field, M. J.; Karplus, M. *J. Comp. Chem.* **1990**, *11*, 700.
(2)   Gao, J. *J. Phys. Chem.* **1992**, *96*, 537.
(3)   Tomasi, J.; Persico, M. *Chem. Rev.* **1994**, *94*, 2027.
(4)   Cramer, C. J.; Truhlar, D. G. In *Reviews in Computational Chemistry*, Vol. 6, Boyd, D. B.; Lipkowitz, K. B., Eds.; VCH: New York, 1995; p. 1.
(5)   Rivail, J.-L.; Rinaldi, D. In *Computational Chemistry: Reviews of Current Trends*, Vol. 1; Leszczynski, J., Ed.; World Scientific: Singapore, 1996; p. 139.
(6)   Cramer, C. J.; Truhlar, D. G. *J. Am. Chem. Soc.* **1991**, *113*, 8305, 9901(E).

(7) Cramer, C. J.; Truhlar, D. G. *Science* **1992**, *256*, 213.
(8) Cramer, C. J.; Truhlar, D. G. *J. Comp. Chem.* **1992**, *13*, 1089.
(9) Cramer, C. J.; Truhlar, D. G. *J. Comput.-Aided Mol. Des.* **1992**, *6*, 629.
(10) Liotard, D. A.; Hawkins, G. D.; Lynch, G. C.; Cramer, C. J.; Truhlar, D. G. *J. Comp. Chem.* **1995**, *16*, 422.
(11) Giesen, D. J.; Storer, J. W.; Cramer, C. J.; Truhlar, D. G. *J. Am. Chem. Soc.* **1995**, *117*, 1057.
(12) Giesen, D. J.; Cramer, C. J.; Truhlar, D. G. *J. Phys. Chem.* **1995**, *99*, 7137.
(13) Hawkins, G. D.; Cramer, C. J.; Truhlar, D. G. *Chem. Phys. Lett.* **1995**, *246*, 122.
(14) Chambers, C. C.; Hawkins, G. D.; Cramer, C. J.; Truhlar, D. G. *J. Phys. Chem.* **1996**, *100*, 16385.
(15) Hawkins, G. D.; Cramer, C. J.; Truhlar, D. G. *J. Phys. Chem.* **1996**, *100*, 19824.
(16) Giesen, D. J.; Gu, M. Z.; Cramer, C. J.; Truhlar, D. G. *J. Org. Chem.* **1996**, *61*, 8720.
(17) Giesen, D. J.; Chambers, C. C.; Cramer, C. J.; Truhlar, D. G. *J. Phys. Chem.* **1997**, *101*, 2061.
(18) Giesen, D. J.; Hawkins, G. D.; Liotard, D. A.; Cramer, C. J.; Truhlar, D. G. *Theor. Chem. Acc.* **1997**, *98*, 85.
(19) (a) Hawkins, G. D.; Cramer, C. J.; Truhlar, D. G. *J. Phys. Chem. B* **1997**, *101*, 7147. (b) Hawkins, G. D.; Cramer, C. J.; Truhlar, D. G. to be published.
(20) Hawkins, G. D.; Cramer, C. J.; Truhlar, D. G. *J. Phys. Chem. B*, to be published.
(21) Zhu, T.; Li, J.; Hawkins, G. D.; Cramer, C. J.; Truhlar, D. G. to be published.
(22) (a) Li, J.; Hawkins, G. D.; Cramer, C. J.; Truhlar, D. G. *Chem. Phys. Lett.* to be published. (b) Hawkins, G. D.; Cramer, C. J.; Truhlar, D. G. unpublished.
(23) Lee, B.; Richards, F. M. *J. Mol. Biol.* **1971**, *55*, 371.
(24) Hermann, R. B. *J. Phys. Chem.* **1972**, *76*, 2754.
(25) Burkert, U.; Allinger, N. L. *Molecular Mechanics*; American Chemical Society: Washington, D.C., 1982.
(26) Weiner, S. J.; Kollman, P. A.; Nguyen, D. T.; Case, D. A. *J. Comp. Chem.* **1986**, *7*, 230.
(27) Brooks, C. R.; Bruccoleri, R. E.; Olafson, B. D.; States, B. J.; Swaminathan, S.; Karplus, M. *J. Comp. Chem.* **1983**, *4*, 187.
(28) Hagler, A. T.; Ewig, C. S. *Comput. Phys. Commun.* **1994**, *84*, 131.
(29) Allinger, N. L.; Zhou, X. F.; Bergsma, J. *J. Mol. Struct. (Theochem)* **1994**, *118*, 69.
(30) (a) Berendsen, H. J. C. In *Computer Simulation of Biomolecular Systems*, Vol. 2; van Gunsteren, W. F., Weiner, P. K., Wilkonson, A. J., Eds.; ESCOM: Leiden, 1993; p. 161. (b) Smith, P. E.; van Gunsteren, W. F. *ibid.*, p. 182.
(31) Smith, P. E.; Pettit, B. M. *J. Chem. Phys.* **1991**, *95*, 8430.
(32) (a) Lee, F. S.; Warshel, A. *J. Chem. Phys.* **1992**, *97*, 3100. (b) Tasaki, K.; McDonald, S.; Brady, J. W. *J. Comp. Chem.* **1993**, *14*, 278. (c) Guernot, J.; Kollman, P. A. *J. Comp. Chem.* **1993**, *14*, 295.
(33) Kirkwood, J. G. *J. Chem. Phys.* **1934**, *351*.
(34) Onsager, L. *J. Am. Chem. Soc.* **1936**, *58*, 1486.
(35) Harrison, S. W.; Nolte, N. J.; Beveridge, D. *J. Phys. Chem.* **1976**, *80*, 2580.
(36) (a) Rivail, J. L.; Terryn, B. *J. Chim. Phys.* **1982**, *79*, 1. (b) Rinaldi, D. *Comput. and Chem.* **1982**, *6*, 155.
(37) Hoijtink, G. J.; de Boer, E.; Van der Meij, P. H.; Weijland, W. P. *Rec. Trav. Chim. Pays-Bas* **1956**, *75*, 487.
(38) Peradejordi, F. *Cah. Phys.* **1963**, *17*, 343.
(39) Jano, I. *Compt. Rendu Acad. Sci. Paris* **1965**, *261*, 103.

(40) Kozaki, T.; Morihasi, M.; Kikuchi, O. *J. Am. Chem. Soc.* **1989**, *111*, 1547.
(41) Tucker, S. C.; Truhlar, D. G. *Chem. Phys. Lett.* **1989**, *157*, 164.
(42) Still, W. C.; Tempczyk, A.; Hawley, R. C.; Hendrickson, T. *J. Am. Chem. Soc.* **1990**, *112*, 6127.
(43) Bucher, M.; Porter, T. L. *J. Phys. Chem.* **1986**, *90*, 3406.
(44) Schaefer, M.; Froemmel, C. *J. Mol. Biol.* **1990**, *216*, 1045.
(45) Dewar, M. J. S.; Zoebisch, E. G.; Healy, E. F.; Stewart, J. J. P. *J. Am. Chem. Soc.* **1985**, *107*, 3902.
(46) Dewar, M. J. S.; Zoebisch, E. G. *J. Mol. Struct. (Theochem)* **1989**, *187*, 1.
(47) Dewar, M. J. S.; Jie, C. *J. Mol. Struct. (Theochem)* **1989**, *187*, 1.
(48) Dewar, M. J. S.; Yate-Ching, Y. *Inorg. Chem.* **1990**, *29*, 3118.
(49) Stewart, J. J. P. *J. Comp. Chem.* **1989**, *10*, 221.
(50) Stewart, J. J. P. *Rev. Comp. Chem.* **1990**, *1*, 45.
(51) Pople, J. A.; Santry, D. P.; Segal, G. A. *J. Chem. Phys.* **1965**, *43*, S129.
(52) Storer, J. W.; Giesen, D. J.; Cramer, C. J.; Truhlar, D. G. *J. Comput.-Aided Mol. Des.* **1995**, *9*, 87.
(53) Simons, J.; Nichols, J. *Quantum Mechanics in Chemistry*; Oxford University Press: New York, 1997.
(54) Parr, R. G.; Yang, W. *Density-Functional Theory of Atoms and Molecules*; Oxford University Press: New York, 1989.
(55) Feller, D.; Davidson, E. R. *Rev. Comp. Chem.* **1990**, *1*, 1.
(56) Mulliken, R. S. *J. Chem. Phys.* **1955**, *23*, 1833.
(57) Löwdin, P.-O. *J. Chem. Phys.* **1950**, *18*, 365.
(58) Chirlian, L. E.; Francl, M. M. *J. Comput. Chem.* **1987**, *8*, 894.
(59) Breneman, C. M.; Wiberg, K. B. *J. Comput. Chem.* **1990**, *11*, 431.
(60) Li, J.; Zhu, T.; Cramer, C. J.; Truhlar, D. G. *J. Phys. Chem.* in press.
(61) Dewar, M. J. S.; Thiel, W. *J. Am. Chem. Soc.* **1977**, *99*, 4899.
(62) Thiel, W.; Voityuk, A. A. *Theor. Chim. Acta* **1992**, *81*, 391, **1996**, *93*, 315(E).
(63) Thiel, W.; Voityuk, A. A. *Int. J. Quantum Chem.* **1993**, *44*, 807.
(64) (a) Thiel, W.; Voityuk, A. A. *J. Mol. Struct. (Theochem)* **1994**, *313*, 141. (b) Thiel, W.; Voityuk, A. A. *J. Phys. Chem.* **1996**, *100*, 616. (c) Thiel, W. *Adv. Chem. Phys.* **1996**, *93*, 703.
(65) Hine, J.; Mookerjee, P. K. *J. Org. Chem.* **1975**, *40*, 287.
(66) Cabani, S.; Gianni, P.; Mollica, V.; Lepori, L. *J. Sol. Chem.* **1981**, *10*, 563.
(67) Pearson, R. G. *J. Am. Chem. Soc.* **1986**, *108*, 6109.
(68) Abraham, M. H.; Whiting, G. S.; Fuchs, R.; Chambers, E. J. *J. Chem. Soc. Perkin Trans. II* **1990**, *291*.
(69) Leo, A. J. *MedChem Database*; BioByte Corp.: Claremont, CA, 1994.
(70) Bondi, A. *J. Phys. Chem.* **1964**, *68*, 441.
(71) Abraham, M. H. *Chem. Soc. Rev.* **1993**, *22*, 73.
(72) (a) Easton, R. E.; Giesen, D. J.; Welch, A.; Cramer, C. J.; Truhlar, D. G. *Theor. Chim. Acta* **1996**, *93*, 281. (b) Li. J.; Cramer, C. J.; Truhlar, D. G. *Theor. Chem. Acc.* in press.
(73) Godbout, N.; Salahub, D. R.; Andzelm, J.; Wimmer, E. *Can. J. Chem.* **1992**, *70*, 560.
(74) Mohamadi, F.; Richards, N. G. J.; Guida, W. C.; Liskamp, R.; Lipton, M.; Caufield, C.; Chang, G.; Hendrickson, T; Still, W. C. *J. Comp. Chem.* **1990**, *11*, 440. We used MACROMODEL–version 6.0.
http://www.columbia.edu/cu/chemistry/mmod/mmod.html
(75) Allinger, N. L.; Yuh, Y. H.; Lii, J.-H. *J. Am. Chem. Soc.* **1989**, *111*, 8551.

(76) Allinger, N. L.; Yuh, Y. H.; Lii, J.-H. *J. Am. Chem. Soc.* **1990**, *112*, 8293.

(77) Allinger, N. L.; Yan, L. *J. Am. Chem. Soc.* **1993**, *115*, 11918.

(78) Hawkins, G. D.; Giesen, D. J.; Lynch, G. C.; Chambers, C. C.; Rossi, I.; Storer, J. W., Rinaldi, D.; Liotard, D. A.; Cramer, C. J.; Truhlar, D. G. AMSOL–version 6.1.1, Oxford Molecular Group: Campbell, CA, 1997.
http://comp.chem.umn.edu/amsol
http://www.oxmol.co.uk/prods/amsol/

(79) AMSOL–version 6.5

(80) G. D. Hawkins, C. J. Cramer, and D. G. Truhlar, OMNISOL–version 1.01, University of Minnesota, Minneapolis, MN 55455
http://comp.chem.umn.edu/omnisol

(81) AMPAC, Semichem. Inc. Shawnee, KS.
http://www.semichem.com/ampac.html
http://comp.chem.umn.edu/ampacm

(82) DGAUSS, a module of UNICHEM, Oxford Molecular Group, Oxford, UK.
http://www.oxmol.com/prods/unichem/cap/dgauss.html
http://comp.chem.umn.edu/dgsol

(83) Frisch, M. J., Trucks, G. W., Schlegel, H. B., Gill, P. M. W., Johnson, B. G., Robb, M. A., Cheeseman, J. R., Keith, T. A., Petersson, G. A., Montgomery, J. A., Raghavachari, K., Al-Laham, M. A., Zakrzewski, V. G., Ortiz, J. V., Foresman, J. B., Peng, , C. Y., Ayala, P. A., Wong, M. W., Andres, J. L., Replogle, E. S., Gomperts, R., Martin, R. L., Fox, D. J., Binkley, J. S., Defrees, D. J., Baker, J., Stewart, J. P., Head-Gordon, M., Gonzalez, C., and Pople, J. A., GAUSSIAN94, Gaussian, Inc., Pittsburgh, PA, 1995.
http://www.gaussian.com/
http://comp.chem.umn.edu/gsm

(84) Schmidt, M. W.; Baldridge, K. K.; Boatz, J. A.; Elbert, S. T.; Gordon, M. S.; Jensen, J. H.; Koseki, S.; Matsunaga, N.; Nguyen, K. A.; Su, S.; Windus, T. L.; Dupuis, M.; Montgomery, J. A. *J. Comput. Chem.* **1993**, *14*, 1347.
http://www.msg.ameslab.gov/GAMESS/GAMESS.html

(85) Li, J.; Cramer, C. J.; Truhlar, D. G. gamesol–version 1.0, University of Minnesota, Minneapolis, MN, 1998.
http://comp.chem.umn.edu/gamesol

Chapter 14

Quantum Mechanical–Molecular Mechanical Calculations of (Hyper-)Polarizabilities with the Direct Reaction Field Approach

Piet Th. van Duijnen, Marcel Swart, and Ferdinand Grozema

Department of Chemistry (OMAC), University of Groningen, Nijenborg 4, 9747AG, Groningen, The Netherlands

In the Direct Reaction Field (DRF) approach to the description of events in the condensed phase, quantum parts (QM) are embedded in a (semi-) classical environment (MM). QM is described with any appropriate wave function, while MM is modeled with point charges and interacting polarizabilities and/or a dielectric continuum, which may have finite ionic strength. The static and response potentials are made part of QM's hamiltonian (hence Direct RF), leading to one- and two-electron contributions. Hence, we obtain also a good estimate of the dispersion. For QM/MM and MM/MM interactions point charges and polarizabilities are treated as belonging to (model) charge distributions. The rest of the short range repulsion is accounted for by a model atom pair potential borrowed from CHARMM. The same model—with a Saltier-Quirked dispersion expression, based on the same interacting polarizabilities—leads to a definition of a classical, polarizable force field used in QM/MM and MM-only Monte Carlo simulations. Here we present calculated static (hyper-)polarizabilities α, β and γ for some molecules in various environments.

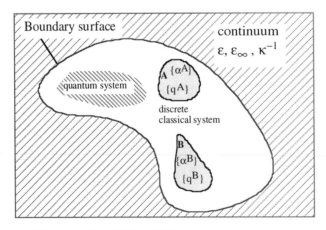

Figure 1. The DRF model for the condensed phase.

The DRF model. Our DRF approach to condensed phase problems(1-3) is summarized in Figure 1, where we see one or more "quantum systems" (QM) with explicit electrons and a classical system (MM) consisting of a number of discrete classical sub-systems: A,B,...characterized by sets of point charges $\{q^A\}$, $\{q^B\}$,... and polarizabilities $\{\alpha^A\}$, $\{\alpha^B\}$,... The discrete part may be embedded in a continuum with total and optical dielectric constants ε and ε_∞, respectively, and a finite ion-strength leading to a Debye screening radius of κ^{-1}. The point charges are generally obtained from appropriate ab initio calculations(4). The polarizabilities come from (ab initio) calculations (5)or from experiment, in either case using Thole's model(6). If required, the discrete classical parts are subject to Monte Carlo sampling using the same parameterizaton in the force field as outlined below(7).

The Energy without Continuum. The total energy of the discrete system without continuum can be written as:

$$\Delta U^{discr} = \Delta U^{QM} + \Delta U^{MM} + U^{QM/MM} \tag{1}$$

with ΔU^{QM} the expectation value of the vacuum hamiltonian of QM over the non-vacuum wave function. With $\mathcal{V}_{sp} = \dfrac{1}{|\mathbf{r}_p - \mathbf{r}_s|}$ the operator for the coulomb potential at

$\mathbf{p}$ from a source at $\mathbf{s}$, and $f_{sp} = -\nabla_p \mathcal{V}_{sp}$ the corresponding electric field we get:

$$\Delta U^{MM} = \sum_{\substack{i,j \\ A \neq B}} q_i^A \, \mathcal{V}_{ij} \, q_j^B + \frac{1}{2} \sum_{\substack{i,j,A,B \\ rs \notin AB}} q_i^A \, f_{ir} \, A_{rs} \, f_{sj} \, q_j^B + \Delta U_{disp}^{MM} + \Delta U_{rep}^{MM} \tag{2}$$

i.e. the classical coulomb interaction, the induction energy (also often called "screening of the coulomb interaction"), the dispersion and short range repulsion energies, respectively.

The A_{rs} are elements of the relay matrix $\mathbb{A}$, constructed from the $\{\alpha^A\}$ (see next section), and which may here be seen as a normal polarizability. In all inductive contributions the polarization 'cost' energy is here included, although we keep it in practice separated in order to make it also possible to deal with non-equilibrium situations.

For the classical dispersion energy we use the Slater-Kirkwood expression:

$$\Delta U_{disp}^{S-K} = \sum_{i<j} \frac{1}{4} \frac{\text{Tr}\left(\alpha_i \, t(i;j)^2 \, \alpha_j\right)}{\left(\sqrt{\alpha_i/n_i} + \sqrt{\alpha_j/n_j}\right)} \tag{3}$$

with $t(i;j)$ the interaction tensor for induced dipoles in i and j.
For the short range repulsion we use the CHARMM(8) expression:

$$\Delta U_{rep}^{MM} = \sum_{i<j} \frac{3}{4} \frac{\alpha_i \, \alpha_j \, (r_i + r_j)^6}{\left(\sqrt{\alpha_i/n_i} + \sqrt{\alpha_j/n_j}\right)} r_{ij}^{-12} \tag{4}$$

in which α_i, n_i, and r_i are the isotropic polarizability, number of valence electrons, and radius of atomic center i, respectively, and r_{ij} is the distance between centers i and j. We use the integral number of valence electrons of an atom, and the same atomic polarizabilities that go into the electrostatic (for the damping function, see later), response, and dispersion terms, leaving only the atomic radii as independent parameters. The latter are obtained from wave functions as local (atomic) contributions to the second moment of the electronic charge distribution.

The QM/MM Interaction Energy. For the QM/MM interaction we have, for a single determinant wave function:

$$\Delta U^{QM/MM} =$$

$$= \sum_{A,i,j} q_i^A \, \mathcal{V}_{ij} \, Z_j + e \sum_{A,i} q_i^A \langle \mathcal{V} \rangle_i \; +$$

$$+ \sum_{A,i,j,r,s} q_i^A f_{ir} \, A_{rs} \, f_{sj} \, Z_j + \, e \sum_{A,k,j,r,s} q_j^A f_{jr} \, A_{rs} \langle f(s;k) \rangle \; +$$

$$+ \frac{1}{2} \sum_{i,j,r,s} Z_i \, f_{ir} \, A_{rs} \, f_{sj} \, Z_j \; + \, e \sum_{i,k,r,s} Z_i \, f_{ir} \, A_{rs} \langle f(s;k) \rangle \; +$$

$$+ \frac{1}{2} e^2 \sum_{k,r,s} \langle \, f(k;r) \, A_{rs} \, f(s;k) \, \rangle + \frac{1}{2} e^2 \sum_{k,\ell,r,s} \langle f(k;r) \rangle_r \, A_{rs} \, (\, 1\text{-}P_{12} \,) \langle f(s;\ell) \rangle + \Delta U_{rep}^{QM/MM} \tag{5}$$

The repulsion term in equation 5 is the same as in equation 4, although we may optionally get the radii of the QM atoms "on the fly" instead of fixing them on their vacuum values.

Further we see in equation 5 interactions between nuclei and electrons with the point charges, and their screening, the screening of the nuclear repulsion, nuclear attraction and electron/electron interactions, respectively. In order to distinguish between source and recipient in the expectation values of the field—e.g. $\langle f(k;s) \rangle$, i.e. the electric field at s due to electron k—we have made explicit the electron labels (k,ℓ) and the electronic charge (e) so as to avoid ambiguity in the signs of the various terms. Also we have separated the screening of the self energy in the electron/electron screening term—which leads to a one-electron operator— from the screening of the two-electron term, which leads to screened coulomb and exchange contributions.

All corresponding operators are added *directly* to the vacuum hamiltonian $\mathcal{H}^0$ of QM:

$$\mathcal{H} = \mathcal{H}^0 + \mathcal{H}^{es} + \mathcal{H}^{rf} \tag{6}$$

with $\mathcal{H}^{es}$ and $\mathcal{H}^{rf}$ the electrostatic and reaction potential operators, respectively. The difference of expectation values

$$\langle \Psi | \mathcal{H} | \Psi \rangle - \langle \Psi^0 | \mathcal{H}^0 | \Psi^0 \rangle + \Delta U_{rep}^{QM/MM} = \Delta U_{int}^{QM/MM} \tag{7}$$

contains all first and second order contributions normally obtained when Ψ refers to a "supermolecule" SCF calculation, and more. The expectation value of $\mathcal{H}^{rf}$, i.e.

$$\langle \mathcal{H}^{rf} \rangle = -\frac{1}{2} \langle f^\dagger \, \mathbb{A} \, f \rangle = -\frac{1}{2} \langle f \rangle^\dagger \, \mathbb{A} \, \langle f \rangle - \frac{1}{2} \gamma \{ \langle f^\dagger \, \mathbb{A} \, f \rangle - \langle f \rangle^\dagger \, \mathbb{A} \, \langle f \rangle \} =$$
$$= \Delta U_{DRF}^{ind} + \Delta U_{DRF}^{disp} \tag{8}$$

contains, in addition to the induction term—which will also be obtained from a supermolecule SCF calculation—a fluctuation term, which was already long ago (1,9) shown (for $\gamma = 1$) to be an upper limit to the dispersion interaction:

$$\Delta U_{DRF}^{disp} = \gamma \{ \langle SELF \rangle - \langle SEX \rangle \} \tag{9}$$

In fact, this term differs a constant factor from the second order perturbation dispersion expression, albeit in its one center expansion and within the Unsöld approximation(7). The first term in braces in equation 9 is the change in one-electron self energy of QM due to $\mathbb{A}$, and the second is the screened two-electron exchange energy. Obviously, the last term is only available separately in the single determinant approximation to the wave function for QM. For these cases we take

$$\gamma = \frac{I_{MM}}{I_{QM} + I_{MM}} \tag{10}$$

with I_{QM} and I_{MM} the ionization energies of QM and MM, respectively. If QM and MM are identical we have $\gamma = 0.5$, while for $\gamma = 0$ the method renders in fact the Average Reaction Field (ARF) rather than the DRF approach. ARF is usually obtained by coupling QM and MM by adding the expectation value of the reaction potential to $\mathcal{H}^0$ and reiterate e.g. the Self Consistent Recation Field (SCRF) procedure(10), in stead of using the reaction potential operator "Directly".

The ARF approach is used by us for non-SCF wave functions, because in constructing Configuration Interaction (CI) or Multi Configuration SCF (MCSCF) wave functions, the explicit control over the exchange interactions is lost. An estimate of the dispersion is then obtained by performing one-determinant DRF calculations for the leading components of the CI expansion and scaling them with the corresponding expansion coefficients.

The Relay Matrix. The relay matrix $\mathbb{A}$ is a (super)matrix representing the total effective polarizability of the complete discrete classical part, in which all particles interact self consistently. Taking a set of points $\{\mathbf{p}\}$ with polarizabilities $\{\alpha_p\}$ in a uniform electric field $\mathbf{F}^0$ we have for the induced dipole moment in point $\mathbf{p}$:

$$\mu_p = \alpha_p \left(\mathbf{F}^0(\mathbf{p}) - \sum_{q \neq p}^{N^{pol}} \mathfrak{t}(\mathbf{p};\mathbf{q}) \, \mu_q \right) \tag{11}$$

A formal solution for the $\{\mu_p\}$ can be found by collecting the N^{pol} equations into a single supermatrix equation of dimension $3N^{pol} \times 3N^{pol}$:

$$\mathbf{M} \left(\alpha^{-1} + \mathbb{T} \right) = \mathbf{F}^0 \tag{12}$$

where $\mathbf{F}^0$ and $\mathbf{M}$ are $3N^{pol}$-dimensional vectors, and α and $\mathbb{T}$ are square $3N^{pol} \times 3N^{pol}$ matrices. The supervectors and matrices are blocked into $3N^{pol}$ and $3N^{pol} \times 3N^{pol}$ elements, respectively: $\mathbf{M}_p = \mu_p$, $\alpha_{pq} = \alpha_p \, \delta_{pq}$, $\mathbb{T}_{pq} = \mathfrak{t}(\mathbf{p};\mathbf{q})(1 - \delta_{pq})$, and δ_{pq} is the Kronecker delta. Then

$$\mathbb{A} = \left(\alpha^{-1} + \mathbb{T} \right)^{-1} \tag{13}$$

may be considered as an ordinary polarizability matrix (but of an N^{pol} membered system):

$$\mathbf{M} = \mathbf{F} \, \mathbb{A} = \mathbf{F} \left(\alpha^{-1} + \mathbb{T} \right)^{-1} \tag{14}$$

$\mathbb{A}$ is obtained either by an exact matrix inversion or by solving the associated linear equations by iteration. We note that (14) is a self consistent solution for any field, e.g. the electric field of QM. The $\{\mathfrak{t}(\mathbf{p};\mathbf{q})\}$ are, when appropriate, screened according to the method described by Thole(6) in which (atomic) polarizabilities are taken as related to (model) charge distributions, the widths of which are related to the $\{\alpha_p\}$. This leads also to a consistent screening of the potentials and fields of interaction for overlapping charge distributions. In general the polarizabilities are constructed following Thole's original recipe and parameterization for obtaining (molecular) polarizabilities with experimental accuracy. If needed this model can be re-parameterized to reflect computed polarizabilities from specific basis sets (Swart, M and Duijnen P.Th. van, submitted)
The advantage of this way of treating the relay matrix is that only atomic polarizabilities are needed as input, while changes in geometry will be automatically reflected in $\mathbb{A}$. Optionally one may reduce parts of $\mathbb{A}$ first to group polarizabilities so as to reduce the dimensionality of the problem.

Extension with a Continuum. The extension with a dielectric continuum around the discrete part(s) is straightforward. For example, one may divide the continuum into an infinite set of finite volumes $\{V_i\}$ and assign a polarizability $\alpha_i = \chi V_i$ to each of them with $\chi = \frac{\varepsilon - 1}{4\pi}$ where χ the electric susceptibility. These polarizabilities may be used directly in equation 13, were it not for the infinite dimensions that result. Using standard mathematical procedures the volume integrations can be reduced to surface integrals. For a discretized surface (boundary element method, BEM(11)) the final result can be expressed in a set of linear equation of finite dimensions:

$$
\begin{pmatrix} \mathbf{M}_p \\ \Omega_I \\ \mathbf{Z}_I \end{pmatrix} = \mathbb{A} \begin{pmatrix} \mathbf{F}_p^0 \\ \dfrac{1}{2\pi(1+\varepsilon)}\mathbf{V}_I^0 \\ \dfrac{-\varepsilon}{2\pi(1+\varepsilon)}\mathbf{F}_I^0 \cdot \mathbf{n}_I \end{pmatrix} \tag{15a}
$$

with

$$
\mathbb{A} = \begin{pmatrix} \alpha_p^{-1} + \mathbb{T}'_{qp} & \nabla_p \mathbb{K}_{Ip}\mathbf{S}_I & \nabla_p \mathbb{L}_{Ip}\mathbf{S}_I \\ \dfrac{-1}{2\pi(1+\varepsilon)}\mathbb{F}_{pI} & 1 - \dfrac{\mathbb{K}'_{JI}\mathbf{S}_J}{2\pi(1+\varepsilon)} & -\dfrac{\mathbb{L}_{JI}\mathbf{S}_J}{2\pi(1+\varepsilon)} \\ \dfrac{-\varepsilon}{2\pi(1+\varepsilon)}\mathbb{T}_{pI}\cdot\mathbf{n}_I & \dfrac{-\varepsilon\,\mathbb{M}_{JI}\mathbf{S}_J}{2\pi(1+\varepsilon)} & 1 - \dfrac{\mathbb{N}'_{JI}\mathbf{S}_J}{2\pi(1+\varepsilon)} \end{pmatrix}^{-1} \tag{15b}
$$

In Eq. (15b) we have added (redundant) indices for clarity: lower case indices for discrete polarizable points and capitals for boundary elements, while 1 is a unit matrix of appropriate dimensions.
In the r.h.s of equation 15a all source fields ($\mathbf{F}$) and potentials ($\mathbf{V}$) are collected, while $\mathbf{n}_I$ is the normal outward unit vector in the representative point of boundary element $\mathbf{S}_I$. In the l.h.s. of equation 15a the vector $\mathbf{M}$ represents the induced dipoles at the classical polarizable points, Ω the set of induced dipoles on the surface $\mathbf{S}$, and $\mathbf{Z}$ (only for $\kappa \neq 0$) the induced charges on $\mathbf{S}$. In the top-left block of equation 15b the matrix equation(13) will be recognized, while the $\mathbb{K}$, $\mathbb{L}$, $\mathbb{M}$ and $\mathbb{N}$ are more or less complicated—potential, field and field gradient like—kernels, depending on ε, κ and the

geometry of S. We note that when leaving out the continuum equation 13 remains, while for a continuum-only the top-left vanishes. For $\kappa = 0$ the continuum related blocks reduce to $(1 - \mathbb{K}\mathbf{S})$, if the source potential is chosen, or to $(1-\mathbb{N}\mathbf{S})$ if the source field is chosen as the inducing agent. Both formulations are equivalent, but in our implementation we use the former because it is numerically more stable (11).

Thus, the general picture remains the same: all information about the reaction potentials are contained in a single relay matrix. The electronic contributions to the source potentials and fields of course have first to be formalized in terms of integrals over basis functions. Such details may be found elsewhere(3).

The wave functions for QM, which may be of any type, are made self consistent with the external (reaction) potentials, thus leading to an overall self consistent solution for the energy and other properties of the complete system. DRF has been implemented(3) in HONDO8.1(12). Obviously, any part of the system may be omitted: leaving out the discrete classical part we obtain something like Tomasi's method(13), or leaving out QM we just have a polarizable classical force field(7).

Influence of an External Electric Field. If a system like described in the previous sections is perturbed by a homogeneous electric field $\mathbf{F}$ the expression for the total energy and the dipole moment of the system can be expanded as:

$$U = U^0 - \mu_i^0 F_i - \frac{1}{2}\alpha_{ij} F_i F_j - \frac{1}{6}\beta_{ijk} F_i F_j F_k - \frac{1}{24}\gamma_{ijk\ell} F_i F_j F_k F_\ell - \dots \tag{16a}$$

$$\mu_i = \mu_i^0 + \alpha_{ij} F_j + \frac{1}{2}\beta_{ijk} F_j F_k + \frac{1}{6}\gamma_{ijk\ell} F_j F_k F_\ell + \dots \tag{16b}$$

with F_i the component of the field in the ith direction, and α, β and γ, respectively, the dipole polarizability and the first and second dipole hyper-polarizabilities. U^0 and μ^0 are the unperturbed energy and dipole moment, respectively.

When in a field $\mathbf{F}$, the total energy of our QM+MM system (equation 1) has to be extended with:

$$\Delta U^F = -\left\{\mu_{class}^{stat}\right\}^\dagger \mathbf{F} - \sum_{A,i,rs} q_i^A f_{ir}^\dagger A_{rs} \mathbf{F} - \frac{1}{2}\mathbf{F}^\dagger \mathbb{A} \mathbf{F} +$$
$$-\left\langle \mu_{QM}^{DRF}(\mathbf{F})\right\rangle^\dagger \mathbf{F} - \sum_{i,rs} Z_i f_{ir}^\dagger A_{rs} \mathbf{F} - \frac{e}{2}\langle f \rangle^\dagger \mathbb{A} \mathbf{F} - \frac{e}{2}\mathbf{F}^\dagger \mathbb{A} \langle f \rangle \tag{17}$$

which again may obtained self consistently, simply by adding the field to the sources in equation 15. In equation 17 we—on purpose—have made explicit the last two terms: the first of them is the interaction of the dipoles induced in MM by the electrons of QM and the external field $\mathbf{F}$, while the last term represents the interaction of the dipoles induced in MM by $\mathbf{F}$ with the electric field of QM. In fact, all terms like these are in general not symmetric and should be written as:

$$\frac{1}{2}\left[V_{source}^\dagger(\mathbf{i};\mathbf{r}) A_{rs} V_{response}(\mathbf{s};\mathbf{j}) + V_{source}^\dagger(\mathbf{j};\mathbf{r}) A_{rs} V_{response}(\mathbf{s};\mathbf{i}) \right] \tag{18}$$

because, e.g. in the presence of a continuum, the source and response operators are different. The symmetric form in equation (17) holds only if $\mathbb{A}$ is constructed from polarizabilities only.

The first derivative of the total energy with respect to the ath component F_a of the field is:

$$\frac{\partial U}{\partial F_a} = - e \langle \mu^a_{QM/DRF} (\mathbf{F}) \rangle - \mu^a_{nuclei} - \mu^a_{class} +$$

$$- \left(\sum_{i,r,s} q_i \, f(i;r)^\dagger \, A_{rs} \right)^a - \frac{1}{2} \left(\mathbf{F}^\dagger \, \mathbb{A} + \mathbb{A}^\dagger \mathbf{F} \right)^a +$$

$$- e \left(\langle f(\mathbf{F}) \rangle^\dagger \, \mathbb{A} \right)^a - \left(\sum_{i,r,s} Z_i \, f(i;r)^\dagger \, A_{rs} \right)^a \tag{19}$$

i.e. the ath components of the static and induced dipole moments of all parts of the system.

The second derivatives look like:

$$\frac{\partial^2 U}{\partial F_{ab}} = - \frac{e \partial \langle \mu^a_{QM/DRF}(\mathbf{F}) \rangle}{\partial F_a} - \frac{e \partial (\langle f(\mathbf{F}) \rangle^\dagger \, \mathbb{A})}{\partial F_a} - A_{ab} =$$

$$= - \left\{ A_{ab} + \alpha^{QM/DRF}_{ab} \right\} \tag{20}$$

where the $\mathbf{F}$-dependence of the expectation values goes obviously through the electron density. The result is simply the sum of the components of the polarizabilities of QM (in its classical environment!) and of MM.

One can use equation 16a or equation 16b directly by applying various field strengths and determining numerically the derivatives of $U(\mathbf{F})$ or $\mu(F)$ to obtain (hyper)polarizabilities. This finite field (FF) approach is applicable independent of the type of wave function for QM.

For closed shell systems the Coupled Perturbed Hartree-Fock (CPHF) method (14) is an option which is in general faster than FF. In CPHF the first order perturbing hamiltonian for polarizabilities for a molecule in vacuum is the set of molecular dipole moment integrals. In DRF the CPHF approach is in principle possible by simply extending this hamiltonian with the integrals related to the moments induced by the QM electrons in MM, because they also interact with $\mathbf{F}$ and with dipoles induced by $\mathbf{F}$.

Here the complete perturbing hamiltonian becomes, expressed in terms of basis set charge distributions:

$$h^a_{mn} = \left(\mu^0_{mn} \right)^a + \left(\sum_{rs} f(mn;r) A_{rs} \right)^a \tag{21}$$

where m,n refer to basis functions. Also here is the character of MM—discrete and/or continuum—of no relevance. In recent work of Cammi et al. (15,16) , using the PCM method(13), the latter contributions seem to be not present, or at least the polarization of the continuum by $\mathbf{F}$ is not shown. Without any further adaptation to the code of HONDO8.1(12) we calculated the (hyper-) polarizabilities of various dimers and compared them with fully SCF results and other work.

Applications.

Water dimers. First we looked into a hydrogen bonded water dimer near its equilibrium geometry as depicted in Figure 2a, using the CPHF method.

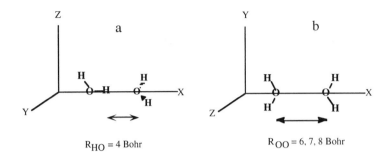

Figure 2. Geometry of the water dimers. a: Hydrogen bonded. b: Repulsive. Monomers have their equilibrium geometries.

Since this QM/MM pair is not symmetric, two calculations are required (QM/MM and MM/QM) and the results should be added in order to make them comparable with the supermolecule SCF results. The latter we corrected for the basis set superposition errors (BSSE) which are significant, despite of using Dunning's DZP(17) basis, in particular in the second hyper-polarizabilities. Results are collected in Table I.

Table I. Polarizabilities and 2nd hyper-polarizabilities of the hydrogen bonded water dimer from CPHF calculations. (Atomic units).

	monomer [a]		*DRF*		*dimer* [b]	
	donor	*acceptor*	*Q : don.*	*Q : acc.*	*Q* [c]	*DRF* [d]
α_{xx}	6.68	5.57	6.90	5.86	12.9	12.9
α_{yy}	3.03	7.37	2.93	7.14	10.0	10.1
α_{zz}	6.25	3.03	6.36	2.90	9.21	9.30
γ_{xxxx}	278	61.2	342	74.7	446	435
γ_{xxyy}	9.35	139	8.53	138	148	149
γ_{xxzz}	1.64	6.45	13.5	6.71	1.06	22.8
γ_{yyyy}	12.1	190	10.6	171	182	181
γ_{xxzz}	8.22	11.1	7.55	9.20	15.9	16.6
γ_{zzzz}	247	12.1	276	10.8	258	295

[a]Monomer in equilibrium geomemetry. [b]Donor in xz-plane; acceptor in xy-plane; R_{HO} = 4.0 Bohr. [c]Fully quantum mechanical, corrected for BSSE. [d]Sum of DRF donor(Q) and acceptor(Q).

The contents of the last two columns of Table I compare well, except for γ_{xxzz} , so far for reasons unknown. The results give some confidence that the separate DRF results indeed represent the influence of the classical partner's polarizability on these properties in the dimer, and that the DRF operators contain the correct physics. Although we have to perform two calculations to obtain an estimate of the properties of the dimer, we note that they require much less computational effort, since only the monomer basis is used.

We went through the same exercise for a symmetric planar dimer in a repulsive geometry, with the oxygen facing each other (See Figure 2b). For the radius of O in the monomer we found $r_O \approx 3.5$ Bohr. We did three dimer calculations for $R_{OO} = 6.0$ 7.0, and 8.0 Bohr, respectively, i.e. inside, at, and just outside the vanderwaals distance. Results are given in Table II.

Table II. Polarizabilities and 2nd hyper-polarizabilities of a planar, repulsive water dimer[a] from CPHF calculations. (Atomic units).

	$R_{OO} = 6$ Bohr		$R_{OO} = 7$ Bohr		$R_{OO} = 8$ Bohr	
	Q [a]	DRF[c]	Q[b]	DRF[c]	Q[a]	DRF[c]
α_{xx}	10.8	11.8	10.7	11.6	10.6	11.4
α_{yy}	14.6	15.1	14.7	15.2	14.7	14.8
α_{zz}	6.00	5.99	6.02	6.04	6.03	6.02
γ_{xxxx}	137	159	135	147	134	135
γ_{xxyy}	308	303	297	300	291	289
γ_{xxzz}	10.2	16.1	12.8	15.6	13.4	13.8
γ_{yyyy}	363	385	367	397	370	379
γ_{xxzz}	22.2	23.8	22.1	24.2	22.2	22.2
γ_{zzzz}	23.2	23.7	23.7	24.5	23.8	23.8

[a] Monomer geometry as in Table I. Dimer is in xy-plane with the O-atoms facing. [b] Fully quantum mechanical, corrected for BSSE. [b]Twice the calculated values because of symmetry.

For $R_{OO} = 8$ the DRF results are in almost perfect agreement with the full quantum calculations, while for the shorter distances they are—regarding the simplicity of the model—still very acceptable.

Formaldehyde in Aqueous Solution. Next we tried to repeat some recent work of Cammi et al.(16), who calculated, among other, the static (hyper)polarizabilities of formaldehyde in vacuum and in water with the Polarizable Continuum Method (PCM)(13). Using the same DZP basis set we obtained practically the same vacuum results for $<\alpha>$, $<\beta>$ and $<\gamma>$, the small deviations being attributed to a slightly different geometry. In our continuum-only approach for modeling water—which is, apart from details, equivalent to PCM—the cavity was defined by the Connolly solvent accessible surface(18).

However, putting the boundary at 1.2 times the vanderwaals radii—the PCM standard(13)—the DRF/CPHF wave functions gave much larger changes in the (hyper)polarizabilities than reported by Cammi et al.. This is most likely a consequence of not "renormalizing" the reaction potential for the "leaking" electronic charge(13), due to the—too small—distance between the electron cloud and the cavity's boundary. We have serious doubts about the physical contents of such uniform corrections (See Ref. 14, Eqs. 62-64), and prefer cavities that are large enough to contain the solute and its first solvation shell(19). By putting the boundary at 1.5 times the vanderwaals radii we obtained numbers comparable to Cammi's. (See Table III).

Our strongest objection to the continuum approach is related to the arbitrariness in the choice of the boundary(19,20), while we think that (changes in) the solute's sensitive electronic properties are dictated by the first solvation shell and its structure. Therefore we turned to the discrete version of DRF and chose randomly a small number of solvent configurations from a 5×10^5 steps Monte Carlo run for formaldehyde surrounded by 52 water molecules (Grozema, F. and Duijnen P.Th. van, submitted) and calculated the average polarizabilities.

Table III. Calculated average (hyper-)polarizabilities for formaldehyde from CPHF in DZP basis.

	vacuum		in water			
	quantum chemical		continuum [b]		discrete [c]	
$<\alpha>$	12.6	12.7[a]	15.6 (14.1)	14.6[a]	11.8±0.2	(11.7)[d]
$<\beta>$	36.5	38.1[a]	81.4 (55.8)	63.2[a]	31 ±1	(29.9)[d]
$<\gamma>$	90.9	86.3[a]	705 (250)	205.7[a]	94 ± 4	(82.8)[d]

[a]Cammi et al. (13). [b] Cavity boundary at 1.2 (1.5)×R_{vdw}. [c] Averaged over 10 configurations randomly chosen from 5×10^5 accepted classical Monte Carlo steps. [d]Smallest value in sample.

The numbers in the last column of Table III are in contrast with those from the continuum calculations: in a discrete environment all polarizabilities remain almost at their vacuum values. If there can be seen a trend at all in our small number of configurations, it is, regarding the smallest values in our sample, that the average polarizabilities are smaller than in vacuum.

Butadienes. Finally, we turned our attention to a more interesting case, viz. the butadiene dimers which show a strong reduction of the longitudinal component of the 2nd hyper-polarizability, $\gamma_{//}$, i.e. along the C-C-C-C chain. For this occasion we re-parameterized Thole's model for getting the calculated polarizability of butadiene in Dunning's DZV(17) basis set, so as enable us to compare SCF and DRF results directly. With the CPHF method we obtained the results as shown in Table IV.

Table IV. Monomer and dimer (hyper-)polarizabilities for a parallel butadiene dimer from CPHF calculations.

	quantum chemical		DRF	
	monomer	dimer [a]	dimer	"solid" [b]
α_{xx}	77	72	26	88
α_{yy}	22	22	16	26
α_{zz}	35	38	23	40
$\gamma_{//}$	7777	4950	$\approx 2 \times 10^6$	13245

[a] Per chain (half of dimer value), BSSE corrected. [b]Monomer surrounded with continuum with $\varepsilon=2.5$. Boundary is a Connolly solvent accessible surface for $R_{probe}=8.4$ Bohr, i.e. approximately 1.2 times the radius of butadiene according to the largest component of the polarizability.

The DRF results make at first sight no sense at all. However, one should realize that the expectation value of the perturbing hamiltonian in the ground state, i.e. the dipole moment induced by the electrons-only in the classical partner in the dimer, is about 14 a.u. The corresponding inducing electric field is about 0.3 a.u., which is more than an order larger than one usually applies in finite field calculations(14), and is much too large to be considered as 'perturbation', in particular with the large polarizability of butadiene. Hence, already the first order derivatives of wave functions and eigenvalues are completely wrong.

Also treating 'solid' butadiene as a continuum in the CPHF method, adding the dipoles induced in the boundary elements to the first order hamiltonian, is not a viable approach. Both α_{xx} and $\gamma_{//}$ increase, in contrast to the fully quantum mechanical dimer results. Here we put a (solute) butadiene molecule in a cavity of such dimensions that at least one other molecule could be put between the solute and the boundary, which was taken as a Connolly surface(18). Even in this large cavity the wave function was not stable, producing a stabilization of more than 10 Hartree, although the SCF procedure converged. No reasonable energy of cavitation (21) can compensate for this. Increasing the radii by 20 % resulted in a stable wave function and the polarizabilities in the last column of Table IV.

Like in the case of formaldehyde, the polarizabilities are increased in the continuum approach. Part of the increase may have been associated with a too large perturbing hamiltonian, but in fact the failure did not come as a surprise to us. We found already earlier (19,20) that a continuum affects the electronic states of the solute in a different way than a discrete environment. The difference is largely due to the lack of structure in the solvent, and the fact that a continuum can only stabilize (or do nothing), while the discrete DRF method balances stabilizing and destabilizing effects. Another part is caused by over polarization, because the continuum is generally too close to the (QM) sources.

Since there is no clear physically sound criterion for where the boundary should be, we did not follow the continuum approach any further, but we adapted the discrete DRF model to the finite field approach, which is straightforward and gives better control over the final states. The (hyper-) polarizabilities were obtained by fitting the dipole moment of QM, as function of the applied field, to the polynomial of equation 16b. We used the dipoles rather than the energies because the former are much more sensitive to small external influences.

It appeared that also this approach is not without problems. If the external field is too strong, the over polarization, mentioned with respect to the CPHF method, obviously occurs also here and may even lead to induced dipoles of the wrong sign. If the external field is too weak, the changes in the dipole are too small to be significant in the fitting procedure.

The results are also sensitive to the number of results used in the fit procedure, and to the starting densities in the SCF procedure. Hence this procedure is not easily automized, but by repeating it carefully several times with varying parameters, we obtained consistently the values of Table V.

Table V. Monomer, dimer and heptomer (hyper-)polarizabilities of butadiene from finite field calculations.

	quantum chemical		DRF	
	monomer	dimer [a]	dimer	"solid" [b]
α_{xx}	77	72	71	42
α_{yy}	22	22	21	25
α_{zz}	35	38	38	37
$\gamma_{//}$	7830	4814	4368	- 3681

[a]Per chain (half of dimer value), BSSE corrected.
[b]Central butadiene (QM) + six nearest neighbors (MM)

First we note that the first two columns of Table V compare very well with the corresponding columns of Table IV, showing that our finite field method is sound. Here one should realize that the "per chain" quantum mechanical values should be different from the DRF values, in particular in the γs, because the former contain also higher order interaction contributions not present in the latter. Because the polarizability of butadiene is much larger than that of water the differences are here more manifest than in the water dimers. The DRF results for the dimer are, like for the water dimers, very satisfying. They show that the larger part of the changes in the static (hyper-)polarizabilities, due to the monomer's environment, can be understood with purely classical arguments, in this case in terms of the classical polarizability only.

The "solid state" result for $\gamma_{//}$ in Table V is negative and looks therefore very unphysical, but one should realize that the external electric field—used in the FF approach—on the central (QM) butadiene is nearly quenched by the field of the induced dipoles at its six first neighbors. Hence, the total field is so small that the non-linear response here will fall into the noise, but we expect the contribution to γ_{xxxx} of the central molecule in the "bundle" to be close to zero. Presently we are investigating series of n-mers of butadiene (n = 1,7) both fully quantum mechanically, and with DRF.

Conclusion. We have briefly summarized the Direct Reaction Field (DRF) approach to QM/MM calculations on condensed phase problems, derived expressions for the energy and the dipole moment for mixed quantum/classical systems in a uniform external electrostatic field, and their derivatives with respect to components of the field. The appropriate expressions were used in Coupled Perturbations Hartree-Fock (CPHF) and Finite Field (FF) calculations of α, β and γ of various molecules in various environments, and we compared some results with those of purely quantum chemical calculations.

For not too polarizable molecules (e.g. water and formaldehyde) both CPHF and FF give DRF results consistent with fully quantum chemical calculations, as long as the classical environment is modeled with discrete molecules. The continuum approach to solvent or solid state effects shows systematically an increase in α, β and γ, while discrete calculations show nothing significant or a decrease, even if averaged over a small number of solvent configurations. The different results from the two models is attributed to the fact that a dielectric continuum can only stabilize and has no structure. In contrast, the discrete model balances stabilizing and destabilizing effects.

For very polarizable molecules only the FF with DRF is able to mimic the fully quantum chemical SCF calculations. CPHF fails in these cases because the DRF contributions to the first order perturbing hamiltonian are too large. For the butadiene dimer we obtained FF/DRF results in good agreement with the SCF values, and probably a good estimate of α in the solid state from calculations on butadiene surrounded with its six first neighbors. The contribution to $\gamma_{//}$ of the central molecule turns out to be negative, because the external field is almost quenched by the field of the induced dipoles in the classical molecules. This introduces large numerical errors in the higher order polarizabilities, but we expect this contribution to be very small indeed. Reports of our present research in this field will be published shortly elsewhere.

Acknowledgments. PTvD thanks Prof. B.Kirtman, of UC Santa Barbara, for his suggestion to try DRF on (hyper-) polarizabilities and for many stimulating discussions. The Computer Center of the University of Groningen is acknowledged for granting ample time on the Cray J932 supercomputer.

References

1. Thole, B.T.; Duijnen, P.Th. van *Theor. Chim. Acta* **1980,** *55,* 307.
2. Duijnen, P.Th. van; Juffer, A.H.; Dijkman, J.P. *J. Mol. Struct. (THEOCHEM)* **1992,** *260,* 195.
3. Vries, A.H. de; Duijnen, P.Th. van; Juffer, A.H.; Rullmann, J.A.C.; Dijkman, J.P.; Merenga, H.; Thole, B.T. *J. Comp. Chem.* **1995,** *16,* 37;1445.
4. Thole, B.T.; Duijnen, P.Th. van *Theor. Chim. Acta* **1983,** *63,* 209.
5. Duijnen, P.Th. van; Swart, M. *J.Phys.Chem.* **1997,** submitted.
6. Thole, B.T. *Chem. Phys.* **1981,** *59,* 341.
7. Duijnen, P.Th. van; Vries, A.H. de *Int. J. Quant. Chem.* **1996,** *60,* 1111.

8. Brooks, B.R.; Bruccoleri, R.E.; Olafson, B.D.; States, D.J.; Swaminathan, S.J.; Karplus, M. *J. Comp. Chem.* **1983,** *4,* 187.
9. Thole, B.T.; Duijnen, P.Th. van *Chem. Phys.* **1982,** *71,* 211.
10. Tapia, O. *J. Mol. Struct. (THEOCHEM)* **1991,** *226,* 59.
11. Juffer, A.H.; Botta, E.F.F.; Keulen, B.A.M. van; Ploeg, A. van der; Berendsen, H.J.C. *J. Comp. Phys.* **1991,** *97,* 144.
12. Dupuis, M.; Farazdel, A.; Karma, S.P.; Maluendes, S.A., In *MOTECC-90;* Clementi, E. Ed; ESCOM, Leiden, 1990; 277.
13. Tomasi, J.; Persico, M. *Chem. Rev.* **1994,** *94,* 2027.
14. Sim, F.; Chin, S.; Dupuis, M.; Rice, J.E. *J.Phys.Chem.* **1993,** *97,* 158.
15. Cammi, R.; COssi, M.; Mennucci, B.; Tomasi, J. *J. Chem. Phys.* **1996,** *105,* 10556.
16. Cammi, R.; Cossi, M.; Tomasi, J. *J.Chem.Phys.* **1996,** *104,* 4611.
17. Dunning Jr., T.H.; Hay, P.J., In *Methods of Electronic Structure Theory;* Schaefer III, H. F. Ed; 1977; 1.
18. Connolly, M.L. *Science* **1983,** *221,* 709.
19. Vries, A.H. de; Duijnen, P.Th. van; Juffer, A.H. *Int. J. Quant. Chem, Quant. Chem. Symp.* **1993,** *27,* 451.
20. Vries, A.H. de; Duijnen, P.Th. van *Int. J. Quant. Chem.* **1996,** *57,* 1067.
21. Pierotti, R.A. *Chem. Rev.* **1976,** *76,* 717.

BIOCHEMICAL APPLICATIONS

Chapter 15

The Local Self-Consistent Field Principles and Applications to Combined Quantum Mechanical– Molecular Mechanical Computations on Biomacromolecular Systems

Xavier Assfeld, Nicolas Ferré, and Jean-Louis Rivail

Laboratoire de Chimie Théorique—UMR 7565, Institut Nancéien de Chimie Moléculaire, Université Henri Poincaré, BP 239, 54506 Nancy-Vandoeuvre

A new combined Quantum mechanics (QM) molecular mechanics (MM) computational scheme, working at the *ab initio*, DFT or semi-empirical levels is developed. The region of the macromolecule which separates the quantum subsystem to the classical one is represented by predetermined strictly localized bond orbitals. The resulting Local Self Consistent Field method (LSCF) allows the fully analytical computation of the forces, giving rise to a consistent Classical-Quantum Force Field (CQFF) which can be used in the same way as a traditional force field in molecular modeling computations.

The method is illustrated by the study of the influence of an α helix of polyalanine on the electronic and conformational properties of a peptidic bond at the C and N terminals of the chain as well as in the middle of the helix. These effects appear to be far from negligible.

The importance of molecular mechanics (MM) for modeling the structure of biomolecules is now well recognized. A major limitation of this modeling tool is that it is built on the concept of well defined chemical bonds and, therefore, it cannot be used to model chemical transformations during which some bonds are broken and some others formed. This can only be reached by considering the electrons and the nuclei which constitute the molecule, *i.e.* by using quantum mechanics (QM). In addition, biomolecules are highly deformable systems with many configurations in similar potential energy and in which the only meaningful energetic quantity is the free energy of the system. Usually those system consist of one (or more) macromolecule(s), a solvent and, very often, a substrate. The usual statistical mechanical methods, such as Monte Carlo integration or molecular dynamics (MD) currently require the computation of one million, or more, different configurations. Therefore, although the performances of computational codes for quantum chemistry and computers, are constantly and rapidly improving, it is still not realistic to consider the possibility of a fully quantum molecular dynamics computation on a biochemical system. Some drastic reduction of the computational requirements are necessary. To fulfill this constraint, one can use the idea of representing the larger part of the system, which is not substantially modified by the chemical reaction, by a simple molecular mechanical model whilst quantum mechanics is used to describe the main modifications of the electronic structure during the course of the chemical reaction. These combined QM/MM approaches are becoming more and more popular (*1-2*) and are relatively straigthforward to set up when the reaction involves small molecules in solution. In this case the quantum system is well defined (the reacting molecule) and the solvent is described classically (*3-4*). When the reaction involves a macromolecule, the necessity of considering only a part of this chemically bonded system in the quantum computation causes a difficult problem. This problem is generally solved by replacing the actual macromolecule by a model molecule containing the atomic arrangements of interest and by introducing, in the Hamiltonian of this molecule, the interaction with the rest of the system. From the electronic point of view, the non-reacting part of the macromolecule is replaced, in this procedure, by a monovalent "link atom", usually hydrogen. This can introduce some severe difficulties when computing the interaction with the neighboring classical atoms and so, to avoid this difficulty, a local modification of the system is required.

The aim of the present paper is to present a method which fulfills the following requirements :
- easy to implement;
- valid at any level of theory : Hartree Fock (*ab initio*, or semi empirical); post Hartree- Fock or Density Functional Theory (DFT);
- fully derivable in order to allow the safe computation of the forces acting on the atoms for any configuration.

This method is expected to provide us with a consistent QM/MM force field which allows full energy minimizations as well as molecular dynamics simulations either by using the traditional Born-Oppenheimer approach or the Car-Parrinello quantum dynamics scheme (*5*).

We shall focus our attention on the method itself and illustrate it with one selected example application.

Principles of the local self consistent field method
The detailed theory has been published previously (*6*).

The method rests upon the division of the system into 3 parts treated at three different levels of theory
- the quantum part which is described as an ensemble of electrons and nuclei. Within the Born-Oppenheimer approximation, the nuclei have fixed positions and the description of the electron cloud requires that a set of molecular orbitals (MO) has to be computed.

- an intermediate part in which the electrons are described by a set of L localized molecular orbitals (LMO) which are assumed to be known. These LMOs are calculated in such a way that they represent, as well as possible, the electronic distribution in the corresponding bonds. Once determined, they remain constant during the computation of the MOs. This region plays the role of a buffer between the quantum part and the classical one;

- the classical part which is described by a classical force-field.

For the computations, a basis set of N atomic orbitals is used. It is defined on all the atoms comprising the quantum as well as the intermediate part of the molecule.

Let $\{..|\mu)...\}$ be the basis set. The LMOs denoted as $|\ell_i\rangle$ (i from 1 to L) are expressed as a linear combination of the atomic orbitals

$$|\ell_i\rangle = \sum_{\mu} a_{\mu i}|\mu\rangle \tag{1}$$

and they are assumed to be mutually orthogonal.

The molecular orbitals are maintained orthogonal to all the localized ones. The procedure consists of expanding the MOs on a set of functions orthogonal to the LMOs. These functions being built on the basis set of N atomic orbitals, assumed to be linearly independent, there will be N-L such linearly independent basis functions which are expected to be obtained from the original basis set by a linear transformation expressed by means of a Nx(N-L) matrix. It is obtained in two steps :

i) a set of N normalized functions, orthogonal to the LMOs is obtained by subtracting, from the atomic orbitals, their projection onto the LMOs. Let $|\tilde{\mu}\rangle$ be one of these functions. It derives from the original AOs by

$$|\tilde{\mu}\rangle = \sum_{\nu} \left[1 - \sum_{i=1}^{L} S_{\nu i}^2 \right]^{1/2} \cdot \left[\delta_{\mu\nu} - \sum_{i=1}^{L} \sum_{\eta=1}^{N} a_{\nu i} a_{\eta i} S_{\eta\mu} \right] |\nu\rangle \tag{2}$$

where $S_{\nu i}$ stands for the overlap integrals $\langle \nu | \ell_i \rangle$, $S_{\eta\mu}$ for $\langle \eta | \mu \rangle$ and $\delta_{\mu\nu}$ is the Kronecker delta. Obviously, there are L linear dependences.

Let **M** be the matrix transforming the AOs into these intermediate functions.

$$\{|\tilde{\mu}\rangle\} = \mathbf{M}\{|\mu\rangle\} \tag{3}$$

ii) the linear independence of the functions we are looking for is obtained by means of a canonical orthogonalization written as

$$\{|\mu'\rangle\} = \mathbf{X}^{-1}\{|\tilde{\mu}\rangle\} \tag{4}$$

where $\{|\mu'\rangle\}$ and $\{|\tilde{\mu}\rangle\}$ are two column matrices of dimension N-L and N, respectively, and **X** is the canonical orthogonalization matrix (7-9). **X** is obtained by diagonalizing the $\widetilde{\mathbf{S}}$ overlap matrix of the $|\widetilde{\mu}\rangle$ functions and eliminating the L eigenvectors corresponding to zero eigenvalues. The eigenvectors are then divided by the square root of the corresponding eigenvalue giving rise to the Nx(N-L) $\mathbf{X}$ matrix. Finally, the required orthogonal basis set is deduced from the original basis set by a (N-L)xL matrix $\mathbf{B}$ defined by

$$\mathbf{B}^+ = \mathbf{X}^+.\mathbf{M} \tag{5}$$

This set of functions is now ready to express the molecular orbitals of the quantum part as

$$|k\rangle = \sum_{\mu} C'_{\mu k} |\mu'\rangle \tag{6}$$

and the expected occupied and virtual MOs $|k\rangle$ are defined by (N-L) sets of (N-L) coefficients $C'_{\mu k}$ which can be arranged in a (N-L)x(N-L) matrix $\mathbf{C'}$. Equivalently, these functions are linear combinations of the N original AOs and the coefficients of this expansion can be found in a rectangular $\mathbf{C}$ matrix which is deduced from the previous one by the linear transformation

$$\mathbf{C} = \mathbf{B}.\mathbf{C'} \tag{7}$$

The electron distribution of the quantum part in the basis of the AOs is expressed by a density matrix $\mathbf{P}^Q$. Similarly, the electrons in the localized orbitals give rise to a density matrix $\mathbf{P}^L$ so that all the electrons considered in this study are represented by a density matrix

$$\mathbf{P} = \mathbf{P}^Q + \mathbf{P}^L \tag{8}$$

The ($\mu\nu$) element of the Hartree-Fock matrix corresponding to this electron distribution is written

$$F_{\mu\nu} = H^t_{\mu\nu} + \sum_{\lambda,\sigma} P_{\lambda\sigma} G_{\mu\nu\lambda\sigma} \tag{9}$$

where $\mathbf{G}$ stands for the electron-electron (Coulomb+exchange) operator and $\mathbf{H}^t$ for the sum of the core Hamiltonian and the electrostatic interaction of the electrons with the charge distribution of the classical part. This term includes Coulomb interactions with localized classical point charges but may also be extended to higher permanent moments as well as the induced ones.

The above linear transformation allows us to express this matrix in the orthogonal basis set leading to

$$\mathbf{F'} = \mathbf{B}^+.\mathbf{F}.\mathbf{B} \tag{10}$$

The C' matrix of the eigenvectors of **F'** is obtained by diagonalizing the former matrix

$$\varepsilon = \mathbf{C'^+ . F'C'} \tag{11}$$

where ε is the diagonal matrix of the eigenvalues. The expansion of the MOs in the original basis set is obtained as usual after the back transformation (equation (7)). After taking into account the occupation number of each orbital, one gets the density matrix $\mathbf{P}^Q$ and the process is repeated until convergence according to the usual SCF procedure. The same procedure holds for solving the Kohn-Sham equations in the DFT formalism in which the electron density is split, similarly, into the self-consistent part ρ^Q and the frozen part ρ^L.

Compared with a standard algorithm, the method requires two slight modifications :

- the splitting of the density matrix into a constant $\mathbf{P}^L$ component and a computed one $\mathbf{P}^Q$;

- the computation of the **B** matrix. This being done, the **B** matrix plays the same role as the orthogonalization matrix in a standard LCAO MO computation and the self-consistent field computation can use the standard codes without further modification.

We call this method *Local Self-Consistent Field* (LSCF).

Energy and forces. The Classical/Quantum Force-Field

After convergence of the LSCF computation, the energy of the system is expressed by

$$
\begin{aligned}
E = &\frac{1}{2}\sum_\mu\sum_\nu \left(P_{\mu\nu}^Q + P_{\mu\nu}^L\right)\left(H_{\mu\nu}^t + F_{\mu\nu}\right) + \frac{1}{2}\sum_M\sum_N \frac{Z_M Z_N}{R_{MN}} \\
&+ \sum_M\sum_K \frac{Z_M Q_K}{R_{MK}} + V_{CC} + V'_{CQ}
\end{aligned}
\tag{12}
$$

In this expression M and N stand for two nuclei of the quantum subsystem, $Z_M Z_N$ for their charges and R_{MN} for the distance between them. Q_K is the net charge of atom K in the classical subsystem, V_{CC} is the configurational energy of the classical part and V'_{CQ} stands for the van der Waals interaction of the classical atoms with the quantum ones. Note that in equation (12), the electrostatic interaction has been reduced to Coulomb terms for sake of simplicity. It is extended to more sophisticated interactions without any difficulty (*10*). If we denote by $\partial_q\Gamma$ the partial derivative of the quantity Γ with respect to the coordinate q, the energy derivatives can be put in the form

$$\partial_q E = \sum_{\mu,\nu} \left(P_{\mu\nu}^Q + P_{\mu\nu}^L \right) \left[\partial_q H_{\mu\nu}^t + \frac{1}{2} \sum_{\lambda,\sigma} \left(P_{\lambda\sigma}^Q + P_{\lambda\sigma}^L \right) \partial_q G_{\mu\nu\lambda\sigma} \right]$$

$$+ \frac{1}{2} \sum_M \sum_N \partial_q \frac{Z_M Z_N}{R_{MN}} + \sum_M \sum_K \partial_q \frac{Z_M Q_K}{R_{MK}} + \partial_q V'_{CQ} + \partial_k V_{CC} \quad (13)$$

$$+ \sum_{\mu,\nu} \partial_q \left(P_{\mu\nu}^Q \right) F_{\mu\nu} + \sum_{\mu,\nu} \partial_q \left(P_{\mu\nu}^L \right) F_{\mu\nu}$$

In this expression, most of the terms are computed in standard QM/MM codes. The only ones which deserve special attention are the last two sums which involve the density matrices derivatives.

The derivatives of the self-consistent density matrix $\mathbf{P}^Q$ can be computed according to the traditional procedure. Indeed, the MOs are eigenfunctions of the Fock operator so that if $c_{\mu i}$ is the coefficient of $|\mu\rangle$ in the expression of MO $|i\rangle$ with an occupation number n_i :

$$\sum_{\mu,\nu} \partial_q \left(P_{\mu\nu}^Q \right) F_{\mu\nu} = \sum_i \sum_{\mu,\nu} 2n_i c_{\mu i} F_{\mu\nu} \partial_q c_{\nu i} = \sum_i \sum_{\mu,\nu} 2n_i \varepsilon_i c_{\mu i} S_{\mu\nu} \partial_q c_{\nu i} \quad (14)$$

if one assumes that the coefficients are real.

Similarly, the orthogonalization of the MO provides us with the equation

$$\sum_{\mu,\nu} \left(2c_{\mu i} S_{\mu\nu} \partial_q S_{\nu i} + c_{\mu i} c_{\nu i} \partial_q S_{\mu\nu} \right) = 0 \quad (15)$$

Equations (14) and (15) allow us to write

$$\sum_{\mu,\nu} \partial_q \left(P_{\mu\nu}^Q \right) F_{\mu\nu} = - \sum_i n_i \varepsilon_i \sum_{\mu,\nu} c_{\mu i} c_{\nu i} \partial_q S_{\mu\nu} = - \sum_{\mu,\nu} W_{\mu\nu} \partial_q S_{\mu\nu} \quad (16)$$

if we introduce the "energy weighted density matrix" $\mathbf{W}$ defined by

$$W_{\mu\nu} = \sum_i n_i \varepsilon_i c_{\mu i} c_{\nu i} \quad (17)$$

This reasoning is no longer valid for the terms containing the density matrix $\mathbf{P}^L$ since the LMO's are not eigenfunctions of the Fock operator. If one considers the projection of the atomic orbitals on the subspace of the LMOs, one can write :

$$\sum_{\mu,\nu} P^L_{\mu\nu} F_{\mu\nu} = \sum_{\mu,\nu}\sum_{\ell} n_\ell a_{\mu\ell} a_{\nu\ell} \langle \mu | \sum_{k} | k \rangle \langle k | \mathbf{F} | \nu \rangle$$

$$= \sum_{\mu,\nu}\sum_{\lambda,\sigma}\sum_{\ell,k} n_\ell a_{\mu\ell} a_{\lambda k} S_{\mu\lambda} a_{\nu\ell} a_{\sigma k} F_{\sigma\nu} \tag{18}$$

where $|\ell\rangle$ and $|k\rangle$ stand for LMOs.

By differentiating the density matrix $\mathbf{P}^L$ with respect to coordinate q, and by taking into account the orthonormalization condition, one gets, after some simple algebra :

$$\sum_{\mu,\nu} \partial_q \left(P^L_{\mu\nu} \right) F_{\mu\nu} = -\sum_{\mu,\nu}\sum_{\lambda,\sigma} \left(P^L_{\mu\nu} P^L_{\lambda\sigma} F_{\sigma\nu} \partial_q S_{\mu\lambda} \right) \tag{19}$$

This equation reduces to equation (16) when the orbitals are eigenfunctions of operator $\mathbf{F}$ and can be considered as a generalization of the former.

Therefore, the energy derivatives *i.e.* the forces have a simple expression and the only derivatives to be computed are those which are required in any QM/MM code.

The computation of the derivatives of the density matrix of the localized orbitals is made necessary by the fact that, in the course of a geometry optimization or a molecular dynamics simulation, the orientation of the LMOs varies. Therefore, their expansion in the basis of the atomic orbital defined in the laboratory frame is modified. It then becomes necessary to get the new expression of the LMOs after each such step. This expression is obtained by the appropriate rotation and renormalization processes.

Treatment of the intermediate region
Although the localized orbitals, in the intermediate region, do not need to be reduced to localized bond orbitals, we chose to limit this region to the bonds connecting the last atoms of the quantum subsystem, which we shall denote by X_ℓ, and the first atoms of the classical one, which we shall denote by Y_ℓ. Therefore, each link $X_\ell Y_\ell$ will be described by a strictly localized bond orbital (SLBO) $|\ell\rangle$.

These SLBOs can be defined at several degrees of approximation. A simple solution can be found in having predetermined localized orbitals in the standard basis sets for the most common bonds occurring in the macromolecules of interest. A more refined, but very convenient procedure, consists in considering a model molecule of the fragment which includes the bond of interest and performing a full quantum chemical computation on this molecule followed by an orbital localization. Among the various possible localization procedures, the Weinstein-Pauncz criterion (11) appears to be very efficient to localize one given bond orbital. A further refinement can be achieved by a subsequent use of the Magnasco-Perico criterion (12). The bond orbital is then reduced to a SLBO by eliminating the contribution of orbitals which are not centered on the pair of atoms defining the bond and normalizing the remaining function.

These localized orbitals are usually defined for equilibrium bond lengths, which may vary slightly in the course of a MD simulation or a geometry optimization. The corresponding energy derivatives may be obtained from an appropriate empirical force constant which has to be rescaled from the classical force field because the electrostatic interaction between the two atoms is evaluated differently.

The last empirical parameter needed in the system is the effective charge Q_Y* on the frontier atom separating the classical subsystem from the localized orbital. In the present study, this charge is simply defined after Q_Y, the charge of this atom Y in the classical force field, by $Q_Y*=Q_Y+1$ since one electron has been withdrawn from this atom to enter the localized bond orbital.

The whole procedure including the various approximations defines a Classical-Quantum Force Field (CQFF).

In the case of semi-empirical quantum chemical methods, very useful for treating large quantum subsystems, the method can be further simplified. Owing to the neglect of diatomic differential overlap, the only parameters of a SLBO which intervene in the computation are the nature of the hybrid orbital of the quantum frontier atom and the coefficient of this orbital in the SLBO. In addition, the neglect of overlap integrals makes the orthogonal basis set very easy to get since the only transformations required are the definition of hybrid orbitals, on the frontier atoms of the quantum subsystem, which have to be orthogonal to the hybrid orbitals of the SLBO's (*13-14*).

Applications

Most of the force fields are parametrized from *ab initio* results obtained on isolated small molecules. In order to investigate to what extent the properties of a molecular fragment may be modified under the influence of its macromolecular surroundings, we decided to study three different fragments of an α-helix composed of nine alanine molecules H$($NH-CHCH$_3$-CO$)_9$OH (Figure 1). All calculations have been performed at the MP2/6-31G**//HF/3-21G level of theory, using the frozen core approximation in the evaluation of the correlation energy. Since the localized orbitals are not eigenfunctions of the Fockian, they need to be considered as frozen with respect to the MP2 perturbation procedure. The conformational studies have been performed by varying a chosen dihedral angle by 30° steps and optimizing all the other coordinates of the quantum part. In addition, in order to analyze the influence of the intramolecular interactions on the properties of this fragment, four different computations have been considered : i) a full QM/MM computation using the LSCF scheme denoted LSCF/MM; ii) a LSCF computation in which all interactions with the surroundings have been set equal to zero (denoted LSCF). In addition, to understand separately the influence of the electrostatic and the van der Waals interactions, two additional computations have been performed; iii) a LSCF computation in which the only active contribution from the classical part is the electrostatic one (LSCF/ELE) and iv) a LSCF computation with the classical van der Waals potential only (LSCF/VDW). The van der Waals contribution (denoted VDW in the tables) is obtained by subtracting the LSCF results from the LSCF/VDW ones.

Similarly, the polarization contribution (POL) is the difference between LSCF/ELE and LSCF. The comparison of the results obtained under these different conditions enables us to separate the various contributions.

Results and discussion
* C-terminal residue (residue number 9). The quantum subsystem has the following linear formula $C-CO-NH-CHCH_3-COOH$. Figure 2 shows the energy variation along the rotation around the NH-CH bond (i.e. the traditional φ diedral angle). One can see that the LSCF/MM curve exhibits a deep minimum around the value $\varphi=-125°$ and a second one, less pronounced for $\varphi=-60°$ which is the dihedral angle for an α-helix structure. The quantum part alone, represented by the LSCF curve, exhibits a minimum close to the value $\varphi=-160°$. This curve appears to be very close to that obtained by a MP2/6-31G**//HF/3-21G computation on the molecule $H-CO-NH-CHCH_3-COOH$. The small differences are due to the fact that in the LSCF case, we model the C-C bond with a localized orbital, whereas in the molecule, this bond is replaced by a C-H bond. Both curves exhibit a maximum for $\varphi\approx-110°$.

The relative van der Waals energy between the quantum and the classical parts does not vary much and its influence on the φ value can be considered as negligible.

Analyzing the geometry of the LSCF/MM minimum, we found that the acidic proton of the carboxylic group appears to be H-bonded to the oxygen atom of the carbonyl group of residue number 5 (1-4 H bond). The distance between these two atoms is displayed on figure 3.

The influence of this hydrogen bond can be seen by considering the polarization energy, obtained by subtracting the LSCF and the van der Waals energies to the LSCF/MM energies, which is minimized for $\varphi=-110°$.
The peptidic bonds are known to be rather polarizable due to an easy electron shift from the nitrogen atom to the carbonyl group, as seen on the two limiting mesomeric forms shown on scheme 1. Due to the fact that the dipole moments of the peptidic bonds are parallel in an α helix, we expect a strong polarization which favors form 2. These electronic modifications are expected to be visible on the variations of the bond orders, and on equilibrium bond lengths as well. Table 1 collects the bond distance and the Mayer bond order (*15*) of the peptidic bond of the quantum part. We observe an increase of the Mayer C-N bond order and a decrease of the C-O bond order, as expected, but both bond lengths increase slightly from the free fragment to the fragment interacting with the rest of the helix. A more detailed analysis made possible by considering the effect of the various contributions to the intramolecular interactions (table 1) shows that the unexpected lengthening of the C-N bond is due to the van der Waals interaction but the electrostatic contribution does have the expected effect.

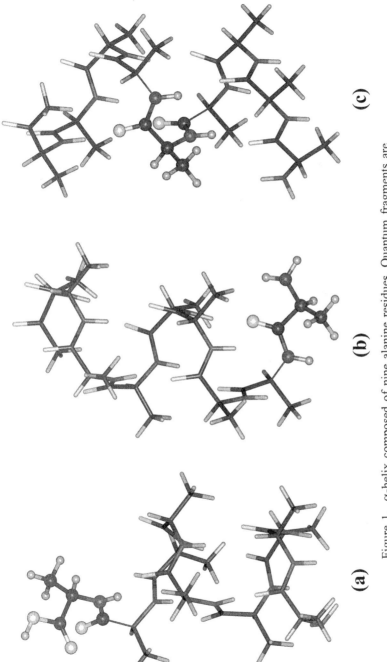

Figure 1. α-helix composed of nine alanine residues. Quantum fragments are shown in balls and sticks representation, classical parts in sticks only. Chemical bonds linking the two subsystems are displayed with a tiny stick. Quantum parts are (a) the C-terminal residue, (b) the N-terminal residue, and (c) the central fragment.

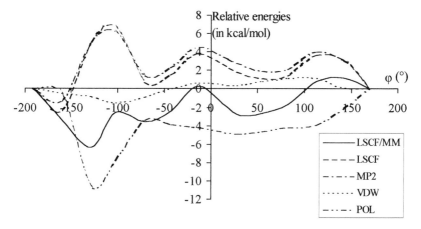

Figure 2. Relative energies of the C-terminal residue for different values of the φ angle (acronyms are defined in the text).

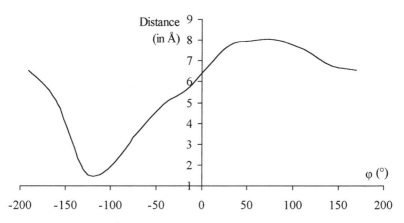

Figure 3. Distance (in Å) between the hydrogen atom of the carboxylic group of the C-terminal residue and the oxygen atom of the carbonyl group of the fifth residue versus φ values.

Table 1. Bond length in Å (left column) and Mayer bond order (right column) corresponding to various calculations (for acronyms, see text).

Bond	LSCF/MM		LSCF/ELE		LSCF/VDW		LSCF	
N-C	1.369	0.926	1.362	0.931	1.371	0.908	1.365	0.911
C-O	1.226	1.687	1.224	1.685	1.223	1.704	1.221	1.705

N-terminal residue (residue number 1). Results concerning this residue, which has the linear formula C-NH-CO-CHCH$_3$-NH$_2$, are shown on figure 4. The rotation is performed around the CO-CH bond. This rotation is related to the values of the

Table 2. Bond length in Å (left column) and Mayer bond order (right column) corresponding to various calculations (for acronyms, see text).

Bond	LSCF/MM		LSCF/ELE		LSCF/VDW		LSCF	
N-C	1.342	1.164	1.340	1.174	1.341	1.150	1.339	1.164
C-O	1.225	1.675	1.228	1.663	1.222	1.705	1.223	1.697

traditional ψ angle. The LSCF/MM curve exhibits a minimum in a region where $-10° \leq \psi \leq 40°$, which, again, does not correspond to the value of ψ for an α-helix ($\psi_\alpha = -40°$). Similar results are obtained without the classical part (LSCF curve). One notices that the van der Waals interaction and the polarization energy do not vary significantly with ψ.

Unlike the C-terminal residue, the N-terminal residue cannot form an H-bond between its NH$_2$ end group and the rest of the helix. For this reason the LSCF and LSCF/MM curves are very close. The influence of electrostatic interaction, which is not very visible on the conformational energy curve, appears when one considers the properties of the peptidic bond (table 2). Again, the polarization of the bond increases the weight of the mesomeric form 2 (scheme 1).

Central residue (residue number 5). In order to gain more information on these cooperative polarization effects in α helices, an additional study has been performed on the central part of the helix i.e. residue n° 5. The two peptidic bonds formed by this amino acid are included in the quantum part which has the following linear formula C-CO-NH-CHCH$_3$-CO-NH-C. Two different computations have been carried out : at fixed geometry, one with the full LSCF/MM scheme and a second one in the absence of the quantum classical interactions (LSCF). The polarization effects are analyzed by considering the dipole moment of this fragment and its components along the three main axes of the helix (table 3). The helicoidal axis is chosen as the Z axis, the X axis is defined so that the C$_\alpha$ of residue 5 is located on this axis (figures 5a,b). All the results are consistent with an electron shift from the NH groups to the CO ones under the influence of the surroundings, leading to an increase of the CN bond orders and a decrease of the C=O one. The largest variation of the dipole moment occurs, as expected, in the Z direction, but one notices that the X component, which should not vary noticeably due to the fact that the induced moments on the peptidic bonds are antiparallel in the X direction, undergoes the second largest variation. An analysis of the Mulliken charges shows that this variation mainly comes from a polarization of the side chain. The sum of the atomic

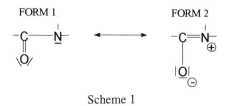

Scheme 1

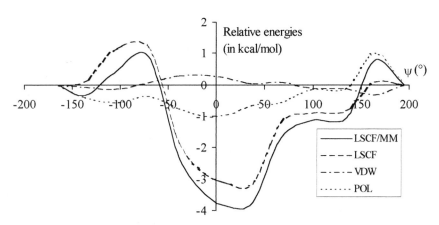

Figure 4. Relative energies of the N-terminal residue for different values of the ψ angles (for acronyms, see text).

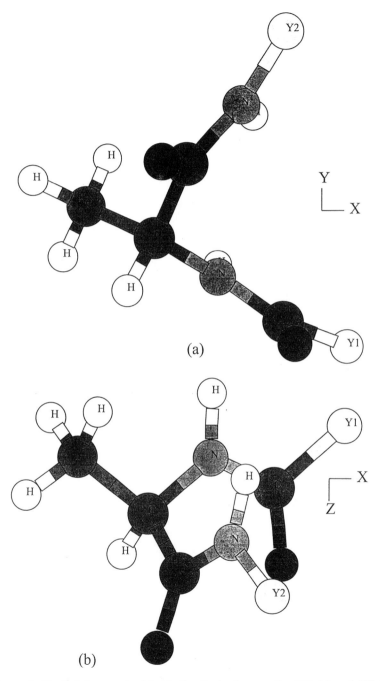

Figure 5. Central fragment of the helix. Projection on the XY (a) and XZ (b) planes. Y1 and Y2 denote the first carbon classical atoms in the LSCF scheme.

charges of the CH$_3$ group increases from +0.021 a.u. to +0.047 a.u. when the interaction with the rest of the helix is turned on.

Table 3. Dipole moment of residue 5 (in Debye) isolated (LSCF) and embedded in the helix (LSCF/MM).

| | **Electric Dipole Moment (in Debye)** | |
	LSCF/MM	LSCF
X	-3.716	-3.397
Y	1.424	1.589
Z	-4.866	-3.552
Total	6.286	5.165

These results show the strong polarizing effect of the helix on each of its constituents. This effect is particularly marked in the center of the helix. One of its most unexpected consequences is the noticeable perturbation that the methyl side chain undergoes in α polyalanine. These observations clearly indicate that the use of a quantum chemical description of a region of a macromolecule may be highly recommendable when one deals with second order effects on the electronic structure or the conformational potential energy surface.

Conclusions

The Local Self Consistent Field method is a viable alternative to the approaches using link atoms in QM/MM computations. It is easy to implement in a standard *ab initio* or DFT code and the computation of the forces by means of a fully analytical algorithm gives rise to a Classical Quantum Force Field which can be used in energy minimization or in molecular dynamics simulations. This methodology makes possible the study of chemical reactions involving large molecular systems, like enzymes, but it can also be used to derive the *in situ* conformational potential energy surface part of a macromolecule, which appears to be quite sensitive to the influence of its surroundings.

In the case of reactions involving a large number of atoms, reference *ab initio* or DFT studies may appear as still too expensive, in particular in molecular dynamics studies. In these cases, a whole semi-empirical study may appear as more appropriate. Unfortunately, these methods are known to give rather poor results, in particular for the energies of the transition states which are often overstimulated. Nevertheless, the

possibility of performing more accurate single point computations by means of non-empirical approaches may appear as very useful to rescale the result and to reduce strongly the arbitrariness of some interpretations arising from pure semi-empirical results.

References
1. Field, M.J., Bash, P.A. and Karplus, M., *J. Comp. Chem.* **1990**, *11*, 700.
2. Gao, J. in *Reviews in Computational Chemistry* K.B. Lipkowitz and D.G. Boys eds. (VCH, New York) **1995**, *7*, 119-185.
3. Tuñón, I., Martins-Costa, M.T.C., Millot, C. and Ruiz-Lopez, M.F. *J. Chem. Phys.* **1997**, *106*, 3633-3642.

4. Strnad, M., Martins-Costa, M.T.C., Millot, C., Ruiz-Lopez, M.F. and Rivail, J.L. *J. Chem. Phys.* **1997**, *106*, 3643-3657.

5. Car, R. and Parrinello, M.P., *Phys. Rev. Lett.*, **1985**, *55*, 2471

6. Assfeld, X. and Rivail, J.L. *Chem. Phys. Lett.* **1996**, *263*, 100-106.

7. Löwdin, P.O. *Int. J. Quantum Chem.* **1967**, *15*, 811.

8. Löwdin, P.O. *Advan. Quantum Chem.* **1970**, *5*, 185.

9. Roby, K.R. *Chem. Phys. Lett.* **1971**, *11*, 6; **1972**, *12*, 579.

10. Szabo, A. and Ostlund, N.S., *Modern Quantum Chemistry* (Mc Graw-Hill, New York 1989) p. 144.

11. Ruiz-Lopez, M.F. and Rivail, J.L. in *Encyclopaedia of Computational Chemistry* (J. Wiley and Sons, New York) to be published.

12. Weinstein, H. and Pauncz, R. *Symp. Faraday Soc.* **1968**, *2*, 23.

13. Weinstein, H. and Pauncz, *Adv. Atomic Molec. Phys.* **1971**, *7*, 97.

14. Magnasco, V. and Perico, A. *J. Chem. Phys.* **1967**, *47*, 971; **1968**, *48*, 800.

15. Thery, V., Rinaldi, D., Rivail, J.L., Maigret, B. and Ferenczy, G.G. *J. Comp. Chem.* **1994**, *15*, 269.

16. Monard, G., Loos, M., Thery, V., Baka, K. and Rivail, J.L. *Int. J. Quantum Chem.* **1996**, *58*, 153.

17. Mayer, I., *Chem. Phys. Lett.*, **1983**, *97*, 270.

Chapter 16

Investigating Enzyme Reaction Mechanisms with Quantum Mechanical–Molecular Mechanical Plus Molecular Dynamics Calculations

Peter L. Cummins and Jill E. Gready

Division of Biochemistry and Molecular Biology, John Curtin School of Medical Research, Australian National University, P.O. Box 334, Canberra, ACT 2601 Australia

In this paper we describe a coupled semiempirical (AM1, MNDO and PM3) quantum mechanical and molecular mechanical (QM/MM) plus molecular dynamics and free energy perturbation (MD/FEP) approach for studying chemical reaction mechanisms in the active sites of enzymes. The feasibility of this approach for studying enzyme catalysis is illustrated in a calculation of the free energy change for the hydride-ion transfer step between NADPH and a novel substrate, 8-methylpterin, in the active site of dihydrofolate reductase.

The coupled quantum mechanical and molecular mechanical (QM/MM) method can be used to study chemical processes in condensed phases, such as biological catalysis (*1-3*). Several QM/MM potentials within the semiempirical molecular orbital approximation have been developed for this purpose (*4-7*). In this paper we describe the implementation of one of these QM/MM potentials (*7*) in the study of enzyme-catalyzed reactions. The molecular dynamics (MD) simulation method is used together with the free energy perturbation (FEP) method (*8*) in order to compute free energy differences between points along the reaction coordinate. We illustrate the application of the methodology by calculating the free energy change for the enzymic hydride-ion transfer step between the cofactor nicotinamide adenine dinucleotide phosphate (NADPH) and a novel substrate, 8-methylpterin (*9*). The enzyme dihydrofolate reductase (DHFR) catalyses this transfer between NADPH and the substrate molecule (S) according to

$$NADPH + SH^+ \rightarrow NADP^+ + SH_2$$

To define the role of the enzyme and the relationships between protein structure and function in catalysis, we decompose the free energies into different components. Consequently, we are able to determine the relative importance of the contributions of various amino acid residues to the free energy change. As the number of QM atoms is an important consideration in QM/MM studies, for reasons of computational efficiency, we also examine the feasibility of partitioning the NADPH molecule into various QM and MM regions. The "best" model is one where all molecules taking part in the chemical reaction are treated by QM. This does not present a problem for the 8-

methylpterin substrate, and remains possible for the larger NADPH molecule. However, it is advantageous to divide the NADPH molecule into QM and MM regions provided this does not adversely affect the accuracy of the results. The cpu timings obtained from the present calculations may be extrapolated in order to estimate increases which will be required for larger QM regions.

Methods

The Combined QM/MM Model. The computations were performed using Molecular Orbital Programs for Simulations (MOPS) (*10*). Although MOPS can perform traditional geometry optimization and transition state searches (*11*), it is clear that MD and Monte Carlo simulations will become increasingly important as these provide a more realistic description of condensed phases, particularly for biological systems. Thus, our emphasis has been in the development of computer codes that combine MD simulations with QM/MM methods. When performing MD simulations using QM/MM potentials, we are generally restricted to semiempirical QM methods for reasons of computational efficiency. Consequently, the current version of MOPS can perform MINDO/3, MNDO, AM1 and PM3 semiempirical QM calculations only.

Molecular dynamics simulations require the use of relatively simple potential energy functions. The QM/MM potentials must be developed with this fact in mind. The total potential energy (E) of a system partitioned into quantum and molecular mechanical groups of atoms may be written in the general form (*3*)

$$E = E_{QM} + E_{MM} + E_{QM/MM} \tag{1}$$

where E_{QM} is the Hartree-Fock SCF energy of the quantum system, E_{MM} is the energy of the molecular mechanics part of the system and $E_{QM/MM}$ is the interaction energy between the quantum and molecular mechanics parts of the system. The QM/MM interaction energy is written in the usual way as the sum of polar (electrostatic) and nonpolar (van der Waals) terms,

$$E_{QM/MM} = E_{ele} + E_{vdW} \tag{2}$$

In equation 2, E_{ele} is a semiempirical approximation for the interaction between the nuclei and electrons of the QM system and the atomic charges of the MM system. Within the neglect of diatomic differential overlap (NDDO) approximation (*12*), the total QM/MM interaction energy may be calculated by the summation of pair-potential energies, i.e. the electrostatic term may be written in the form

$$E_{ele} = \sum_{ij} q_i V_j(R_i) \tag{3}$$

where $V_j(R_i)$ is the electrostatic potential (EP) at the position of molecular mechanics atom i, R_i, produced by an atom j of the QM region. The EP can be expressed solely in terms of atom-centered contributions (all equations given in atomic units),

$$V_j = \sum_{\mu\nu} P_{\mu\nu} I_{\mu\nu} + Z_j/R_{ij} \tag{4}$$

where Z_j is the core charge of atom j, R_{ij} is the distance from QM atom j to MM atom i, $P_{\mu\nu}$ are elements of the first order density matrix, and $I_{\mu\nu}$ are the one-electron integrals of the coulomb operator involving only those orbitals ϕ_ν and ϕ_μ centered on the j'th atom. All the semiempirical methods (MNDO, AM1 and PM3) make use of two-electron repulsion integrals of the form (ss,μv) in order to approximate one electron integrals $I_{\mu v}$ (12,13). As this approximation underestimates the core-electron attractions, the core-core repulsions must likewise be reduced from their coulomb values in order to maintain the balance between the attractions and repulsions in the molecule. In terms of the two-electron integrals, (ss,μv), equation 4 may be written in the form

$$V_j = \sum_{\mu v} P_{\mu v}(ss,\mu v)_m + Z_j(ss,ss) \tag{5}$$

where the subscript m indicates the transformation of (ss,μv) from the molecular cartesian to local (diatomic) coordinate system. For the EP, the one electron integrals simplify to the following expressions ($p = p_x$ or p_y, $z = p_z$ orbitals) (14):

$$(ss,ss) = 1/R_i \tag{6a}$$

$$(ss,sz) = \frac{1}{2}\left[(R_i - D_1)^{-1} - (R_i + D_1)^{-1}\right] \tag{6b}$$

$$(ss,zz) = \frac{1}{2}\left[\frac{1}{2}\left\{(R_i + 2D_2)^{-1} + (R_i - 2D_2)^{-1}\right\} - (ss,ss)\right] \tag{6c}$$

$$(ss,pp) = \frac{1}{2}\left[(R_i^2 + 4D_2^2)^{-1/2} - (ss,ss)\right] \tag{6d}$$

where the parameters D_1 and D_2 are determined by the condition that the dipole and quadrupole moments of each atomic orbital product $\phi_\mu \phi_v$ are equal to those for the corresponding point-charge distributions. Values for D_1 and D_2 are readily calculated using the formulae (13)

$$D_1 = \frac{(2^n+1)(4\zeta_{ns}\zeta_{np})^{n+1/2}}{3^{1/2}(\zeta_{ns} + \zeta_{np})^{2n+2}} \tag{7a}$$

$$D_2 = \left[(2n+1)(n+1)/10\right]^{1/2}(\zeta_{np})^{-1} \tag{7b}$$

where ζ_{ns} and ζ_{np} are, respectively, the exponents of the Slater s and p atomic orbitals for principal quantum number n.

The charge distribution of the quantum system is polarized by the electrostatic potential produced by the atom-centered charges q_i of the molecular mechanics system by performing the self-consistent field (SCF) calculation using the appropriately perturbed one-electron Fock matrix elements, i.e.

$$F_{\mu v} = F_{\mu v}^{\circ} + \delta\sum_i q_i P_{\mu v}(ss,\mu v)_m \tag{8}$$

where $F_{\mu v}^{\circ}$ is the matrix element for the unperturbed (i.e. gas phase) system, and $\delta = 1$ if v and μ are on the same atom and $\delta = 0$ if v and μ are on different atoms, as required by the NDDO approximation. Note, however, that we do not include terms that describe explicitly the polarization of the MM region.

The E_{vdW} term in equation 2 is expressed simply as the sum over pair-potential functions from the MM part of the force field. The attractive part of the potential is given by the usual R^{-6} dispersion term. Within MOPS there are several options for the repulsive part of the vdW term

$$V(R) = A/R^n - B/R^6 \qquad (9)$$

where $n = 12$ gives the usual Lennard-Jones potential, or $n = 9$. Alternatively, the repulsions may be described by an exponential to give

$$V(R) = C.exp(-Ar) - B/R^6 \qquad (10)$$

for the vdW terms. Note that usually only one of the above options is applied to all the non-bonded interactions. However, it is often desirable to allow for greater flexibility and to be able to treat particular interactions as special cases. Thus, two types of nonbonded interactions may be defined each described by a different function. The special case interactions are given by

$$V(R) = A/R^{12} - B/R^{10} \qquad (11)$$

although other functional forms could be included as options. These special non-bonded terms are important for the correct description of H bonding between QM and MM regions (7).

Free energy perturbation methods. The free energy differences between reactant and product states were calculated using the free energy perturbation (FEP) approach (8). Thus, we are interested in the energy as a function of some predefined reaction coordinate vector **r**. The Helmholtz free energy change ΔA from **r** to $\mathbf{r} \pm \Delta \mathbf{r}$ is then given by the perturbation formula

$$\Delta A_k(\mathbf{r}, \mathbf{r} \pm \Delta \mathbf{r}) = -(RT)^{-1} \ln \left\langle \exp\left[-\left(E_k(\mathbf{r} \pm \Delta \mathbf{r}) - E_k(\mathbf{r})\right)/RT\right] \right\rangle_\mathbf{r} \qquad (12)$$

where the ensemble average is over the state defined at some point **r** on the reaction coordinate. The subscript k denotes a total free energy or a free energy component as defined, for example, in equations 1 and 2. As implied in equation 12, the reaction path is divided into "windows" of width $\Delta \mathbf{r}$, the total free energy change being the sum over all such windows. We estimate the free energy change for each window by taking the mean of plus and minus perturbations, i.e.

$$\Delta A_{ki} = \{\Delta A_k(\mathbf{r}_i, \mathbf{r}_i + \Delta \mathbf{r}) - \Delta A_k(\mathbf{r}_i, \mathbf{r}_i - \Delta \mathbf{r})\}/2 \qquad (13)$$

gives the free energy change for the i'th window.

Molecular dynamics simulations and FEP calculations. The system is initially built in the standard way for an MM/MD calculation from a number of relatively small fragments or "residues" (15). All residues are assumed to be MM unless specified as QM, i.e. the QM region in MOPS can only be defined on a residue by residue basis. In the present study, the 8-methylpterin molecule and the four residues of the NADPH molecule, as shown in Figure 1, were included in the QM

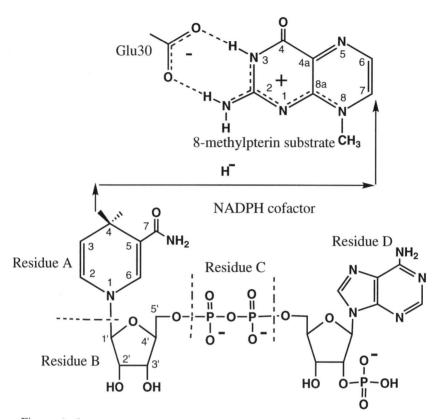

Figure 1. 8-methylpterin substrate and NADPH cofactor. The cofactor is partitioned into residues A (nicotinamide), B (ribose), C (phosphate) and D (adenine) for the purposes of defining the QM/MM boundary. The hydride-ion transfer takes place between C4 of the nicotinamide ring of NADPH and C7 of the substrate.

region. Several QM/MM models for the cofactor were also defined by restricting the QM region to contain only residues A, A and B (AB) or A, B and C (ABC). The cofactor residues that do not form part of the QM region, all of the protein residues, and all water molecules, form the MM region. Note that if covalent bonds cross the boundary between the QM and MM regions, so-called "link" atoms (4) must be introduced to satisfy valence requirements and effectively model the actual system. Where a covalent bond in the NADPH cofactor crosses the boundary between the QM and MM regions, "link" atoms F, CH_3 and H were used. Thus, we denote the possible QM parts of the cofactor as A-L, AB-L, ABC-L or ABCD, where the link atom L may be F, CH_3 or H. Note that link atoms do not interact directly with the MM atoms through explicit terms in the Hamiltonian, but are required only to satisfy valency requirements of the QM fragment. Terms in the MM force field describing bonds, angles and dihedrals that include both QM and MM atoms are retained, while the nonbonded electrostatic and vdW terms are treated in the usual way as described in equation 2. In the present MD simulations the link atoms are treated in the same way as all other atoms, i.e. their dynamical behaviour is determined by the forces acting on them according to classical mechanics. Although this approach introduces artificial degrees of freedom, it is easy to implement and provides a consistent treatment of the total energy and gradient that is required for solving the classical equations of motion in MD. To minimise the risk of any adverse effects on the calculation of reaction free energies, it is important that the link atoms are not located too close to the reaction center.

The force field of Weiner et al. (16,17) (i.e. the AMBER force field) was used for the protein, while the TIP3P model parameters were used for water (18). For the QM systems the AM1 method (19) was used and the vdW parameters were chosen from an optimized set obtained from a previous QM/MM study of solute-solvent interactions (7). The starting coordinates for the simulations were taken from a previous study of solvated DHFR ternary complexes (20). The two inter-atomic r(C4-H) and r(C7-H) distances were chosen for the reaction coordinates and were maintained at fixed values by application of the SHAKE algorithm (21). The algorithm of Berendsen et al. (22) was used to perform the MD simulation at a temperature of 300K and a time step of 0.001 ps. In the FEP calculations the reaction coordinates were divided into windows of width Δr = 0.01 or 0.02 Å requiring 160 windows to obtain the total free energy difference between reactant and product states. With each window consisting of 800 steps of equilibration followed by 800 steps of data collection for calculation of the free energies, the overall simulation time was close to the minimum used by Ho et al. (23) in FEP calculations of hydride-transfer reactions in solution.

Results and Discussion

The free energy differences between reactant and product states were computed along the path shown in Figure 2a. Figure 2b shows the free energies as a function of the reaction coordinate for the calculation in which all cofactor residues (ABCD) were treated by QM. Based on previous AM1 studies (24) of gas-phase analogue reactions, the r(C4-H) and r(C7-H) bond lengths at the TS are expected to be between 1.3 and 1.5 Å. Thus, the path closely follows the minimum free energy path and consequently approaches the transition state. The free energies for the hydride-ion transfer from the cofactor to the substrate are summarized in Table I. The activation free energies are given in parentheses. Calculations were also performed by carrying out the FEP calculations along different pathways, the results suggesting relatively small path dependence (25).

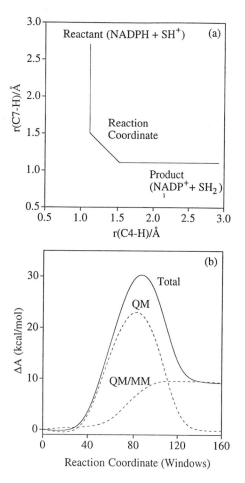

Figure 2. (a) Reaction pathway used in the MD/FEP calculations. The pathway is defined by allowing increments (windows) Δr of absolute value 0.00, 0.01 or 0.02 Å such that $|\Delta r(\text{C4-H})| + |\Delta r(\text{C7-H})| = 0.02$ Å. The reactant state is defined at $r(\text{C4-H}) = 1.125$ Å and $r(\text{C7-H}) = 2.705$ Å, while the product state is defined at $r(\text{C4-H}) = 2.704$ Å and $r(\text{C7-H}) = 1.105$ Å. (b) Free energy changes (kcal/mol) as a function of the reaction coordinate (number of windows in the MD/FEP calculation) for QM species ABCD (Figure 1). The total, QM and QM/MM component free energies were calculated using equation 12.

The change in the QM/MM component is dominated by the electrostatic as opposed to vdW type of interactions (Table I). In simulations with all NADPH residues treated by QM, the interactions with the enzyme plus solvent (i.e. the QM/MM component) contribute 10 kcal/mol to the free energy change. This is an interesting result as it suggests that the TS for the hydride-ion transfer step is in fact destabilized by electrostatic interactions with the enzyme (Figure 2b). However, the activation barrier of ca. 30 kcal/mol remains similar to that calculated for analogue gas-phase reactions (24). To understand the origin of the 10 kcal/mol free energy change we require a knowledge of the contributions from individual residues. As the current version of MOPS does not perform such a complex free energy decomposition, we have used a computationally simpler method of comparing the thermally averaged potential energy differences between reactant and product states. On the basis of the mean differences in the interaction energies, the increase in the QM/MM component of the free energy was rationalized largely in terms of the loss of electrostatic binding energy in the H-bond interaction (Figure 1) between the charged Glu30 side-chain and the substrate (25).

Table I. Free energy changes ΔA (kcal/mol) for the hydride-ion transfer path shown in Figure 2a.

QM species[a]	ΔA (kcal/mol)				
	QM	QM/MM	Ele[b]	Total[c]	Total[d]
A-F	21.66	-4.36	-4.29	17.30	17.30 (39.2)
A-CH$_3$	8.91	-7.41	-7.33	1.50	1.50 (34.0)
A-H	11.42	-4.67	-4.42	6.75	6.76 (30.6)
AB-H	-20.06	0.71	0.81	-19.35	-19.35 (21.9)
ABC-H	5.26	9.24	9.52	14.50	14.54 (33.0)
ABCD	-0.15	9.31	9.56	9.16	9.20 (29.3)

[a]QM parts of the cofactor A-L, AB-L, ABC-L or ABCD, where the link atom L may be F, CH$_3$ or H. Note the entire substrate molecule is treated by QM.
[b]Electrostatic/polarization component of the free energy (see equation 2).
[c]Total obtained by summing the QM and QM/MM components.
[d]Total calculated directly from equation 12. The results in parentheses are the estimates of the activation free energy.

The relative timings for different parts of the MD/FEP calculations are given in Table II. Note the SCF plus QM/MM terms account for more than 70% of the total cpu time. For the smallest QM system (A-H) the QM/MM terms account for nearly 52% of the cpu time. However, as the number of QM atoms is increased the proportion of time spent on the SCF part increases. After about 50 to 60 atoms the SCF calculation accounts for the largest proportion of cpu time. The percentages for QM gradients remain relatively static with increasing size of the QM system, while the importance of the MM part diminishes from 15.5% (A-H) to 2.8% (ABCD).

Figure 3 illustrates the effect of increasing the number of QM atoms on the proportion of the calculation taken up by the SCF part. On the basis of the present MD/FEP calculations, we would expect the SCF part to account for more than 90% of the total cpu time after ~150 atoms. The total cpu times as a function of the number of QM atoms for the hydride-ion transfer calculations are illustrated in Figure 4. The cpu time rises rapidly after 50 QM atoms, which coincides with the predominance of the SCF part of the calculation over the calculation of the QM/MM terms (Table II). For the the entire NADPH molecule treated by QM (96 atoms including the substrate), the total cpu time to compute the free energy change from reactant to product would be ~1700 cpu hours on an SGI Power Challenge R10000 processor. Beyond 100 atoms the calculations rapidly become impractical on a single processor, due largely to the computational expense of performing the SCF part. Although we have access to powerful vector hardware, a Fujitsu VPP300, improvements in performance on this machine will require much improved vectorization of the SCF calculations than we can currently achieve.

Table II. Relative (%) timings on a single SGI PowerChallenge (R10000) cpu for components of the MD/FEP calculations using MOPS.

Component	A-H[a]	AB-H[a]	ABC-H[a]	ABCD[a]
QM (Hamiltonian)	2.0	2.4	2.4	2.4
QM (SCF)[b]	21.8	34.1	43.4	62.8
QM (gradients)	4.9	5.8	5.8	5.5
QM/MM terms	51.5	44.9	38.9	25.1
MM (non-bonded)	15.5	9.6	6.9	2.8
MM (bonded)	0.5	0.3	0.2	0.1
Pair lists	0.5	0.3	0.2	0.1
Other	3.2	2.7	2.1	1.3

[a]QM fragment of NADPH. See Figure 1 for NADPH residues. H is a hydrogen "link" atom. The substrate molecule is treated by QM in all calculations.
[b]Fock matrix construction and diagonalization.

Clearly, parallelization would allow for the efficient treatment of larger QM systems. The current version of MOPS does not perform parallel computations. Fortunately, however, as the structure of the FEP calculation allows for any number of independent calculations to be performed simultaneously, the problem is well suited to parallelization (23). The availability of machines with multiple processors can be exploited by performing a number of independent FEP calculations concurrently on separate processors. The reaction path can be divided into segments each containing equal numbers of windows, the number of segments being equal to the number of processors. The total free energy change is then the sum over free energies obtained from the individual FEP calculations. To ensure that the structures at the beginning of each segment are properly equilibrated a period of MD is allowed before starting the FEP calculations. This procedure could be fully automated in a parallel version of MOPS. In the present calculations in which the QM region includes all cofactor residues and which would have taken ~1700 cpu hours on a single SGI Power Challenge processor, we divided the pathway into 8 segments, each containing 20 windows. An additional 20 ps of MD equilibration was allowed before starting the FEP calculations. The whole calculation took less than 10 days to complete.

Although reducing the number of QM atoms in the calculations is desirable on efficiency grounds, not all of the reactions involving cofactor fragments terminated with link atoms provide an accurate model of the reaction where the entire NADPH cofactor is treated quantum mechanically. There are often large differences in the QM component free energies. These may be rationalized in terms of the electronic

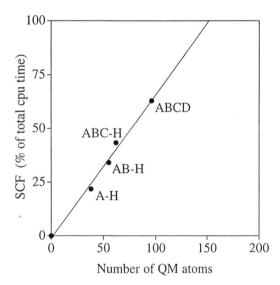

Figure 3. SCF calculation as a percentage of total cpu time for the free energy calculation of the hydride-ion transfer in DHFR (results in Table II) plotted as a function of the number of QM atoms. All timings are referenced to a single SGI Power Challenge (R10000) processor. A-H, AB-H, ABC-H and ABCD indicate the QM parts of the cofactor as in Figure 1 (with H link atoms). Note the entire substrate molecule is treated by QM.

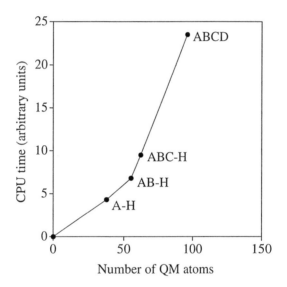

Figure 4. Relative total cpu times for the free energy calculation of the hydride-ion transfer in DHFR plotted as a function of the number of QM atoms. All timings are referenced to a single SGI Power Challenge (R10000) processor. A-H, AB-H, ABC-H and ABCD indicate the QM parts of the cofactor as in Figure 1 (with H link atoms). Note the entire substrate molecule is treated by QM.

structures of the fragments. The charges obtained by summing the atomic partial charges on the QM residues are given in Table III. As both F and CH_3 are strongly electron withdrawing in the field of the MM atoms, the nicotinamide residue (A) is left with a net positive partial charge in A-F and A-CH_3, whereas a net negative partial charge is obtained in A-H and when all the NADPH residues (i.e. ABCD) are treated by QM. On oxidation of the cofactor (product state) the positive charge on residue A in A-L is overestimated compared with residue A in ABCD, but this overestimation is least when the link atom is H. Thus, an electron donating group (e.g. H) would be the preferred choice for the link in A-L. In contrast, residue B (ribose) in AB-H has a partial charge opposite in sign to the one in ABCD. Although the charge on C is grossly overestimated in ABC-H, its value does not significantly change on oxidation of the cofactor. Note that the QM species with charge distributions similar to corresponding residues in ABCD also tend to yield similar free energies (Table II).

Table III. Net charges (a.u.) on QM residues in reactants and products.

QM species[a]	Residue	Reactants[b]	Products[b]
A-F	Substrate	1.00	-0.09
	A (Nicotinamide)	0.25	1.32
	F (link)	-0.25	-0.23
A-CH_3	Substrate	1.00	-0.10
	A (Nicotinamide)	0.21	1.36
	CH_3 (link)	-0.21	-0.26
A-H	Substrate	1.00	-0.06
	A (Nicotinamide)	-0.09	0.98
	H (link)	0.09	0.08
AB-H	Substrate	1.00	-0.10
	A (Nicotinamide)	-0.20	0.10
	B (Ribose)	0.33	1.32
	H (link)	-0.13	-0.32
ABC-H	Substrate	1.00	-0.04
	A (Nicotinamide)	-0.12	0.77
	B (Ribose)	-0.30	-0.14
	C (Phosphate)	-2.43	-2.47
	H (link)	0.84	0.88
NADPH	Substrate	1.00	-0.05
	A (Nicotinamide)	-0.13	0.74
	B (Ribose)	-0.47	-0.31
	C (Phosphate)	-0.74	-0.74
	D (Adenine)	-1.65	-1.63

[a]QM parts of the cofactor A-L, AB-L, ABC-L or ABCD, where the link atom L may be F, CH_3 or H. Note the entire substrate molecule is treated by QM.
[b]Charges obtained over 5 ps of coordinate averaging.

While analysis of the QM fragment charge distributions may be useful for determining the appropriate QM species and link atoms as a model for a particular reaction, it should be remembered that different polar environments may significantly alter the distribution of electronic charge. It is known from other QM/MM studies (26) that charges on MM atoms close to the QM/MM boundary tend to induce large polarization between QM residues and the link atoms. In the present study, also, electronic polarization appears much larger where link atoms are used. This is particularly noticeable where the QM/MM boundary divides highly polar groups. However, the link atom charges are similar in reactant and product states of A-L and ABC-H, and consequently the polarization would not be expected to have a major effect on the free energy.

Conclusions

We have described a combined semiempirical quantum mechanics and molecular mechanics (QM/MM) plus molecular dynamics and free energy perturbation (MD/FEP) approach for the study of enzymic reaction mechanisms. The approach was illustrated by a study of the enzyme-catalysed hydride-ion transfer step between NADPH cofactor and 8-methyl-pterin substrate. The NADPH cofactor was partitioned into a number of subunits (residues) to be treated by either QM or MM force-fields. With appropriate choices of QM cofactor residues and link atoms we have demonstrated the feasibility of modeling the full QM treatment of the NADPH cofactor in terms of both the electronic distributions of the QM residues and the free energies calculated using the MD/FEP method. For the smallest QM fragment (nicotinamide) used in the present study, we found that a H atom was the most appropriate choice for the link due to its electron donating properties. However, we predict that if the system requires more than 100 QM atoms to accurately describe the reaction, it would be highly desirable to use a parallel MD/FEP algorithm.

The total free energy change of 10 kcal/mol for the transfer was attributed largely to the QM/MM contribution arising from the specific H-bond interaction of the negatively-charged side-chain of Glu30 with the protonated substrate molecule. Thus, the hydride ion is transferred at the expense of a reduction in the binding energy between substrate and enzyme (25).

Acknowledgments

We gratefully acknowledge large grants of computer time from the Australian National University Supercomputer Facility on the SGI Power Challenge and Fujitsu VPP300 systems. The research was funded by an ANU Strategic Development Grant.

Literature Cited

1   Aqvist, J.; Warshel, A. *Chem. Rev.* **1993**, *93*, 2523.
2   Merz, K.M. *Curr. Opin. Struct. Biol.* **1993**, *3*, 234.
3   Gao, J. *Rev. Comput. Chem.* **1996**, *7*, 119.
4   Field, M.J.; Bash, P.A.; Karplus, M. *J. Comput. Chem.* **1990**, *11*, 700.
5   Vasilyev, V.V.; Bliznyuk, A.A.; Voityuk, A.A. *Int. J. Quant. Chem.* **1992**, *44*, 897.
6   Bash, P.A.; Ho, L.L.; MacKerell, A.D.; Levine, D.; Hallstrom, P. *Proc. Natl. Acad. Sci. USA* **1996**, *93*, 3698.
7   Cummins, P.L.; Gready, J.E. *J. Comput. Chem.* **1997**, *18*, 1496.
8   Mezei, M.; Beveridge, D.L. *Ann. N.Y. Acad. Sci.* **1986**, *1*, 482.
9   Thibault, V.; Koen, M.J.; Gready, J.E. *Biochemistry* **1989**, *28*, 6042.
10   Cummins, P.L. *Molecular Orbital Programs for Simulations (MOPS)*, **1996**.
11   Cummins, P.L.; Gready, J.E. *J. Comput. Chem.* **1989**, *10*, 939.
12   Dewar, M.J.S.; Thiel, W. *J. Am. Chem. Soc.* **1977**, *99*, 4899.
13   Dewar, M.J.S.; Thiel, W. *Theoret. Chim. Acta* **1977**, *46*, 89.
14   Cummins, P.L.; Gready, J.E. *Chem. Phys. Lett.* **1994**, *225*, 11.

15   Weiner, P.K.; Kollman, P.A. *J. Comput. Chem.* **1981**, *2*, 287.
16   Weiner, S.J.; Kollman, P.A.; Case, D.A.; Singh, U.C.; Ghio, C.; Alagona, G.; Profeta, S.; Weiner, P. *J. Am. Chem. Soc.* **1984**, *106*, 76.
17   Weiner, S.J.; Kollman, P.A.; Nguyen, D.T.; Case, D.A. *J. Comput. Chem.* **1986**, *7*, 230.
18   Jorgensen, W.L.; Chandrasekhar, J.; Madura, J.D.; Impey, R.W.; Klein, M.L. *J Chem. Phys.* **1983**, *79*, 926.
19   Dewar, M.J.S.; Zoebisch, E.G.; Healy, E.F.; Stewart, J.J.P. *J. Am. Chem. Soc.* **1985**, *107*, 3902.
20   Cummins, P.L.; Gready, J.E. *J. Comput. Chem.* **1996**, *17*, 1598.
21   van Gunsteren, W.F.; Berendsen, H.J.C. *Mol. Phys.* **1977**, *34*, 1311.
22   Berendsen, H.J.C.; Postma, J.P.M.; van Gunsteren, W.F.; DiNola, A.; Haak, J.R. *J. Chem. Phys.* **1984**, *81*, 3684.
23   Ho, L.L.; MacKerell, A.D.; Bash, P.A. *J. Phys. Chem.* **1996**, *100*, 4466.
24   Cummins, P.L.; Gready, J.E. *J. Comput. Chem.* **1990**, *11*, 791.
25   Cummins, P.L.; Gready, J.E. *J. Comput. Chem.* submitted.
26   Bakowies, D.; Thiel, W. *J. Phys. Chem.* **1996**, *100*, 10580.

Chapter 17

Ab Initio and Hybrid Molecular Dynamics Simulations of the Active Site of Human Carbonic Anhydrase II: A Test Case Study

Ursula Röthlisberger

Inorganic Chemistry, ETH Zentrum, ETH Zurich, CH-8092 Zurich, Switzerland

Ab initio molecular dynamics (AIMD) and combined Hybrid/AIMD simulations appear as promising candidates for an *in situ* modeling of enzymatic reactions. In order to probe the capabilities of these methods for the characterization of enzymatic processes we have chosen the enzyme Human Carbonic Anhydrase II (HCAII) as a test case system. Several models of the active site have been studied in this work including *ab initio* cluster models of different sizes, ranging from about 30-90 atoms. We have also extended these quantum mechanical (QM) models by taking into account the electrostatic effects of the surrounding protein through a hybrid scheme where the external field is described via charges from a classical MD force field. By comparing the structural and electronic properties of these different models we can probe the importance of size effects of the QM part as well as the influence of the protein environment. For the largest size cluster model we have been able to directly observe part of the enzymatic reaction cycle, namely the initial proton transfer steps from the zinc-bound water towards the Nδ-atom of the proton acceptor group His 64.

Understanding enzymatic processes at the molecular level is one of the most challenging problems in (bio)chemistry. A modeling of such phenomena however, is a formidable task: an appropriate method for the description of enzymatic reactions should be able (i) to treat fairly large systems of several thousand of atoms; (ii) take into account dynamical effects at finite temperature; and (iii) provide an adequate description of chemical reactions.

Classical molecular dynamics simulations are able to provide a finite temperature description, and nowadays, systems with several thousands of particles can be treated for periods in the ns time scale. There are however limits to this approach of renouncing an explicit treatment of the electronic structure problem. One of the most severe is the lack of an adequate description of chemical reactions. Phenomena, such as the making and breaking of chemical bonds require the use of an electronic structure calculation. On the other hand, traditional *ab initio* quantum chemical calculations that are able to fulfil such a task are only applicable to relatively small systems.

Moreover, static *ab initio* and semiempirical quantum calculations are not able to describe the dynamical effects known to be crucial for protein function (*1*). Clearly, a modeling of enzymatic reactions necessitates a combined approach as realized e.g. in the Hybrid QM/MM methods pioneered by Warshel and others (*2-9*). An alternative way of directly combining an electronic structure calculation with a molecular dynamics scheme was presented in 1985 with the development of the Car-Parrinello method (*10*). This method enables *ab initio* molecular dynamics simulations on a potential energy surface that is calculated on the fly via a first principle method. Most of the existing *ab initio* molecular dynamics schemes are based on density functional (DFT) methods (*11,12*). It is well known that the treatment of electron correlation effects is an indispensable prerequisite for an accurate description of chemical reactions barriers. At difference with standard quantum chemical methods, DFT calculations allow to take account of electronic correlation effects at a relatively modest computational cost. Therefore, AIMD simulations based on DFT represent a promising tool for the investigation of chemical reactions in the gas phase and in condensed phase systems. Currently, it is possible to treat systems of a few hundred atoms and by combining AIMD simulations with a classical MD force field in a QM/MM fashion (Hybrid-AIMD) the system size can be further extended.

We have chosen the enzyme Human Carbonic Anhydrase II (HCAII) as a test case system to explore the capabilities of AIMD and combined Hybrid/AIMD simulations for the modeling of enzymes. HCAII is one of the best characterized enzymes. A considerable amount of experimental data, e.g. high resolution crystal structures of the wild type enzyme and several mutants as well as many theoretical studies are available for comparison.

Here, we report our preliminary results from the investigation of three models of the active site: two *ab initio* cluster models of different size: MOD-A (consisting of $\sim$30 atoms), MOD-B (consisting of $\sim$90 atoms) and a hybrid version of MOD-B, named MOD-C. By comparing the structural, electronic and dynamical properties of these different models we can get indications of the size effects of the quantum part and the influence of the (classical) environment.

Our calculations show that the smallest size quantum model (MOD-A) does not provide an adequate description for neither structural, nor electronic or dynamical properties. In contrast, a cluster model of the size of MOD-B is able to reproduce the structural properties of the real system quite accurately and provides also a qualitative description of the electronic and dynamical features. In fact, during the room temperature MD simulations of MOD-B several proton transfer reaction steps are observed that are part of the proposed enzymatic cycle (*13*). However, for an accurate quantitative description of the reaction barriers the inclusion of the electrostatic effects of the protein surrounding is indispensable.

Computational Method

The Car-Parrinello method for *ab initio* molecular dynamics simulations is well described in a number of publications (*10,14-17*). In the following, we only give the relevant computational details for this particular study.

The electronic structure is described in the framework of density functional (*11,12*) and pseudo potential (pp) theory. The first series of test calculations were performed using the local-density approximation (LDA) with the exchange-correlation functional described by the Perdew-Zunger parametrization (*18*) of the Ceperley-Alder (*19*) results for the homogeneous electron gas.

Instead of aiming at a very detailed quantitative description, the main motivation of the current work is to probe the general capabilities of AIMD simulations for a modeling of enzymatic reactions. We expect that the LDA calculations will provide us with this first qualitative view that can be refined later on. Due to the known deficiencies of LDA to overestimate hydrogen-bonding and underestimate barriers for proton tunneling a subsequent quantitative description has to apply generalized gradient corrections (GC), such as the gradient corrected functionals of Becke (*20*) for the exchange and the one of Lee, Yang and Parr (*21*) for the correlation.

In our AIMD calculations, the valence orbitals are expanded in a basis of plane waves using periodic boundary conditions with a cubic super cell of edge a = 18 Å. The Poisson equation is solved for isolated systems using the algorithm of Hockney (*22,23*) enabling a treatment of charged systems. The effect of the ionic cores is described via nonlocal *ab initio* pseudo potentials of the supersoft Vanderbilt (*24*) type with an energy cutoff E_{cut} for the one-electron wave functions of 25 Ry. We have constructed supersoft LDA pps with cutoff radii r_{cs}= 2.1 a.u., r_{cp}= 2.4 a.u. and r_{cd}= 2.0 a.u. for zinc, r_{cs}= r_{cp}= 1.4 a.u. for oxygen, and nitrogen, r_{cs}= r_{cp}= 1.25 a.u. for carbon and r_{cs}= 0.80 a.u. for hydrogen. In the case of zinc, the 3d and the 4s valence electrons have been explicitly included in the calculations.

To test the quality of our supersoft pps (pseudo potential set PP-I), we have performed test calculations with a corresponding set of normconserving soft pseudo potentials (PP-II) and a plane wave cutoff of E_{cut}= 70 Ry. This second set has been generated according to the scheme of Troullier and Martins (*25*) with cutoff radii of r_{cs}= 1.8 a.u., r_{cp}= 2.0 a.u. and r_{cd}= 1.8 a.u. for zinc.

Molecular Dynamics Runs.   For the molecular dynamics runs a fictitious mass of the electronic degrees of freedom of μ = 600 a.u. was used. Standard constant energy MD simulations were performed and the equations of motion were integrated using a Velocity Verlet algorithm with a time step of 6 a.u. ($\sim$0.15 fs).

Inclusion of the External Electrostatic Field.   All the calculations in this paper have been performed with the program CPMD V2.5-V3.0 (*26*). In order to take into account the electrostatic effects of the surrounding protein we have implemented an extension of the code where the quantum system can be treated in an external electrostatic field described by a set of Gaussian charge distributions n_J^{ex}:

$$n_J^{ex}(\vec{r} - \vec{R}_J^{ex}) = -\frac{q_J^{ex}}{R_c^3}\pi^{-3/2}e^{-\frac{|\vec{r} - \vec{R}_J^{ex}|^2}{R_c^2}}$$

Where q_J^{ex} is the external charge J located at the position $\vec{R}_J^{ex}$ and R_c is the width of the Gaussian distribution. The external charges q_J^{ex} are treated in full analogy to the ionic core charges of the quantum system. The total charge density of the system at any point $\vec{r}$ is given by the sum of the electron density $n^{el}(\vec{r})$, the Gaussian charge distributions of the ionic cores $n_I(\vec{r} - \vec{R}_I)$ and the Gaussian distributed external charges $n_J^{ex}(\vec{r} - \vec{R}_J^{ex})$:

$$n_{tot}(\vec{r}) = n^{el}(\vec{r}) + \sum_I n_I(\vec{r} - \vec{R}_I) + \sum_J n_J^{ex}(\vec{r} - \vec{R}_J^{ex})$$

or in reciprocal space:

$$n_{tot}(\vec{g}) = n^{el}(\vec{g}) + \sum_I n_I(\vec{g})e^{i\vec{g}\vec{R}_I} + \sum_J n_J^{ex}(\vec{g})e^{i\vec{g}\vec{R}_J^{ex}}$$

Using a method developed by Hockney (*22*) the Poisson equation is solved for this total charge distribution subject to boundary conditions for an isolated system (*23,26*). This enables a treatment of charged systems within a periodic super cell approach such as the one used in our AIMD simulations. In this study, we have represented the electrostatic field of the surrounding protein using the point charges of the all atom force field of AMBER 4.0 (*27*) smeared out with Gaussian distributions of width $R_c = 1.0$ a.u.. Only point charges that are located more than three bonds away from any QM atom have been included.

The System: Human Carbonic Anhydrase II (HCAII)

Human carbonic anhydrase II is a zinc metallo enzyme that catalyses the reversible hydration of CO_2 to bicarbonate HCO_3^-. It consists of a single polypeptide chain of 260 amino acids with a molecular weight of $\sim$29 kD. The active site sits at the bottom of $\sim$15 Å deep conical cavity that is open towards the solvent. With a turn over rate at room temperature of $\sim 10^6 s^{-1}$ it is one of the fastest enzymes known. X-ray structures show that the zinc ion is coordinated to three histidine residues (His94, His96 and His119) and that a water molecule is bound to the zinc ion in an approximately tetrahedral arrangement. This water molecule has a pK_a around 7-8 and can thus be deprotonated to OH^- under physiological conditions. The zinc bound H_2O/OH^- is connected via a hydrogen-bonded network to the rest of the protein ($H_2O/OH^- \rightarrow$Thr199 $\rightarrow$Glu106). Another hydrogen-bonded network extends from the zinc bound hydroxide/water via two solvent molecules to a histidine group (His 64) located in the upper channel ($H_2O/OH^- \rightarrow$HOH 318 $\rightarrow$HOH 292 $\rightarrow$His64). The direct zinc ligands (His94, His96 and His119) and the two residues involved in the hydrogen-bonded network around the zinc-bound water (Thr199 and Glu106) are conserved in all animal carbonic anhydrases (*28*) and site specific mutagenesis experiments have indicated the crucial importance of these residues for the activity of the enzyme by controlling a precise coordination geometry at the zinc center (*29*). The three direct histidine ligands form hydrogen-bonds with three other residues (Gln92, Asn244 and Glu117) which have also a (smaller) effect on the catalytic activity of the enzyme (*30*). The strength of these additional hydrogen-bonds is crucial in determining the imidazolate character of the three histidines around the zinc which in turn influences the charge on the metal ion and therefore the acidity of the zinc-bound water. The whole system of direct and indirect zinc-ligands together with the electrostatic environment created by charged groups close to the active site are thus involved in the subtle fine tuning of the pK_a of the catalytic water molecule.

A rather clear picture of the reaction mechanism has emerged during the last years (*13*). Although some points are still controversial, it is generally accepted that the catalytic reaction involves the steps of binding the CO_2 via a nucleophilic attack of the zinc-bound OH^-, binding of an additional water molecule, release of HCO_3^- and regeneration of the Zn-OH^- through deprotonation of the zinc bound water molecule. The latter step constitutes the rate determining step (*31*) and most probably involves the histidine residue (His64) as a proton shuttle. Experiments estimate a free energy barrier of $\sim$10 kcal/mol for the overall proton transfer reaction originating mainly from solvent reorganization or conformational changes while the intrinsic barrier for proton transfer could be as low as 1.25 kcal/mol (*31*).

Models of the Active Site. Many theoretical investigations have been performed on models of HCAII (*32-40*) at various levels of sophistication. *Ab initio* studies have mainly been limited to minimum cluster models, such as $[Zn(NH_3)_3]^{2+}$

(*32-36*). The real situation is by far more complex due to the environmental effects of the surrounding protein and the presence of water molecules.

We have investigated three different models of the active site (MOD-A, MOD-B and MOD-C). The starting point for all three models was the high resolution (1.54 Å) crystal structure of Liljas et al. (*41*) from the Brookhaven data bank (PDB entry: 2CBA). We have hydrogenated and hydrated the original crystal structure using the classical MD package AMBER 4.0 (*27*). The initial positions of the hydrogens were relaxed by keeping the heavy atoms fixed and the resulting structure was used to generate the starting coordinates for models A-C:

MOD-A. Model A is a zinc-trisimidazole complex with a water or a hydroxide group as forth ligand.

MOD-B. Model B consists of the tetrahedrally coordinated zinc center (Zn^{2+}-H_2O/OH^-, His94, His96, His119) and the essential residues involved in the hydrogen bonding network (Thr199 and Glu106). The eight ordered water molecules resolved in the crystal structure that are within a distance of 7Å from the zinc ion have also been included. By using Connolly surfaces we have checked that there is not enough space to fit in an additional water molecule close to the active site. The residues were fixed at a position close to the backbone and otherwise left free. Where covalent bonds had to be broken to cut out the cluster model from the entire protein structure dummy hydrogen atoms have been used to saturate the QM model. Figure 1 shows a graphical representation of MOD-B. Model B contains a total of 86 atoms.

MOD-C. Model C is an extension of MOD-B that takes the electrostatic external field of the protein into account represented by Gaussian broadened point charges located within a distance of 7.5-9 Å from the centre of the simulation box.

Results

In this sections, we are comparing the structural, dynamic and electronic properties of the three different models. The differences that occur between them can serve to gain insight which level of detail is necessary for an adequate description of a given property. In particular we try to estimate the size effects of the QM part and the influence of the environment.

Structural Properties. As an example for a characteristic structural property, the zinc-oxygen bond distance for models A and B is compared in Table I. To give an idea of the overall accuracy of our computational scheme we have also included the corresponding results for a hexaquo zinc complex for which values from other LDA calculations and experiments are available for comparison. Table I shows that both pseudo potential sets (LDA-I and LDA-II) result in very similar values for the zinc-oxygen distance (2.05 and 2.07 Å, respectively). Both values are also in very good agreement with a LDA value reported in the literature (LDA-III) (*42*). This result convinced us to make use of the more economic Vanderbilt pseudo potentials (PP-I) in our subsequent calculations.

As a further test of the computational scheme we have investigated the influence of the gradient corrected exchange-correlation functional BLYP (*20, 21*) on the same system. The zinc-oxygen distance is roughly 0.1 Å larger than in the LDA case (2.17 with respect to 2.07 Å). However, both, LDA and BLYP values are close to the range of experimental data.

MOD-A. In the case of the zinc-trisimidazole complex, the zinc-oxygen

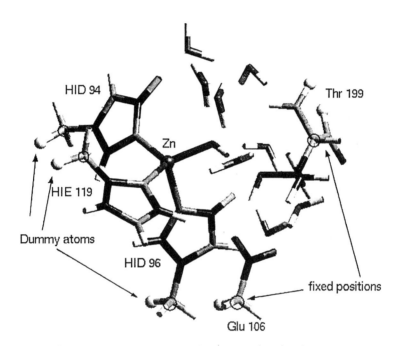

Figure 1. Graphical representation of model B (see text). Atoms that are kept fixed during the simulations are indicated with a circle. Dummy atoms are indicated with white balls.

distance changes distinctly upon deprotonation of the zinc-bound water ($\Delta d =$ 0.22 Å). Such a drastic change of the zinc-oxygen distance is not observed when comparing the crystal structures at low and high pH (41). Apparently, in the real enzyme the protein environment helps to stabilize the zinc-oxygen distance during protonation/deprotonation. This shows clearly that such a simplified model is not able to capture the main structural features of the real enzyme.

Table I. Zinc-Oxygen Distances of Different Model Complexes

	LDA-I	LDA-II	LDA-III	BLYP-II	exp
$Zn(H_2O)_6^{2+}$	2.05	2.07	2.06^1	2.16	2.05-2.14 [1]
MOD A OH/HOH	1.82/2.04				$2.05/2.05^2$
MOD B OH/HOH	1.90/1.91				$2.05/2.05^2$

LDA-I: LDA results with pseudo potential set I (supersoft Vanderbilt); LDA-II: LDA results with pseudo potential set II (Troullier-Martins); LDA-III: LDA results from a calculation with a basis set of Gaussian type orbitals (42); BLYP-II: BLYP results with pp set II; exp: experimental values. All distances are given in Angstrom
[1] ref. (42) and references therein.
[2] values of the experimental crystal structures of the high and low pH forms of the enzyme (41).

MOD-B. Starting from the initially prepared coordinates of the high resolution crystal structure the geometry of this model has been optimized with local relaxation techniques and with the help of molecular dynamics runs at low temperature. During this optimization runs, only relatively small deviations from the experimental structure occur. These consist mainly of a reorientation of the water molecules for which the hydrogen positions had to be assumed in the beginning. Thus, it appears that a model of this size is able to retain the appropriate structure of the active site. In particular the zinc-oxygen distance in the hydroxide and in the water form are now essentially identical. The absolute values are significantly smaller than the ones obtained for the experimental crystal structures. The deviation of (Δd=-0.15 Å) is however similar to the one observed between the LDA and BLYP results for the hexaquo zinc complex indicating that this deficiency could probably be cured when using the BLYP-functional.

Dynamical Properties

MOD-A. As for the structural properties, also the dynamical properties of model A differ clearly from the real system. We have performed a short MD simulation (1 ps) at room temperature and found that the zinc-bound hydroxide can rotate around the Zn-O axis. This is in contrast to the real protein where the zinc-bound nucleophile is kept by the hydrogen-bonded network of Thr199 and Glu106 in a defined orientation (38) appropriate for the binding of the CO_2. Furthermore, the mobility of the imdazole rings is much higher than the ones of the histidine residue that are kept quite rigidly in place as indicated by T-factors of 5-10 reported for the crystal structure (41).

MOD-B: The Proton Transfer Reaction. To investigate the dynamical properties of model B, we have performed a 1 ps MD simulation at body temperature. Being aware of the known deficiencies of the LDA in significantly underestimating proton transfer barriers we tried to use this simulation to make an efficient scan of the potential energy surface of the system 'with reduced barriers'. During these MD runs a spontaneous proton transfer reaction is observed. Starting from the hydroxide form of the enzyme a proton from a neighboring water molecule (HOH 318) is transferred to the zinc bound OH$^-$ and the charged defect can be transferred to the next water molecule by a further switch of a proton. In these proton transfer reactions a simultaneous shortening of several hydrogen-bonds (between the zinc-bound water, HOH 318 and HOH 292) occurs and protons can be exchanged easily back and forth via these three solvent molecules that form a kind of proton-exchange pathway. The water molecules involved in this process are indeed the ones connected in the real protein via a hydrogen-bonded network to the hypothetical proton shuttle group His64. Figure II shows the temporal evolution of the four oxygen-hydrogen distances (Zn-HO...H-O-H(318)...OH$_2$(292)) that form the proton relay pathway. Two of these OH-distances correspond to covalent O-H bonds (as indicated in Figure II by OH distances of 1.0-1.2 Å) and two to hydrogen-bonded O...H distances in the range of 1.3-1.6 Å (the hydrogen-bonded O...H distances are somewhat shorter than could be expected due to the overestimation of hydrogen-bonding within the LDA). It is apparent in Figure II that the monitored pairs of oxygen and hydrogen atoms can change their mutual distance from covalent to hydrogen-bonded and vice versa, i.e. the protons can be exchanged between two neighboring oxygen atoms. Prior to such a proton transfer, the OH distances involved in the relay adjust simultaneously to a similar value around 1.2-1.3Å (corresponding to the symmetric position of the hydrogen between two oxygen atoms). Such a concerted change can be seen around 200, 340 and 440 fs. During our simulation, we have only observed that the zinc bound water molecule exchanges its proton via this pathway and no other proton transfers were observed along different hydrogen-bonded networks. The findings of our simulations are thus in very good agreement with the proposed role of His64 as proton shuttle group.

The fact that it is possible to observe directly part of the enzymatic reaction cycle is very encouraging. Our approach is completely bias-free in the sense that no knowledge about likely reactions or reaction coordinates is necessary. Such an unbiased approach seems particularly promising for the study of systems where the enzymatic reaction is not yet known in detail.

Electronic Properties: Effects of the Surrounding.

The proton affinity of the zinc-bound water molecule is the result of a subtle fine tuning via hydrogen-bonded networks and electrostatic environment effects. This quantity can thus serve as a sensitive indicator of differences in the electronic structure that will have a critical influence on the enzymatic reaction. As a first attempt to quantify the effect of the electrostatic environment and the varying size of the cluster model we have therefore calculated the proton affinities for the three models. We obtained a value of 177 kcal/mol for model A. In model B the proton affinity is distinctly higher (240 kcal/mol). The inclusion of the electrostatic environment enhances this value even more. To the best of our knowledge, the proton affinity of HCAII is not known experimentally. The only experimental values available for a rough bracketing are the gas phase proton affinities of water (166.7 kcal/mol) (43) and OH$^-$ (390.8 kcal/mol) (44). Even though our preliminary results might not be quantitatively accurate, they nevertheless indicate the strong effects of the size of the quantum part and the electrostatic environment. More calculations are however necessary to establish if this approach can describe such a delicate property quantitatively. In particular, more tests are needed to quantify the effects of gradient-corrected exchange-correlation functionals. Furthermore, different ways of describing the external electrostatic field such as the

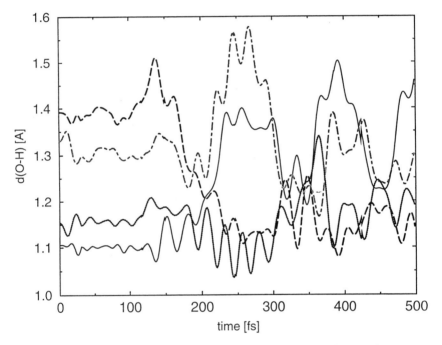

Figure 2. Temporal evolution of 4 characteristic oxygen-hydrogen distances involved in the proton relay. Note the simultaneous contraction of the O-H distances around 200, 340 and 440 fs prior to a proton transfer.

particular set of classical point charges, the Gaussian broadening and the charge exclusion scheme have to be tested thoroughly.

Summary and Outlook

We have studied three models of the active site of human carbonic anhydrase II using *ab initio* molecular dynamics simulations. The different systems include gas phase cluster models ranging from 30-90 atoms and an extended model that takes into account the effects of the electrostatic field of the surrounding protein. The three models show distinctly different structural, dynamic and electronic properties. With the largest cluster model we were able to observe spontaneous proton transfer reactions that are part of the catalytic cycle. Our preliminary results show that this approach may provide a very promising possibility to find likely reaction pathways for systems where the reaction mechanism is not known a priori.

Acknowledgments.   I like to thank Paolo Carloni for his help in the preparation of the initial structure of the active site and Michael L. Klein for his generous and continuous support of this project. This work is part of a Profile 2 fellowship of the Swiss National Science Foundation (Profile2-2124-39 558.93).

Literature Cited

1. Kraut, E.G.J. *Science* **1988**, *242*, 533
2. Dewar, M.J.S.; Haselbach, E. *J. Am. Chem. Soc.* **1970**, *92*, 590
3. Wang, I.S.; Karplus, M. *J. Am. Chem. Soc.* **1973**, *95*, 8160
4. Warshel, A.; Levitt, M. *J. Mol. Biol.* **1976**, *103*, 227
5. Singh, U.C.; Kollman, P.A.; *J. Comp. Chem.* **1986**, *7*, 718
6. Field, M.J.; Bash, P.A.; Karplus, M. *J. Comp. Chem.* **1990**, *11*, 700
7. Gao, J. *J. Phys. Chem.* **1992**, *96*, 537
8. Bala, P.; Leysyng B.; McCammon, J.A. *Chem. Phys.* **1994**, *180*, 271
9. Merz, K. *J. Comp. Chem.* **1995**, *16*, 113
10. Car, R.; Parrinello, M. *Phys. Rev. Lett.* **1985**, *55* 2471
11. Hohenberg, P.; Kohn, W. *Phys. Rev.* **1964**, *136*, B864
12. Kohn, W.; Sham, L.J. *Phys. Rev.* **1965**, *140*, A1133
13. *Bioinorganic Chemistry*; Bertini, I.; Gray, H.; Lippard, S.J.; Valentine, J.S., Eds.; University Science Books, California (1994)
14. Car, R.; Parrinello, M. *Proceedings of the NATO ARW: Simple Molecular Systems at very High Density*, Les Houches (France), NATO ASI Series, New York, Plenum Press 1988
15. Galli, G.; Parrinello, M. In *Computer Simulations in Material Science*, Meyer, M.; Pontikis, V., Eds.; Kluwer, Dordrecht, 1991; pp 283
16. Payne, M.C.; Teter, M.P.; Allan, D.C.; Arias, T.A.; Joannopoulos, J.D. *Rev. Mod. Phys.* **1992**, *64*, 1045
17. Remler, D.K.; Madden, P.A. *Mol. Phys.* **1990**, *70*, 691
18. Perdew, J.; Zunger, A. *Phys. Rev.B* **1981**, *23*, 5048ff
19. Ceperley, D.M.; Alder, B.J. *Phys. Rev. Lett.* **1980**, *45*, 566ff
20. Becke, A.D. *Phys. Rev. B* **1988**, *38*, 3098ff
21. Lee, C.; Yang, W.; Parr, R.G. *Phys. Rev. B* **1988**, *37*, 785
22. Hockney, R.W. *Methods Comput. Phys.* **1070**, *9*, 136
23. Barnett, R.N.; Landman, U. *Phys. Rev. B* **1993**, *48*, 2081
24. Vanderbilt, D. *Phys. Rev. B* **1990**, *41*, 7892

25. Troullier, N.; Martins, J.L.; *Phys. Rev. B* **1991**, *43*, 1993
26. CPMD Version 2.5-3.0, Hutter, J,; Ballone, P.; Bernasconi, M.; Focher, P.; Fois, E.; Goedecker, S.; Parrinello, M.; Tuckerman, M. MPI für Festkörperforschung and IBM Zurich Research Laboratory 1995-98.
27. Cieplak, P.; Bayly, C.I.; Gould, I.R.; Merz, K.M. Jr; Ferguson, D.M.; Spellmeyer, D.C.; Fox, T.; Caldwell, J.W.; Kollman, P.A. *J. Am. Chem. Soc.* **1995**, *117*, 5179-5197
28. Tashian, R.E. *BioEssays* **1989**, *10*, 186
29. Xue, Y.; Liljas, A.; Jonsson, B.-H.; Lindskog, S. *Proteins* **1993**, *17*, 93
30. Lesburg, C.A.; Christianson, D.W *J. Am. Chem. Soc.* **1995**, *117*, 6838
31. Silverman, D.N.; Tu, C.; Chen, X.; Tanhauser, S.M.; Kresge, A.J.; Laipis, P.J. *Biochemistry* **1993**, *32*, 10757
32. Pullman, A.; Demoulin, D. *Int.J. Quant. Chem.* **1979**, *16*, 641
33. Liang, J.-Y.; Lipscomb, W.N. *Biochemistry* **1989**, *28*, 9724
34. Jacob, O.; Cardenas, R.; Taipa, O.; *J. Am. Chem. Soc.* **1990**, *112*, 8692
35. Sola, M.; Lledos, A.; Duran, M.; Bertran, J. *J. Am. Chem. Soc.* **1992**, *114*, 869
36. Garmer, D.R.; Krauss, M. *J. Am. Chem. Soc.* **1992**, *114*, 6487
37. Merz, K.M. *Mol. Biol.* **1990**, *214*, 799
38. Merz, K.M. *J. Am. Chem. Soc.* **1991**, *133*, 406
39. Aqvist, J.; Fothergrill, M.; Warshel, A. *J. Am. Chem. Soc.* **1993**, *115*, 631-635
40. Hwang, J.-K.; Warshel, A. *J. Am. Chem. Soc.* **1996**, *118*, 11745-11751
41. Hakansson, K.; Carlsson, M.; Svensson, L.A.; Liljas, A. *J. Mol. Biol.* **1992**, *227*, 1192
42. Andzelm, J. In *Density Functional Methods in Chemistry*, Labanowski, J.K.; Andzelm, J., Eds.; Springer Verlag (1991), p.155 ff
43. Collyer, S.M.;R McMahon, T.B. *J. Phys. Chem.* **1983**, *87*, 909
44. Lias, S.G.; Bartmess, J.E.; Liebman, J.F.; Holmes, J.L.; Mallard, W.G. *J. Phys. Chem. Ref. Data* **1988**, *17*, Suppl. 1

Chapter 18

Application of Linear-Scaling Electronic Structure Methods to the Study of Polarization of Proteins and DNA in Solution

Darrin M. York[1]

Department of Chemistry and Chemical Biology, Harvard University, 12 Oxford Street, Cambridge, MA 02138

Semiempirical quantum calculations of biomolecular systems in solution were performed using recently developed linear-scaling methods to examine the role of solute polarization in the process of solvation. The solvation free energy of several protein and DNA molecules and complexes were computed and decomposed to asses the relative magnitude of electrostatic and polarization contributions.   The effect of solvation and complex formation on the electronic density of states was also studied.

Over the past several decades, the theoretical treatment of large biomolecular systems in solution has been restricted almost exclusively to use of molecular mechanical models[1]. These models typically neglect explicit electronic polarization terms due to the need for very rapid energy and force evaluations in, for example, molecular dynamics calculations. Complementary to molecular dynamics calculations with explicit solvent have been the application of implicit solvation methods for estimation of free energies of hydration of biomolecules[2]. These methods often employ a dielectric continuum approximation and involve solving, for example, the Poisson or Poisson-Boltzmann equation. As with molecular mechanics models, these methods do not treat explicitly the electronic degrees of freedom.

In the present study, we apply recently developed linear-scaling electronic structure methods to the calculation of solvation free energies and electronic density of states (DOS) distributions of several protein and DNA systems in solution.   In this way, quantum mechanical many-body polarization effects are assessed directly.

THEORY

The development of new methods for computing the electronic structure of molecular systems with computational effort that scales

[1]Laboratoire de Chimie Biophysique, Institut le Bel, Université Louis Pasteur, 4, rue Blaise Pascal, Strasbourg 67000, France

approximately linearly with system size[3-7] has recently allowed very large molecules (10^3-10^4 atoms) to be considered for the first time. Conventional single-determinant non-perturbative wave-function theories such as Hartree-Fock and density-functional methods scale as the cube of the system size due to the orthonormality constraint on the molecular orbitals that make up the wave function, or equivalently the idempotency condition for the single-particle density matrix. Here we employ a linear-scaling semiempirical approach, summarized below, to electronic structure, the details of which have been presented elsewhere[5-7].

In Hartree-Fock molecular orbital and Kohn-Sham density functional methods, the electronic energy of the system is governed by the single-particle density matrix

$$\rho_{ij} = \langle \varphi_i | \hat{\rho} | \varphi_j \rangle \tag{1}$$

where φ_i are basis functions for the expansion of the molecular orbitals, and $\hat{\rho}$ is the density operator defined as

$$\hat{\rho} = \sum_m n_m | \psi_m \rangle \langle \psi_m | = f_\beta(\hat{H} - \mu) \tag{2}$$

where n_m are the occupation numbers of the molecular orbitals ψ_m that are solutions of

$$\hat{H} | \psi_m \rangle = \varepsilon_m | \psi_m \rangle \tag{3}$$

In equations (2) and (3), $\hat{H}$ refers generically to either the Fock or Kohn-Sham Hamiltonian operators. The second equality in equation (2) follows from the assumption that the orbital occupation numbers are taken from a Fermi distribution $f_\beta(\varepsilon)$ with inverse temperature β (taken here to correspond to 300K), and Fermi level μ. For localized basis set methods, the density matrix can be partitioned using a set of normalized, symmetric weight matrices W^α that are localized in real space. A convenient choice is to employ a Mulliken-type partition[8]

$$W_{ij}^\alpha = w_i^\alpha + w_j^\alpha \tag{4}$$

where

$$w_i^\alpha = \tfrac{1}{2} \; \forall \, i \in \alpha \tag{5}$$
$$= 0 \; otherwise$$

The global density matrix can then be approximated by a superposition of partitioned elements:

$$\rho_{ij} = \sum_\alpha W_{ij}^\alpha \rho_{ij} \approx \sum_\alpha W_{ij}^\alpha \rho_{ij}^\alpha = \tilde{\rho}_{ij} \tag{6}$$

and

$$\rho_{ij}^\alpha = \langle \varphi_i | f_\beta(\hat{H}^\alpha - \mu) | \varphi_j \rangle \tag{7}$$

where $\hat{H}^\alpha$ is a local projection of $\hat{H}$ in a set of basis functions that are localized in the region of the subsystem α. This local basis set typically is made up of basis functions centered on the atoms contained in the subsystem α in addition to basis functions centered on nearby *buffer* atoms contained in other subsystems. Normalization of the total electron density at each step of the self-consistent field procedure is enforced through adjustment of the chemical potential μ in equation (7). In the case of the Hartree-Fock methods, the total energy is given by

$$E = \tfrac{1}{2}\sum_{ij} \tilde{\rho}_{ij}\left(F_{ij}(\tilde{\rho}) + H_{ij}^{core}\right) \tag{8}$$

where F is the Fock matrix and H^{core} is the one-electron core Hamiltonian matrix. In the above formulation, there is no need for construction or diagonalization of the global Fock matrix, and hence the cubic scaling bottleneck associated with orthogonalization of the molecular orbitals is avoided. With proper choice of buffer region, the method has been demonstrated to be highly accurate and efficient[5-7].

Solvent effects are critical to the behavior and stability of biomolecules in solution; hence, inclusion of solvent effects in electronic structure calculations is important. A recently developed model for high dielectric solvents in quantum mechanical calculations is the conductorlike screening model[9]. This model employs a variational principle based on a conductor, the results of which are subsequently corrected for finite dielectric media. The model has been shown to give accurate results for small polar and ionic solutes in high dielectric media such as water.

In this model, the classical electrostatic energy of a charge distribution q contained in a cavity ($\varepsilon_{cav}=1$) surrounded by a continuum of constant dielectric ε can be written as:

$$E_{el} = \tfrac{1}{2}\iint_{\Omega\,\Omega} \frac{\sigma(\mathbf{r})\sigma(\mathbf{r}')}{|\mathbf{r}-\mathbf{r}'|}d\Omega d\Omega' + \iint_{V\,\Omega} \frac{q(\mathbf{r})\sigma(\mathbf{r}')}{|\mathbf{r}-\mathbf{r}'|}dVd\Omega' + \tfrac{1}{2}\iint_{V\,V} \frac{q(\mathbf{r})q(\mathbf{r}')}{|\mathbf{r}-\mathbf{r}'|}dVdV' \tag{9}$$

where σ is the reaction field surface charge at the boundary Ω between different dielectric regions. For a conductor, the surface charge distribution can be determined by minimization of the total electrostatic energy with q fixed. In matrix notation, this leads to the solution

$$E_{el} = \tfrac{1}{2}\mathbf{q}^T \cdot (\mathbf{C} - \mathbf{B}^T \cdot \mathbf{A}^{-1} \cdot \mathbf{B}) \cdot \mathbf{q} = \tfrac{1}{2}\mathbf{q}^T \cdot \mathbf{G} \cdot \mathbf{q} \tag{10}$$

where σ and $\mathbf{q}$ are column vectors of the reaction field and solute charge distributions, respectively, the matrices $\mathbf{A}$, $\mathbf{B}$, and $\mathbf{C}$ represent Coulombic interactions between $\sigma{:}\sigma$, $\sigma{:}\mathbf{q}$, and $\mathbf{q}{:}\mathbf{q}$, respectively, and $\mathbf{G}$ is the Green's function matrix for the problem. For a finite dielectric, the surface charge distribution is scaled by a factor of $(\varepsilon-1)/\varepsilon$ in accord with the Gauss theorem, and leads to an error on the order of $1/(2\varepsilon)^9$, which is small for a high dielectric medium such as water ($\varepsilon=80$). The main advantages of the method is that direct computation of the Green's function matrix makes analytic derivatives facile, and affords high efficiency with conventional methods that incorporate **G** directly into the Hamiltonian at the beginning the self-consistent field procedure. A disadvantage of the method is the reduced reliability of results for low dielectric media compared to exact solution of the Poisson equation.

The solution equation (10) cannot be applied as written for large molecules since the process of matrix inversion scales as M^3, where M is the dimension of the surface charge vector. However, this problem can be overcome by use of a preconditioned conjugate gradient minimization technique and fast multipole method[10] for linear-scaling evaluation of electrostatic interactions[7]. The method is an iterative minimization technique that requires evaluation of matrix-vector products of the form $\mathbf{A}\cdot\mathbf{x}=\mathbf{b}$ (an M^2 procedure) at each iteration. This operation corresponds to evaluation of the electrostatic potential of the surface charge vector $\mathbf{x}$, and can be realized with order $M\cdot log(M)$ effort using fast multipole methods[10]. The number of iterations can be decreased using a preconditioner matrix $\mathbf{A}_{pc}$ such that $\mathbf{A}\cdot\mathbf{A}_{pc}\approx 1$. We have chosen $\mathbf{A}_{pc}$ as the inverse of the block diagonal matrix constructed by the elements of **A** corresponding to common atomic surface area patches. With this preconditioner, the number of iterations to reach a fixed convergence level does not appear to grow with system size for the molecules considered here. The method has been shown to be accurate and efficient for biological macromolecules[5-7].

METHODS

Structures for DNA $(CG)_8$ helices in canonical A, B, and Z-forms were generated from ideal monomer subunits obtained from fiber diffraction experiments[11]. Initial structures for proteins, and protein-protein and protein-DNA complexes were obtained from nuclear magnetic resonance data in solution, and refined with 50 steps of steepest descents energy minimization to relax the structures on the quantum mechanical energy surface. Quantum mechanical calculations were performed using the self-consistent linear-scaling electronic structure methods described previously[5-7,12] with the semiempirical AM1 Hamiltonian[13]. Subsystem partitions were chosen to be the amino- and nucleic acid biopolymer subunits. Buffer regions were determined using a 8Å distance criterion R_b, and core Hamiltonian, Fock, and density matrix elements were evaluated using a 9Å cutoff R_m (see below). Solvent effects were included self-consistently using a linear-scaling solvation method for macromolecules with atomic radii parameterized to reproduce solvation free energies of amino acid backbone and side-chain homologues and modified nucleic acid bases[7].

Table I. Convergence of energetic quantities (eV) in solution with buffer and matrix element cutoffs R_b/R_m (Å)[*].

	ΔH_f	$\Delta G_{el,sol}$	ε_{homo}	ΔE_{gap}
B-DNA (1006 atoms, total charge -30)				
4/7	-506.092	-444.647	-5.768	5.747
6/7	-506.707	-444.259	-6.755	6.541
8/9	-506.764	-444.255	-7.619	7.405
10/11	-506.765	-444.255	-7.619	7.405
12/13	-506.765	-444.255	-7.619	7.405
crambin (642 atoms, total charge 0)				
4/7	-124.951	-11.853	-5.964	1.830
6/7	-116.040	-11.536	-7.470	5.682
8/9	-116.101	-11.534	-8.692	6.910
10/11	-116.103	-11.534	-8.704	6.923
12/13	-116.103	-11.534	-8.705	6.923

[*]Quantities are the heat of formation ΔH_f, electrostatic component of the solvation free energy $\Delta G_{el,sol}$, highest occupied molecular orbital eigenvalue ε_{homo} and energy gap ΔE_{gap} (see text).

RESULTS

Convergence

Table I summarizes the convergence of energetic quantities with buffer size and matrix cutoff for a canonical $(CG)_8$ B-DNA helix, and crambin in solution. The matrix element cutoff (R_m) was chosen to be slightly larger than the buffer cutoff (R_b) to insure inclusion of all off-diagonal subsystem-buffer matrix elements. A matrix element cutoff below 7Å was observed to lead to larger errors (data not shown). All energetic quantities are well converged using the 8/9Å (R_b/R_m) scheme. Energetic quantities converge slightly faster in the case of B-DNA (~10^{-3} eV with the 8/9Å scheme) than for crambin, particularly for the energy gap (~10^{-2} eV with the 8/9Å scheme),

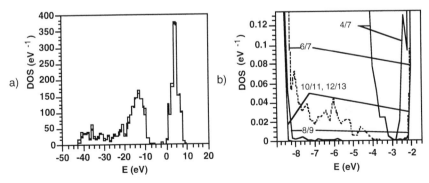

Fig. 1. Convergence of the electronic DOS of crambin in solution with different R_b/R_m schemes: a) overall DOS, b) DOS near the Fermi level (Note: the area of the graph corresponds to 1 electron).

although the difference might not be significant. Figures 1(a,b) illustrate the variation of the DOS for crambin as a function of (R_b/R_m). The overall DOS profile is similar in all cases. A detailed examination of the DOS in the region of the energy gap reveals the low order 4/7Å scheme results in "leakage" of electron density into the region of the gap (Figure 1b). This effect is significantly reduced using the 6/7Å scheme, and becomes almost negligible with larger cutoffs. In what follows, we employ the 8/9Å scheme for determination of electronic properties of protein and DNA systems in solution.

Solvation energies

The process of solvation can be decomposed into different paths, two of which are considered here (Figure 2). The first path involves 1) freezing the electronic degrees of freedom of the solute in the gas phase and allowing this charge distribution to induce an electrostatic solvent reaction field ($\Delta G_{el,gas}$), followed by 2) electronic relaxation (polarization) of the gas phase charge distribution to the final solution phase charge distribution, at the same time allowing the induced reaction field to adjust accordingly (ΔG_{pol}). Both $\Delta G_{el,gas}$ and ΔG_{pol} involve relaxation processes and are stabilizing (lower the total energy). The second path in Figure 2 involves 1) an internal perturbation of the gas phase electronic charge distribution to the solution phase distribution in the absence of a solvation effect (ΔG_{int}), followed by 2) adding the reaction field ($\Delta G_{el,sol}$) that is induced by the solution phase charge distribution. The former is a destabilizing internal energy reorganization contribution, whereas the latter consists of a stabilizing solvent reaction field response to the solution phase charge distribution (analogous to $\Delta G_{el,gas}$ for the gas phase charge distribution). In this study, only electronic energy contributions are evaluated (no cavitation terms) with fixed geometries, and the effect of solvation is approximated by a

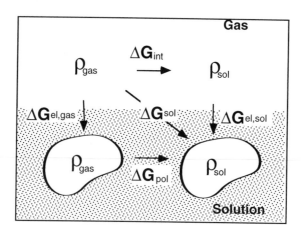

Fig. 2. The solvation process decomposed into different paths.

classical electrostatic dielectric continuum model as described above. The main advance in this study consists of the explicit consideration of the contribution of quantum mechanical many-body effects of the macromolecular solute in the process of solvation.

Table II shows the solvation free energy components for different canonical forms of DNA and several proteins derived from NMR. By their definition, the electrostatic components of the solvation free energy ($\Delta G_{el,gas}$ and $\Delta G_{el,sol}$) bracket the total free energy of solvation ΔG_{sol}; i.e., $-\Delta G_{el,gas} \leq -\Delta G_{sol} \leq -\Delta G_{el,sol}$, or equivalently $\Delta G_{pol} \leq 0 \leq \Delta G_{int}$. The magnitude of the values of $\Delta G_{el,gas}$ and $\Delta G_{el,sol}$ differ from the values for ΔG_{pol} and ΔG_{int} by an order of magnitude or more for the molecules studied here. The ΔG_{pol} contributes approximately 10% for the proteins and 2% for the DNA. This is consistent with results of hybrid quantum mechanical/molecular mechanical simulations of peptides in solution[14]. Although the magnitude of the polarization term is greater (per atom) for DNA than for the proteins, the relative percentage is significantly less due to the dominant $\Delta G_{el,gas}$ term.

The ΔG_{int} term reflects the internal energy penalty the solute pays in order to adopt the ideal solution phase charge distribution. Within the linear response regime, this term is equal to minus the *solute polarization* energy (see below).

The contributions due to solvation and solute polarization to the free energy of binding of myosin to calmodulin[15] and the DNA binding domain of Myb with DNA[16] are summarized in Table III. In both cases, the process of binding has a neutralizing effect as oppositely charged species come together. Consequently, the change in the solvation free energy strongly disfavors complexation, although the overall enthalpy of formation is predicted to be favorable. In these examples $\Delta\Delta G_{pol}$ is positive. This arises from the separated species being more polarized due to both greater exposed surface area and a larger charge induced reaction field than in the

TABLE II: Solvation free energies, enthalpies of formation, and Fermi gap energies (eV) of proteins and DNA*.

Molecule	atoms	ΔG_{sol}	$\Delta G_{el,gas}$	ΔG_{pol}	ΔG_{int}	$\Delta G_{el,sol}$	ΔH_f	ΔE_{gap}
Proteins								
crambin	642	-10.0	-8.7	-1.3	1.6	-11.6	-116.1 *(-106.1)*	6.91 *(5.14)*
bpti	892	-44.2	-41.2	-2.9	3.5	-47.6	-115.0 *(-70.9)*	6.48 *(2.26)*
lysozyme	1960	-68.3	-62.8	-5.6	6.7	-75.0	-296.8 *(-228.5)*	5.99 *(3.46)*
DNA								
A-DNA	1006	-447.6	-440.2	-7.4	8.4	-455.9	-503.1 *(-55.5)*	7.75 *(2.75)*
B-DNA	1006	-437.3	-430.9	-6.4	7.0	-444.3	-506.8 *(-69.5)*	7.41 *(0.40)*
Z-DNA	1006	-457.7	-446.6	-11.1	13.3	-471.0	-507.3 *(-49.6)*	7.78 *(1.06)*

*Unrefined protein coordinates (see text) were obtained from solution NMR data taken from the Brookhaven Protein Data Bank for crambin (1CCN), bovine pancreatic trypsin inhibitor (1PIT), and lysozyme (2LYM). $(CG)_8$ sequences of duplex A, B, and Z-form DNA were constructed from idealized subunits derived from fiber diffraction data (see text). Values for ΔH_f and ΔE_{gap} are given for calculations in solution (plain text) and in the gas phase *(italics)*.

TABLE III: Solvation free energies, enthalpies of formation, and Fermi gap energies (eV) of biomolecular complexes in solution[*].

Molecule	atoms	ΔG_{sol}	$\Delta G_{el,gas}$	ΔG_{pol}	ΔG_{int}	$\Delta G_{el,sol}$	ΔH_f	ΔE_{gap}
Myb-DNA								
complex	2512	-101.2	-86.8	-14.4	12.1	-113.3	-459.3	6.95
Myb	1815	-125.3	-115.1	-10.2	9.6	-134.8	-159.4	7.77
DNA	697	-248.2	-242.7	-5.5	6.1	-254.3	-319.7	7.79
ΔΔ		*272.3*	*271.0*	*1.3*	*-3.6*	*275.8*	*-19.8*	*-0.38*
myosin-calmodulin								
complex	2700	-101.2	-86.8	-14.4	12.1	-113.3	-580.1	6.49
myosin	2259	-65.6	-52.1	-13.6	8.5	-74.1	-546.0	7.33
calmodulin	441	-310.8	-285.8	-24.9	19.1	-329.8	-28.5	8.21
ΔΔ		*168.8*	*153.7*	*15.1*	*-9.7*	*178.4*	*-5.5*	*0.17*

[*]Differences between values for the complexed and separated molecules (ΔΔ) are shown in *italics*. The ΔE_{gap} values for ΔΔ were obtained from a DOS distribution that was a superposition of DOS distributions of the separated molecules. Unrefined coordinates were obtained from solution NMR data taken from the Brookhaven Protein Data Bank for calmodulin-myosin (2BBM), and Myb-DNA (1MSE).

complexed forms. Conversely, $\Delta\Delta G_{int}$ is negative, indicating there is less of an internal energy penalty for adopting the solution phase charge distribution in the absence of a reaction field for the complexes than in the separated species.

It is of interest to determine whether the electronic response of the solute in the process of solvation is a linear response[17]. This can be addressed by considering the process of perturbing the gas phase system by an applied field, and calculating the energetic stabilization that results from the electronic relaxation of the perturbed system. In this case, the external field is taken as the solvent reaction field v_{RF} for the solution phase charge distribution. The process we are interested in is thus:

$$[\rho_{gas}]_o \xrightarrow{v_{RF}(\mathbf{r})} [\rho_{gas}]_{v_{RF}(\mathbf{r})} \xrightarrow{\Delta E_{pol,solute}} [\rho_{sol}]_{v_{RF}(\mathbf{r})} \tag{11}$$

If the electronic response of the solute is a linear response, then the solute polarization energy is given by

$$\Delta E_{LR,solute} = \tfrac{1}{2} \int \delta\rho(\mathbf{r}) v_{RF}(\mathbf{r}) d^3 r \tag{12}$$

where $\delta\rho = \rho_{sol} - \rho_{gas}$ is the *solute polarization density*. The solute polarization energy $\Delta E_{pol,solute}$ of equation (11) is calculated as

$$\Delta E_{pol,solute} = \Delta G_{sol} - \Delta G_{el,sol} + 2\Delta E_{LR,solute} = \Delta E_{LR,solute} + (\Delta G_{int} + \Delta E_{LR,solute}) \tag{13}$$

From equation (13) the quantity ($\Delta G_{int} + \Delta E_{LR,solute}$) is the difference between the solute polarization energy and its ideal linear response value. Figure 3

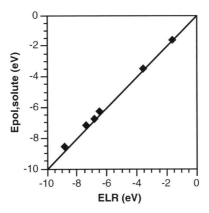

Fig. 3. Regression of the solute polarization energy and ideal linear-response energy (see text).

shows a linear regression of the calculated values of $\Delta E_{pol,solute}$ and $\Delta E_{LR,solute}$. For the systems considered here, the macromolecular response of the solutes is very nearly a linear response. This lends further credence to the use of linear response models for modeling polarization in biomolecular simulations in solution[17].

Energy gaps and electronic DOS

In the present method, the number of electrons for a given set of subsystem molecular orbitals is determined from the Fermi level ε, and given by

$$N(\varepsilon) = \sum_{ij}^{\substack{sub- \\ systems}} \sum_{\alpha} W_{ij}^{\alpha} S_{ij}^{\alpha} \sum_{m} f_{\beta}(\varepsilon_m^{\alpha} - \varepsilon) C_{im}^* C_{jm} \tag{14}$$

where $\mathbf{S}^{\alpha}$ is the subsystem overlap matrix, and $\mathbf{C}^{\alpha}$ is the matrix of molecular orbital expansion coefficients, and $f_{\beta}(\varepsilon)$ and $\mathbf{W}^{\alpha}$ are the Fermi function and partition weight matrix, respectively, as defined earlier. The electronic density of states is defined as

$$g(\varepsilon) = \left(\frac{\partial N}{\partial \varepsilon} \right)\Big|_{\varepsilon} = -\sum_{ij}^{\substack{sub- \\ systems}} \sum_{\alpha} W_{ij}^{\alpha} S_{ij}^{\alpha} \sum_{m} f_{\beta}'(\varepsilon_m^{\alpha} - \varepsilon) C_{im}^* C_{jm} \tag{15}$$

where the derivative of the Fermi function is $f'_\beta(\varepsilon)=-\beta exp(\beta\varepsilon)/(1+exp(\beta\varepsilon))^2$. It is often convenient, and more numerically stable, to calculate the DOS by finite differences, especially with a large value of the inverse temperature β. Use of a finite β establishes a unique mapping between the number of electrons and corresponding Fermi level, i.e. $\varepsilon \leftrightarrow N$ and therefore $\varepsilon=\varepsilon(N)$.

The function $\varepsilon(N)$ can be easily solved numerically. This allows a definition of electronic structural quantities analogous to conventional molecular orbital methods, in particular we define:

$$\varepsilon_{homo} = \varepsilon(N - \delta), \quad \varepsilon_{lumo} = \varepsilon(N + \delta), \quad and \quad \Delta E_{gap} = \varepsilon_{homo} - \varepsilon_{lumo} \tag{16}$$

where δ is a small number, taken here to be 0.05. Here δ is used to avoid numerical instabilities associated with the Fermi energy in the region of energy gaps, especially in the low temperature (large β) limit. To illustrate this, we note that if one uses the standard orbital population conventions (occupation numbers 1 for the N lowest lying orbitals and 0 otherwise, corresponding to the $\beta \to \infty$ limit), the "Fermi level" for an insulator, as defined by the normalization condition, is not unique; i.e. the Fermi level can take on any value between the highest occupied and lowest unoccupied molecular orbital eigenvalues. As values of β become large, determination of the Fermi level becomes numerically unstable. The definitions (16) avoid this instability, and result in rapidly convergent quantities as shown earlier in Table I.

The electronic density of states and energy gap at the Fermi level are useful quantities for describing molecular electronic structure. Here we consider the electronic DOS of the biomolecules and complexes discussed previously in the gas phase and in solution (Tables II & III). In the case of $(CG)_8$ DNA in canonical A, B, and Z-forms, the energy gap ranges from 7.41-7.78 eV. Recently, there has been experimental evidence that long-range electron transfer can occur through the DNA base stack[18]; however, the mechanism of this process is not yet understood, and the subject remains controversial. The energy gap results here suggest that a free electron conduction mechanism is unlikely. The gas phase energy gap values are listed for comparison to illustrate the dramatic effect of solvent stabilization on the electronic structure of these molecules. In the case of the protein systems examined here (all of which are neutral or cationic at neutral pH), the energy gaps in solution are slightly smaller than for the DNA (5.99-6.91 eV), and the electronic DOS are shifted toward more negative values in solution[12]. We note that the energy gaps for both DNA and proteins increase in solution resulting from preferential solvent stabilization of the occupied valence states relative to the virtual states; the smallest change occurs in the case of the hydrophobic protein crambin (5.14 eV to 6.91 eV).

Figure 4 compares the electronic DOS for the myosin-calmodulin and Myb-DNA complexes relative to the uncomplexed species in solution. The value of $\Delta N(E)$ is seen to be non-positive in these calculations. This results from a slight shift in the electronic levels toward more negative values in the uncomplexed molecules, and is more pronounced in the case of Myb-DNA complex. The energy gaps at the Fermi level are similar to those of the proteins, and do not change significantly upon complex formation (Table III).

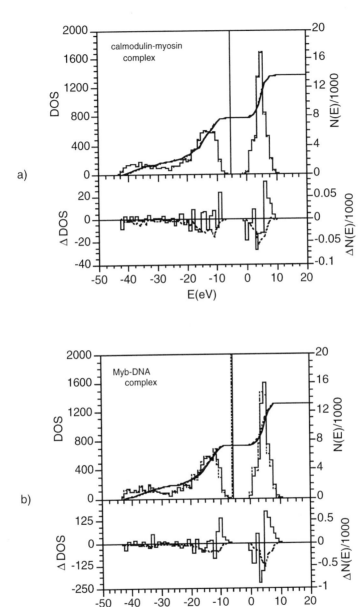

Fig. 4. Electronic DOS (eV⁻¹) and *N(ε)* (see text) in solution for a) calmodulin-myosin, and b) Myb-DNA. Shown are values for the complexes (solid line), superposition of states of the isolated species (dotted line), and the difference (shown immediately below). Vertical lines indicate the Fermi levels.

CONCLUSION

Linear-scaling electronic structure calculations have been performed for several biomolecules in solution at the semiempirical level to investigate the effects of solute polarization on solvation free energies and electronic density of state distributions. Results in the gas phase and in solution are compared. It is demonstrated that polarization contributes on the order of 10% for proteins and 2% for DNA of the total solvation free energy. The electronic response of the solute in the process of solvation is well approximated by a linear response model. In the case of binding between highly charged protein and DNA molecules, the overall $\Delta\Delta G_{sol}$ strongly disfavors binding, as expected, with $\Delta\Delta G_{pol}$ and $\Delta\Delta G_{int}$ making positive and negative contributions, respectively. Solvation has a pronounced effect on the electronic DOS, especially of highly charged biomolecules, causing a shift and broadening of the spectrum. The energy gaps at the Fermi level are observed to significantly increase upon solvation. These results are a first step toward the study of biological macromolecules in solution using self-consistent field methods to treat explicitly quantum mechanical many-body effects of the solute.

ACKNOWLEDGMENTS

The author acknowledges financial support through and NIH postdoctoral fellowship at Harvard University, Cambridge, MA, USA and an EMBO postdoctoral fellowship at Université Louis Pasteur, Strasbourg, France. The author thanks Prof. Martin Karplus for helpful discussions and a critical reading of the manuscript.

REFERENCES AND NOTES

1. C. L. Brooks III, M. Karplus and B. M. Pettitt *Proteins: A theoretical perspective of dynamics, structure, and thermodynamics* (John Wiley & Sons, New York, 1988). A. Warshel, *Computer Simulation of Chemical Reactions in Enzymes and Solutions* (John Wiley & Sons, New York, 1991).

2. Gilson, M. K. *Curr. Opinion in Struct. Biol.* **1995**, *5*, 216. Honig, B.; Nicholls, A. *Science* **1995**, *268*, 1144.

3. Yang, W. *Phys. Rev. Lett.* **1991**, *66*, 438. Yang, W.; Lee, T.-S. *J. Chem. Phys.* **1995**, *103*, 5674.

4. Cortona, P. *Phys. Rev. B* **1991**, *44*, 8454. Baroni, S.; Giannozzi, P. *Europhys. Lett.* 1992, *17*, 547. Galli, G.; Parinello, M. *Phys. Rev. Lett.* **1992**, *69*, 3547. Li, X.-P.; Nunes, W.; Vanderbilt, D. *Phys. Rev. B.* **1993**, *47*, 10891. Mauri, F.; Galli, G.; Car, R. *Phys. Rev. B* **1993**, *47*, 9973. Ordejón, P.; Drabold, D. A.; Grumback, M. P.; Martin, R. M. *Phys. Rev. B* **1993**, *48*, 14646. Stechel, E. B.; Williams, A. P.; Feibelman, P. J. *Phys. Rev. B* **1993**, *49*, 3898. Daw, M. S.*Phys. Rev. B* **1993**, *47*, 10899. Drabold, D. A.; Sankey, O. F. *Phys. Rev. Lett.* **1993,** *70*, 3631. Kohn, W. *Chem. Phys. Lett.* **1993**, *208*, 167. Kohn, W. *Phys. Rev. Lett.* **1996**, *76*, 3168. Stewart, J. J. P. *Int. J. Quant. Chem.* **1996**, *58*, 133. S. L. Dixon, S. L.; Merz, K. M.; *J. Chem. Phys.* **1996**, *107*, 6643. Dixon, S. L.; Merz, K. M., Jr. *J. Chem. Phys.* **1997**, *107*, 879. Millam, J. M.; Scuseria, G. E. *J. Chem. Phys.* **1997**, *106*,

5569. Daniels, A. D.; Millam, J. M.; Scuseria, G. E. *J. Chem. Phys.* **1997**, *107*, 425.

5. York, D. M.; Lee, T.-S.; Yang, W. *J. Am. Chem. Soc.* **1996**, *118*, 10940.

6. Lee, T.-S.; York, D. M.; Yang, W. *J. Chem. Phys.* **1996**, *105*, 2744.

7. York, D. M.; Lee, T.-S.; Yang, W. *Chem. Phys. Lett.* **1996**, *263*, 297.

8. Mulliken, R. S.; Politzer, P. *J. Chem. Phys.* **1971**, *55*, 5135.

9. Klamt, A.; Schüürmann, G. *J. Chem. Soc. Perkin Trans.* **1993**, *2*, 799.

10. Pérez-Jordá, J.; Yang, W. *Chem. Phys. Lett.* **1995**, *247*, 484. Pérez-Jordá, J.; Yang, W. *J. Chem. Phys.* **1995**, *104*, 8003.

11. Arnott, S.; Hukins, D. W. L. *Biochem. Biophys. Res. Commun.* **1972**, *47*, 1504.

12. York, D. M.; Lee, T.-S.; Yang, W. *Phys. Rev. Lett.*, in press.

13. Dewar, M. J. S.; Zoebisch, E. G.; Healy, E. F.; Stewart, J. J. P. *J. Am. Chem. Soc.* **1985**, *107*, 3902. Dewar, M. J. S.; Healy, E. F.; Holder, A. J.; Yuan, Y. C.; *J. Comput. Chem.* **1990**, *11*, 541.

14. Gao, J.; Xia, X.*Science* **1992**, *258,* 631. Gao, J. *Biophys. Chem.* **1994**, *51*, 253. Orozco, M.; Luque, F. J.; Habibollahzadeh, D.; Gao, J. *J. Chem. Phys.* **1995**, *102*, 6145. Thompson, M. A.; Schenter, G. K. *J. Phys. Chem.* **1995**, *99*, 6374.

15. Ikura, M.; Clore, G. M.; Gronenborn, A. M.; Zhu, G.; Klee, C. B.; Bax, A. *Science* 1992, *256*, 632.

16. Ogata, K.; Morikawa, S.; Nakamura, H.; Sekikawa, A.; Inoue, T.; Kanai, H.; Sarai, A.; Ishit S.; Nishimura, Y. *Cell* **1994**, *79*, 639.

17. Rappé, A. K.; Goddard, W. A. *J. Phys. Chem.* **1990**, *94*, 4732. Rick, S. E.; Stuart, S. J.; Berne, B. J. *J. Chem. Phys.* **1994**, *101*, 6141. Bernardo, D. N.; Ding, Y.; Krogh-Jespersen, K.; Levy, R. M. *J. Comput. Chem.* **1995**, *16*, 1141. Gao, J *J. Phys. Chem. B* **1997**, 101, 657. York, D. M.; Yang, W. *J. Chem. Phys.* **1995**, *104*, 159.

18. Murphy, C. J.; Arkin, M. R.; Jenkins, Y.; Ghatlia, N. D.; Bossmann, S. H.; Turro, N. J.; Barton, J. K. *Science* **1993**, *262*, 1023. Arkin, M. R.; Stemp, E. D. A.; Holmlin, R. E.; Barton, J. K.; Hörmann,m A.; Olson, E. J. C.; Barbara, P. F. *Science* **1996**, *273*, 475.

Chapter 19

Exciting Green Flourescent Protein

Volkhard Helms[1,4], Erik F. Y. Hom[1], T. P. Straatsma[2], J. Andrew McCammon[1], and
Peter Langhoff[3]

[1]Department of Chemistry and Biochemistry, University of California at San Diego,
La Jolla, CA 92093–0365
[2]Pacific Northwest Laboratory, Richland, WA
[3]Department of Chemistry, Indiana University, Bloomington, IN 47405

Dynamical and luminescent properties of the Green Fluorescent Protein
(GFP) are being explored via molecular dynamics (MD) simulations and
quantum chemistry calculations with the aim of facilitating the rational development of GFP as a probe for cellular functions. Results from an MD simulation of wild type GFP demonstrate the rigidity of the structural framework
of GFP, and a very stable hydrogen bond network around the chromophore.
Furthermore, excited state calculations have been performed on the chromophore in vacuum, and we report about our work in progress here.

Interfacing Quantum Mechanical and Molecular Mechanics methods has become a
popular device to study properties of a small quantum regime embedded in a solvent
or protein environment that is described classically. Here, we discuss preliminary
results of parallel quantum and classical calculations. Not only do we want to
understand the energetic and dynamic features of an organic chromophore and its
surroundings in detail, we would also like to use its absorption and fluorescent
properties as a detector to study structural and dynamic behaviour of its protein
environment.

Green Fluorescent Protein

Green Fluorescent Protein (GFP) is a spontaneously fluorescent protein isolated
from the Pacific Northwest jellyfish *Aequorea victoria*. Its apparent role in the jellyfish is to transduce, via fluorescence resonance energy transfer, the blue chemilumi-
nescence of another protein, aequorin, as green fluorescent light (1). Since its first

---

[4]email: vhelms@ucsd.edu.

identification, cell biologists have recognized the biotechnological potential of GFP, and have very successfully used GFP as a noninvasive marker in living cells, e.g. as a reporter of gene expression and protein localization (2), and in the detection of protein-protein interactions (3). Recently, three-dimensional crystal structures of the wild type protein, of a Ser65 → Thr mutant, and of a number of other mutants, were determined (4–7).

GFP consists of a single chain of 238 residues that forms an 11-stranded β−barrel, wrapped around a central helix. The barrel is a nearly perfect cylinder 42 Å long and 24 Å in diameter. At the center of this β−barrel 'canister' lies the central helix and the chromophore that is responsible for GFP's luminescence, see Figure 1. This unusual p-hydroxylbenzylideneimidazolidinone chromophore results from the auto-catalytic cyclization of the polypetide backbone between residues Ser65 and Gly67, followed by oxidation of the $\alpha - \beta$ bond of Tyr66. It is completely protected from bulk solvent, although there are a number of bound water molecules in the protein interior.

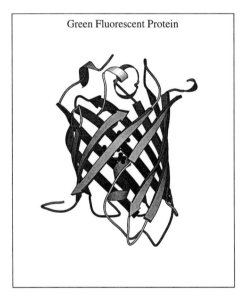

Green Fluorescent Protein

Figure 1 Schematic diagram of the cylindrical shape of GFP. Only elements of secondary structure and the chromophore (in black) are shown.

Wild type GFP has two absorption maxima at 395 and 475 nm that are believed to be due to a neutral and anionic form of the chromophore, respectively (8). A ca. 80% neutral : 20% anionic equilibrium appears to be governed by a hydrogen bonding network that permits proton transfer between the chromophore and neighboring side chains (6). Figure 2 shows simplified drawings of the proposed hydrogen bonding networks around the two ionization states of the chromophore (4, 7).

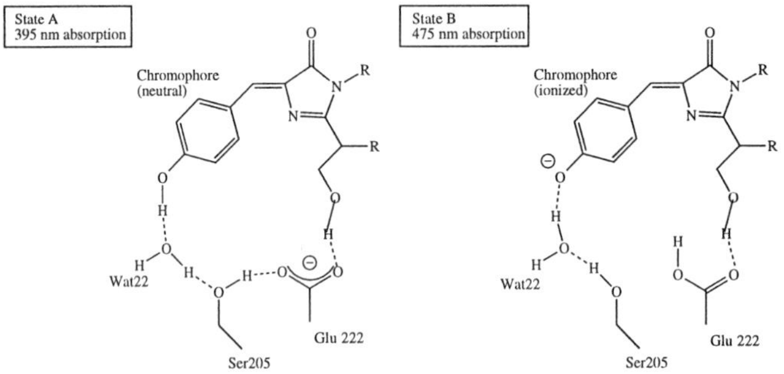

Figure 2 Hydrogen bond networks close to neutral and anionic chromophore.

Mutant versions of GFP with significantly different spectral properties have been constructed, and permit, in principle, simultaneous tagging and localization of proteins and the monitoring of association events *in vivo*. Our present work in progress aims at elucidating the structural, dynamical, and electronic factors responsible for the remarkable properties of GFP.

In spectroscopic single molecule experiments, two mutants exhibited a fast on/off blinking behaviour under constant illumination, and a conversion into a long-lived dark state (9). From here, the molecule could only be switched back to the original emissive state by continuosly irradiating at 405 nm. The authors noted that, presumably, the fast blinking occurs with an anionic chromophore, while the slow conversion reflects the transition to a form with a neutral chromophore. Certainly, these mutants contain different solvent networks in the chromophore surrounding, and the results are not directly transferable to the present study, but they give an indication about the slow timescales involved with these transitions.

Relation to Other Work

This work follows previous theoretical work on the photoisomerization of retinal in the membrane protein bacteriorhodopsin (10) and on metal-containing enzymes such as plastocyanin (11) involved in electron transport. The study of GFP is facilitated compared to these proteins by the availability of excellent crystallographic data with well resolved internal water positions and by the fact that GFP does not contain complex co-factors such as a heme-moiety or metal atoms.

Based on INDO/S calculations of different forms of a model chromophore *in vacuo* and in solvent, it was suggested (12) that the bare nitrogen in the imidazolidinone ring is protonated. This was based on the fact that the results for the protonated chromophore were in best agreement with the experimental absorption frequencies. In our work we do not account for this possibility, because we feel that protonation of the ring nitrogen would cause polar and charged protein groups in its vicinity to move closer to this site than is observed in the crystal structures.

Molecular Dynamics Simulation of GFP

An MD simulation of wild type GFP with a neutral chromophore was performed with the parallel NWChem program (13). The simulation system consisted of one GFP molecule solvated in a cubic, periodic solvent box with dimensions of 69 Å. The AMBER94 all-atom force field (14) was employed, resulting in 3594 solute atoms and 9381 solvent molecules, with a total of 31726 dynamic atoms. Six Na^+-ions were positioned in low energy positions on the protein surface to make the system overall neutral, and long-range electrostatic interactions were treated by the particle-mesh-Ewald technique (15). The partial atomic charges of the ring atoms of the chromophore were obtained from an electrostatic potential fit with the CHELPG method (16) to the HF/6-31G(d) optimized electron density. The partial atomic charges of the atoms of the chromophore's Serine side chain and those of the connecting peptide bonds were taken from the AMBER94 force field. The remaining charge of 0.2 e was assigned to atom CA3. Parameters for covalent and Lennard-Jones interactions were also adapted from AMBER94. The positions of aromatic hydrogens and hydrogens of crystal water molecules were optimized with the WHAT IF program (17). Starting from the crystal structure (6), the system was heated and equilibrated by 100 steps of steepest descent energy minimization of the whole system, by subsequent 20 ps MD equilibration of the solvent only at 300 K, and by short MD simulations of 5 ps length each of the whole system at 50 K, 100 K, ..., 300 K under constant volume and constant energy conditions. A 1 ns MD simulation was then performed at a constant temperature of 300 K and at constant pressure. The individual energy components were well equilibrated after 100 ps, and the last 900 ps were considered for analysis.

The Molecular Dynamics Simulation was Performed to Characterize :

- *How rigid the GFP β-barrel structure is*
  After the crystal structure of GFP became known, it was obvious that the encapsulation of the chromophore is probably responsible for the high quantum yield of fluorescence, for the inability of O_2 to quench the excited state, and for the resistance of the chromophore to titration of the external pH (4, 18).

We investigated the dynamical behavior of the protein during the simulation by an essential dynamics analysis (19) that is implemented in the WHAT IF program (17). Figure 3 shows maximum and minimum projections of the trajectory onto the eigenvector with the largest eigenvalue.

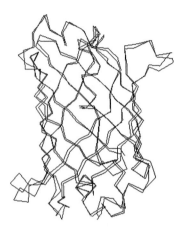

Figure 3 Projections of the MD simulation of GFP wild type onto the eigenvector with the largest eigenvalue.

Detectable motions can only be seen in the loop segments, and the β-barrel is almost unaffected. Apparently, its unusual cylindrical architecture makes GFP a rock-solid protein compared to other proteins folds. Also, the RMS-deviation of the C_α atoms from the energy minimized crystal structure remained at around 0.1 nm during the production phase. This is a very low value since proteins often show RMS-deviations of 0.2 - 0.3 nm during MD simulations of this length.

- *The permeability of the space inside the β-barrel to bulk solvent molecules.*

No water molecules were observed to penetrate the interior of GFP during the simulation.

- *The hydrogen bonding pattern central to the conversion between the neutral and anionic chromophore forms.*

Occupancies were calculated for the hydrogen bond network in the chromophore surrounding. All important hydrogen bonds had occupancies between 80 and 100 % during the simulation. The six water molecules in the immediate vicinity of the chromophore behave like in ice.

A more detailed analysis of the simulations will be presented elsewhere. Not surprisingly, the GFP structure provides a very stable framework for the enclosed chromophore. It is noteworthy that this rigidity extends to the immediate surrounding of the chromophore where a very stable arrangement of hydrogen bonds and water positions was found. The cylindrical β-barrel thus provides a natural protection of the fluorescent signal against environmental factors such as pH, O_2, and metal ions.

Contrary to our original plans that included a statistical sampling of the protein dynamics by performing quantum calculations on a large number of snapshots from the molecular dynamics trajectory, performing single quantum calculations on a small number of selected protein conformations seems perfectly justified.

Quantum Calculations of the GFP Chromophore

The *ab initio* QM calculations aim at characterizing the chromophore absorption spectrum under the influence of the protein environment and thermal fluctuations. Although good quantitative agreement with experimental data is desirable, we are initially most interested in identifying key protein groups which could be modified to shift the wavelength of excitation, improve absorption intensity, and/or narrow the bandwidth of absorption. These calculations will serve as an important base to derive simplified models for predicting spectral changes due to mutations in GFP.

In Vacuo Calculations

The geometries of neutral and anionic model chromophore have been optimized at Hartree Fock 6-31G(d) level (264 basis functions) with the GAMESS program (20). Fifty seven molecular orbitals are occupied in each case. Planar symmetry was employed since the chromophore is found in almost planar conformations in the various crystal structures. Vertical singlet excitations into the 10 lowest virtual IVO orbitals were performed separately for A" → A", A' → A", A' → A', and for A" → A' transitions. The transition dipole moments for mixed A' → A" and A" → A' excitations are close to zero, and all A' → A' excitations are very high energy (> 12 eV). To understand the low-lying spectrum of the chromophore, it is therefore sufficient to focus on A" → A" excitations. The results for IVO calculations for the A" → A" excitations are shown in table I.

Table I. Vertical excitation energies (eV) for A" → A" excitations into virtual IVO orbitals.

from MO # to	state 2	state 3	state 4	state 5	state 6
57	5.534	7.411	8.830	9.654	12.093
56	7.305	8.716	9.990	12.415	13.959
55	7.572	9.613	10.664	11.652	13.658
53	8.889	11.783	12.119	12.763	14.683
51	10.916	13.455	14.235	14.855	17.032
49	11.195	13.489	14.348	15.696	17.962

Only a few vertical excitations have energies below 10 eV. Accordingly, the active space for CASSCF-type calculations may be chosen rather small unless very accurate results are desired. The result from an MCSCF calculation (21) with an active space including the 3 highest occupied π-orbitals and the 3 lowest unoccupied π^*-orbitals is 5.628 eV, in good agreement with the IVO result for the HOMO $\rightarrow$ LUMO excitation. The excitation energies for the anionic chromophore were ca. 1 eV lower, in agreement with the experimentally observed red-shift for the anionic form. Still, the excitation energies are rather high compared to the expected values of ca. 3 eV. We expect a significant drop of these energies by also optimizing the excited state geometries and by including the effects of electron correlation.

Inclusion of the Protein Environment

An important challenge and difficulty arises from the representation of the protein environment. Work on bacteriorhodopsin by the groups of Schulten, Warshel, and others has shown that it is essential to incorporate the protein surroundings for a correct description of the electronic structure of the chromophore. Experimental spectra of a model GFP chromophore in water and in various organic solvents (22) were blue-shifted by 30 - 50 nm compared to spectra in the protein environment. Yet, these shifts are rather small and demonstrate that the chromophore environment in GFP is not of unusual nature.

We plan to represent the protein environment by effective fragment potentials (23) that are being developed in combination with the GAMESS program.

Acknowledgments

We would like to acknowledge the help of Kim Baldridge with using the GAMESS program. Carl Winstead was of critical help with the single excitation calculations. We also thank Roger Tsien for valuable discussions. NWChem Version 3.0, as developed and distributed by Pacific Northwest National Laboratory, P. O. Box 999, Richland, Washington 99352 USA, and funded by the U. S. Department of Energy, was used to obtain some of these results. This work was supported by a postdoctoral NATO fellowship to VH by the Deutscher Akademischer Austauschdienst, and by grants from NSF, the San Diego Supercomputing Center and the NSF Supercomputer Centers MetaCenter program. VH is also a fellow of the Program in Mathematics and Molecular Biology and of the La Jolla Interfaces in Sciences Program.

Literature Cited

(1) Ward, W.; Cody, C.; Hart, R.; Cormier, M. *Photochem. Photobiol.* **1980**, *31*, 611 – 615.

(2) Cubitt, A.; Heim, R.; Adams, S.; Boyd, A.; Gross, L.; Tsien, R. *Trends Biochem. Sci.* **1995**, *20*, 448–455.

(3) Rizzuto, R.; Brini, M.; Giorgi, F. D.; Rossi, R.; Heim, R.; Tsien, R.; Pozzan, T. *Curr. Biol.* **1996**, *6*, 183–188.

(4) Ormö, M.; Cubitt, A.; Kallio, K.; Gross, L.; Tsien, R.; Remington, S. *Science* **1996**, *273*, 1392–1395.

(5) Yang, F.; Moss, G.; Jr, G. P. *Nat. Biotech.* **1996**, *14*, 1246–1251.

(6) Brejc, K.; Sixma, T.; Kitts, P.; Kain, S.; Tsien, R.; Ormö, M.; Remington, S. *Proc. Natl. Acad. Sci. USA* **1997**, *94*, 2306–2311.

(7) Palm, G.; Zdanov, A.; Gaitanaris, G.; Stauber, R.; Pavlakis, G.; Wlodawer, A. *Nat. Struct. Biol.* **1997**, *4*, 361–365.

(8) Chattoraj, M.; King, B.; Bublitz, G.; Boxer, S. *Proc. Natl. Acad. Sci. USA* **1996**, *93*, 8362–8367.

(9) Dickson, R.; Cubitt, A.; Tsien, R.; Moerner, W. *Nature* **1997**, *388*, 355–358.

(10) Logunov, I.; Schulten, K. *J. Am. Chem. Soc.* **1996**, *118*, 9727–9735.

(11) Pierloot, K.; Kerpel, J. D.; Ryde, U.; Roos, B. *J. Am. Chem. Soc.* **1997**, *119*, 218–226.

(12) Voityuk, A.; Michel-Beyerle, M.; Rösch, N. *Chem. Phys. Lett.* **1997**, *272*, 162–167.

(13) *NWChem, A Computational Chemistry Package for Parallel Computers, Version 3.0.* High Performance Computational Chemistry Group, Pacific Northwest National Laboratory, Richland, Washington 99352, USA, 1997.

(14) Cornell, W.; Cieplak, P.; Bayly, C.; Gould, I.; Jr., K. M.; Ferguson, D.; Spellmeyer, D.; Fox, T.; Caldwell, J.; Kollman, P. *J. Am. Chem. Soc.* **1995**, *117*, 5179–5197.

(15) Darden, T.; York, D.; Pedersen, L. *J. Chem. Phys.* **98**, *98*, 10089–10093.

(16) Breneman, C.; Wiberg, K. *J. Comput. Chem.* **1990**, *11*, 361–373.

(17) Vriend, G. *J. Mol. Graph.* **1990**, *8*, 52–56.

(18) Phillips, G. *Curr. Op. Struct. Biol.* **1997**, *7*, 821–827.

(19) Amadei, A.; Linssen, A.; Berendsen, H. *Proteins* **1993**, *17*, 412–425.

(20) Schmidt, M.; Baldridge, K.; Boatz, J.; Elbert, S.; Gordon, M.; Koseki, J. J. S.; Matsunaga, N.; Nguyen, K.; Su, S.; Windus, T. *J. Comp. Chem.* **1993**, *14*, 1347–1363.

(21) Roos, B.; Taylor, P.; Siegbahn, P. *Chem. Phys.* **1980**, *48*, 157–173.

(22) Niwa, H.; Inouye, S.; Hirano, T.; Matsuno, T.; Kojima, S.; Kubota, M.; Ohashi, M.; Tsuji, F. *Proc. Natl. Acad. Sci. USA* **1996**, *93*, 13617–13622.

(23) Day, P.; Jensen, J.; Gordon, M.; Webb, S.; Stevens, W.; Krauss, J.; Cohen, D. G. H. B. D. *J. Chem. Phys.* **1996**, *105*, 1968–1986.

Author Index

Subject Index